Hazardous Materials For First Responders

Second Edition

HAZ MAT

Photo Courtesy of Scott D. Christiansen, Minot, N.D.

Published by
Fire Protection Publications
Oklahoma State University
Stillwater, Oklahoma

Edited by
Michael Wieder
Carol Smith
Cynthia Brakhage

Cover Photo Courtesy of:
Ron Jeffers,
New Jersey Metro Fire Photographers Assoc.

ISBN 0-87939-112-X
Library of Congress 94-61532

Second Edition

Printed in the United States of America 6 7 8 9 10

Dedication

This manual is dedicated to the members of that unselfish organization of men and women who hold devotion to duty above personal risk, who count on sincerity of service above personal comfort and convenience, who strive unceasingly to find better ways of protecting the lives, homes and property of their fellow citizens from the ravages of fire and other disasters . . . ***The Firefighters of All Nations.***

Dear Firefighter:

The International Fire Service Training Association (IFSTA) is an organization that exists for the purpose of serving firefighters' training needs.Fire Protection Publications is the publisher of IFSTA materials. Fire Protection Publications staff members participate in the National Fire Protection Association and the International Association of Fire Chiefs.

If you need additional information concerning our organization or assistance with manual orders, contact:

Customer Services
Fire Protection Publications
Oklahoma State University
Stillwater, OK 74078-8045
1-800-654-4055 Fax 1-405-744-8204

For assistance with training materials, recommended material for inclusion in a manual, or questions on manual content, contact:

Technical Services
Fire Protection Publications
Oklahoma State University
Stillwater, OK 74078-8045
1-405-744-4111 Fax 1-405-744-4112
Email: editors@osufpp.org

THE INTERNATIONAL FIRE SERVICE TRAINING ASSOCIATION

The International Fire Service Training Association (IFSTA) was established as a "nonprofit educational association of fire fighting personnel who are dedicated to upgrading fire fighting techniques and safety through training." This training association was formed in November 1934, when the Western Actuarial Bureau sponsored a conference in Kansas City, Missouri. The meeting was held to determine how all the agencies interested in publishing fire service training material could coordinate their efforts. Four states were represented at this initial conference. Because the representatives from Oklahoma had done some pioneering in fire training manual development, it was decided that other interested states should join forces with them. This merger made it possible to develop training materials broader in scope than those published by individual agencies. This merger further made possible a reduction in publication costs, because it enabled each state or agency to benefit from the economy of relatively large printing orders. These savings would not be possible if each individual state or department developed and published its own training material.

To carry out the mission of IFSTA, Fire Protection Publications was established as an entity of Oklahoma State University. Fire Protection Publications' primary function is to publish and disseminate training texts as proposed and validated by IFSTA. As a secondary function, Fire Protection Publications researches, acquires, produces, and markets high-quality learning and teaching aids as consistent with IFSTA's mission. The IFSTA Executive Director is officed at Fire Protection Publications.

IFSTA's purpose is to validate training materials for publication, develop training materials for publication, check proposed rough drafts for errors, add new techniques and developments, and delete obsolete and outmoded methods. This work is carried out at the annual Validation Conference.

The IFSTA Validation Conference is held the second full week in July, at Oklahoma State University or in the vicinity. Fire Protection Publications, the IFSTA publisher, establishes the revision schedule for manuals and introduces new manuscripts. Manual committee members are selected for technical input by Fire Protection Publications and the IFSTA Executive Secretary. Committees meet and work at the conference addressing the current standards of the National Fire Protection Association and other standard-making groups as applicable.

Most of the committee members are affiliated with other international fire protection organizations. The Validation Conference brings together individuals from several related and allied fields, such as:

- Key fire department executives and training officers
- Educators from colleges and universities
- Representatives from governmental agencies
- Delegates of firefighter associations and industrial organizations
- Engineers from the fire insurance industry

Committee members are not paid nor are they reimbursed for their expenses by IFSTA or Fire Protection Publications. They come because of commitment to the fire service and its future through training. Being on a committee is prestigious in the fire service community, and committee members are acknowledged leaders in their fields. This unique feature provides a close relationship between the International Fire Service Training Association and other fire protection agencies, which helps to correlate the efforts of all concerned.

IFSTA manuals are now the official teaching texts of most of the states and provinces of North America. Additionally, numerous U.S. and Canadian government agencies as well as other English-speaking countries have officially accepted the IFSTA manuals.

Table of Contents

List Of Tables

Preface

The second edition of **Hazardous Materials for First Responders** is written to aid firefighters and other emergency personnel who respond to hazardous materials emergencies. This manual assists personnel in meeting the requirements for First Responders at the Awareness and Operational levels in NFPA 472, *Standard for Professional Competence of Responders to Hazardous Materials Incidents* (1992 edition).

Acknowledgment and special thanks are extended to the members of the validating committee. The following committee members contributed their time, wisdom, and knowledge to this manual:

Chairman
John Eversole
Chicago Fire Department
Chicago, IL

Roy Burton
U.S. Dept. of Transportation
Washington, D.C.

Ron Cody
Redondo Beach Fire Dept.
Redondo Beach, CA

Scott Kerwood
Sni-Valley Fire Prot. Dist.
Oak Grove, MO

Max McRae
Houston Fire Department
Houston, TX

Peter McMahon
Grand Island Fire Co.
Grand Island, NY

Richard Pippenger
Grand Junction, CO

Christopher Plantenga
Lafayette Fire Department
Lafayette, IN

Harold Richardson
Hubbards Fire Dept.
Hubbards, Nova Scotia

Special thanks are given to the Houston (TX) Fire Department Hazardous Materials Response Team and to District Chief Max McRae. Most of the photographs in this manual were obtained with their assistance. Assisting with the project were the following personnel:

District Chief Daniel Snell
Haz Mat 22(D)
Engine 31(D)
Haz Mat 22(C)
  Capt. Bill Wilcox
  Bill Hand
  Stan Ford
  Keith Tolson
  Daniel Banda
  Steven Brooks
  Mike Falco
  David Fletcher

We would also like to extend our thanks to the following individuals and organizations that provided information, photographs, or other assistance toward the completion of this manual.

Oklahoma State University Environmental Health Services
  Emory Moseby
  Kevin Jones

Stillwater (OK) Fire Department
    Tom Bradley
    Jim Morgan
    Tom Low
Lyondell Oil Refinery, Houston, Texas
J.B. Kelly Trucking, Houston, Texas
Mission Petroleum Carriers, Inc., Houston, Texas
Bob Esposito, Warrior Run, Pennsylvania
Joseph J. Marino, New Britain, Connecticut
Linda Gheen, Dallas, Texas
Ron Jeffers, Union City, New Jersey
Harvey Eisner, Tenafly, New Jersey
Sally C. McCann, Gahanna, Ohio
Martin Grube, Virginia Beach, Virginia
Phoenix (AZ) Fire Department
    Kevin Roche, Fire Protection Engineer
    Paul Albertson, PIO Intern
John O'Gorman, Shell Oil Co.
Charles J. Wright, Union Pacific Railroad
Tulsa (OK) Fire Department
Mine Safety Appliances (MSA), Inc.
Scott Aviation, Inc.

Gratitude is also extended to the following members of the Fire Protection Publications staff whose contributions made the final publication of this manual possible.

Barbara Adams, Senior Publications Specialist
Marsha Sneed, Senior Publications Specialist
Susan Walker, Coordinator of Instructional Development
Don Davis, Publications Production Coordinator
Ann Moffat, Graphic Designer Analyst
Desa Porter, Senior Graphic Designer
Shari Downs, Graphic Designer
Kimberly Edwards, Photographic Technician
Ward Barnett, Research Technician
Jason Goetz, Research Technician
Nancy Logue, Research Technician
Matthew Manfredi, Research Technician

Lynne C. Murnane
Lynne Murnane
Managing Editor

Glossary

A

Absolute Pressure
Gauge pressure plus atmospheric pressure.

Absorbents
Inert materials; that is, they have no active properties. They can be used to pick up a liquid contaminant. Some examples of absorbents are soil, diatomaceous earth, vermiculite, sand, and other commercially available products.

Absorption
(1) To take in and make part of an existent whole. (2) Passage of toxic materials through some body surface into body fluids and tissue. (3) Process of picking up a liquid contaminant with an absorbent.

ACGIH
Abbreviation for American Conference of Governmental Industrial Hygienists.

Acid
Compound containing hydrogen that reacts with water to produce hydrogen ions; a proton donor; a liquid compound with a pH less than 2. Acidic chemicals are corrosive.

Acute
(1) Characterized by sharpness or severity; having rapid onset and a relatively short duration. (2) Single exposure (dose) or several repeated exposures to a substance within a short time period.

Acute Health Effects
Health effects that occur or develop rapidly after exposure to a substance.

Aerator
Device for introducing air into dry bulk solids to improve flow ability.

Air Bill
Shipping document prepared from a bill of lading that accompanies each piece or each lot of air cargo.

Air Inversion
Meteorological condition in which the temperature of the air some distance above the earth's surface is higher than the temperature of the air at the surface. Normally, air temperatures decrease as altitude increases. An air inversion traps air, releases gases and vapors near the surface, and impedes their dispersion.

Air Lift Axle
Single air-operated axle that when lowered converts a vehicle into a multiaxle unit, providing the vehicle with a greater load carrying capacity.

Air-Reactive Materials
Substances that ignite when exposed to air at normal temperatures. Also called Pyrophoric.

Air Spring
Flexible, air-inflated chamber on a truck or trailer in which the air pressure is controlled and varied to support the load and absorb road shocks.

Allergen
Material that can cause an allergic reaction of the skin or respiratory system. Also called Sensitizer.

Alpha Radiation
Consists of particles having a large mass and a positive electrical charge; least penetrating of the three common forms of radioactive substances. It is normally not considered dangerous to plants, animals, or people unless it gets into the body.

Ambient Temperature
Temperature of the surrounding environment.

Asphyxia
Suffocation.

Asphyxiant
Any substance that prevents oxygen from combining in sufficient quantities with the blood or from being used by body tissues.

Asphyxiating Materials
Substances that can cause death by displacing the oxygen in the air.

Asphyxiation
Condition that causes death because of a deficient amount of oxygen and an excessive amount of carbon monoxide and/or other gases in the blood.

Assessment Stop
Distant location that is safe for the first responders to stop and evaluate the situation, to complete donning their protective clothing and SCBA, and to report conditions to the communications center.

Atmospheric Pressure
Pressure exerted by the atmosphere at the surface of the earth because of the weight of air. Atmospheric pressure at sea level is about 14.7 psi (101 kPa). Atmospheric pressure increases as elevation decreases below sea level and decreases as elevation increases above sea level.

Atmospheric Storage Tank
Class of fixed facility storage tanks. Pressures range from 0 to 0.5 psig (0 kPa to 4 kPa).

Autoignition
Ignition that occurs when a substance in air, whether solid, liquid, or gaseous, is heated sufficiently to initiate or cause self-sustained combustion independently of the heat source.

Autoignition Temperature
Same as ignition temperature except that no external ignition source is required for ignition because the material itself has been heated to ignition temperature; the spontaneous ignition of the gases or vapor given off by a heated material.

Avulsion
Forcible separation or detachment; the tearing away of a body part.

B

Baffle
Intermediate partial bulkhead that reduces the surge effect in a partially loaded liquid tank.

Base
Substance-containing group forming hydroxide ions in water solution; an alkaline (caustic) substance.

Beta Particle
Particle that is about 1/7000 the size of an alpha particle but has more penetrating power. The beta particle has a negative electrical charge.

Beta Radiation
Type of radiation that can cause skin burns.

Bill Of Lading
Shipping paper used by the trucking industry indicating origin, destination, route, and product. There is a bill of lading in the cab of every truck tractor.

Biochemical
Involving chemical reactions in living organisms.

Bird Box
See Connection Box.

Blasting Cap
See Detonator.

BLEVE
Acronym for Boiling Liquid Expanding Vapor Explosion.

Blood Poisons
See Chemical Asphyxiant.

Blow-Down Valve
Manually operated valve that has the function of quickly reducing tank pressure to atmospheric pressure.

Boiling Liquid Expanding Vapor Explosion (BLEVE)
Major failure of a closed liquid container into two or more pieces when the temperature of the liquid is well above its boiling point at normal atmospheric pressure.

Boiling Point
Temperature of a substance when the vapor pressure exceeds atmospheric pressure. At this temperature, the rate of evaporation exceeds the rate of condensation. At this point, more liquid is turning into gas than gas is turning back into a liquid.

Bulk Container
Cargo tank container attached to a flatbed truck or rail flatcar used to transport materials in bulk. This container may carry liquids or gases.

Bulk Packaging
Packaging, other than a vessel or a barge, including transport vehicle or freight container, in which hazardous materials are loaded with no intermediate form of containment and which:

- For a liquid, the container has a minimum capacity of 119 gallons (450 L).
- For a solid, the containers have a minimum mass of 882 pounds (400 kg).

- For a gas, the container has a minimum water capacity of 1,000 pounds (454 kg). (This information is taken from 49 CFR 171.8.)

Bumper
Structure designed to provide front- and rear-end protection of a vehicle.

C

CANUTEC
Canadian Transport Emergency Centre operated by Transport Canada to assist emergency response personnel in handling dangerous goods emergencies.

Capacity Indicators
Device installed on a tank to indicate capacity at a specific level.

Carboy
A cylindrical container of about 5 to 15 gallons (20 L to 60 L) capacity for corrosive or pure liquids; made of glass, plastic, or metal with a neck and sometimes a pouring tip and cushioned in a wooden box, wicker basket, or special drum.

Carcinogens
Cancer-producing substances.

Cargo Container
See Container.

CAS Number
Number assigned by the American Chemical Society's Chemical Abstract Service that uniquely identifies a specific compound.

CFR
Abbreviation for Code of Federal Regulations.

CGA
Abbreviation for Compressed Gas Association.

Chemical Asphyxiant
Substances that react to keep the body from being able to use oxygen. Also called Blood Poisons.

Chemical Properties
Relating to the way a substance is able to change into other substances. These properties reflect the ability to burn, react, explode, or produce toxic substances hazardous to people or the environment.

CHEMTREC
Manufacturing Chemists Association's name for its Chemical Transportation Emergency Center. The center provides immediate information about handling haz mat incidents. The toll-free number is 1-800-424-9300 (1-202-483-7616 in Washington, D.C., Alaska, and Hawaii).

Chronic
Of long duration or recurring over a period of time.

Chronic Health Hazards
Long-term effects from either a one-time or repeated exposure to a substance.

Class A Fire
Fires involving ordinary combustibles such as wood, paper, cloth, and so on.

Class B Fire
Fires of flammable and combustible liquids and gases such as gasoline, kerosene, and propane.

Cleanout Fitting
Fitting installed in the top of a tank to facilitate washing the tank's interior.

Code of Federal Regulations (CFR)
Formal name given to the books or documents containing the specific United States regulations provided for by law.

COFC
Abbreviation for Container-On-Flatcar.

Colorimetric Tube
Small tubes that change colors when air that is contaminated with a particular substance is drawn through them. Also known as Detector Tube.

Combustible Gas Detector
Indicates the explosive levels of combustible gases.

Combustible Liquids
Liquid having a flash point at or above 100°F (37.8°C) and below 200°F (93.3°C).

Combustion
Self-sustaining process of rapid oxidation of a fuel, which produces heat and light.

Compound
Substance consisting of two or more elements that have been united chemically.

Compressed Gas
Gas that, at normal temperature, exists solely as a gas when pressurized in a container.

Compressed Gas Association (CGA)
Trade association that, among other things, writes standards pertaining to the use, storage, and transportation of compressed gases.

Compressed Gas Trailer
Cargo truck that carries gases under pressure; may be a large single container, an intermodal shipping unit, or several horizontal tubes. Also called Tube Trailer.

Condensation
Process of going from the gaseous to the liquid state.

Cone Roof Storage Tank
Atmospheric storage tank that has a cone-shaped pointed roof with weak roof-to-shell seams that are intended to break when excessive overpressure results inside.

Congenital
Existing at or dating from birth.

Connection Box
Contains fittings for trailer emergency and service brake connections and electrical connector to which the lines from the towing vehicle may be connected. Formerly called Junction Box, Light Box, or Bird Box.

Consignee
Person who is to receive a shipment.

Consist
Rail shipping paper that lists by order the cars in the train. The cars containing hazardous materials are indicated. Also included with the consist may be information on emergency operations for the hazardous materials on board. Also called Train Consist.

Container
Article of transport equipment that is: (a) of a permanent character and strong enough for repeated use; (b) specifically designed to facilitate the carriage of goods by one or more modes of transport without intermediate reloading; and (c) fitted with devices permitting its ready handling, particularly its transfer from one mode to another. The term "container" does not include vehicles. Also referred to as Freight Container, Cargo Container, and Intermodal Tank Container.

Container Chassis
Trailer chassis consisting of a frame with locking devices for securing and transporting a container as a wheeled vehicle.

Container-On-Flatcar (COFC)
Rail flatcar used to transport highway transport containers.

Container Ship
Ship specially equipped to transport large freight containers in horizontal or, more commonly, vertical container cells. Containers are usually loaded and unloaded by special cranes.

Container Specification Number
Shipping container number preceded by "DOT" that indicates the container has been built to federal specifications.

Contaminants
Any foreign substance that compromises the purity of a given substance.

Control Agents
Materials used to contain, confine, neutralize, or extinguish a hazardous material or its vapor.

Convulsant
Poison that causes an exposed individual to have convulsions.

Cooperating Agency
Agency supplying assistance other than direct suppression, rescue, support, or service functions to the incident control effort (Red Cross, law enforcement agency, telephone company, etc.).

Corner Fittings
Strong metal devices located at the corners of a container having several apertures that normally provide the means for handling, stacking, and securing the freight container.

Corner Structures
Vertical frame components located at the corners of a container; integral with the corner fittings.

Corrosive Materials
Liquids or solids that can destroy human skin, or liquids that can severely corrode steel.

Crossover Line
Installed in tank piping systems to allow unloading from either side of the tank.

Cryogens
Gases that are cooled to a very low temperature, usually below -130°F (-90°C), to change to a liquid. Also called Refrigerated Liquid.

Curbside
Side of a trailer nearest the curb when trailer is traveling in a normal forward direction (right-hand side); opposite to roadside.

D

Dangerous Cargo Manifest
Invoice of cargo used on ships containing a list of all hazardous materials on board and their location on the ship.

Dangerous Goods
(1) Any product, substance, or organism included by its nature or by the regulation in any of the nine United Nations classifications of hazardous materials. (2) Term used to describe hazardous materials in Canada. (3) Term used in the U.S. and Canada for hazardous materials aboard aircraft.

Dangerous Goods Guide To Initial Emergency Response (IERG)
Canada's equivalent of the U.S. DOT *Emergency Response Guidebook.*

Dedicated Railcars
Car set aside by the product manufacturer to transport a specific product. The name of the product is painted on the car.

Deflagration
Chemical reaction producing vigorous heat and sparks or flame and moving through the material (as black or smokeless powder) at less than the speed of sound. A major difference among explosives is the speed of the reaction. It can also refer to intense burning; a characteristic of Class B explosives.

Dehydration
Process of removing water or other fluids.

Department Of Transportation (DOT)
Administrative body of the executive branch of the U.S. Federal Government responsible for transportation policy, regulation, and enforcement.

Detector Tube
See Colorimetric Tubes.

Detonator
A device or small quantity of explosive used to trigger an explosion in explosives. Also called Blasting Cap.

Diatomaceous Earth
A light siliceous material consisting chiefly of the skeletons (minute unicellular algae) and used especially as an absorbent or filter. Also called Diatomite.

Diatomite
See Diatomaceous Earth.

Dip Tube
Installed for pressure unloading of product out of the top of the tank.

Dissipate
To cause to spread out or spread thin to the point of vanishing.

Diuretic
Product that tends to increase the flow of urine.

DOE
Abbreviation for the U.S. Department of Energy. *See* Appendix A.

Dollies
See Supports.

DOT
Abbreviation for the U.S. Department of Transportation. *See* Appendix A.

Double-Bottom
See Doubles.

Doubles
Truck combination consisting of a truck tractor and two semitrailers coupled together. Formerly called Double-Trailer or Double-Bottom.

Double-Trailer
See Doubles.

Drop Frame
Two-level frame section of a trailer that provides proper coupler height at the forward end and a lower height for the remainder of the length.

Dry Bulk Carrier
Cargo tanks that carry small, granulated solid materials. They generally do not carry hazardous

materials; however, in some cases they may carry fertilizers or plastic products that can burn and release toxic products of combustion.

Dummy Coupler
Fitting used to seal the opening in an air brake hose connection (gladhands) when the connection is not in use; a dust cap.

Dyspnea
Painful or difficult breathing; rapid, shallow respirations.

E

Element
Most simple substance that cannot be separated into more simple parts by ordinary means.

Elevated Temperature Material
Materials that when offered for transportation or transported in bulk packaging are:

- In a liquid phase and at temperatures at or above 212°F (100°C)
- Intentionally heated at or above their liquid phase flash points of 100°F (38°C)
- In a solid phase and at a temperature at or above 464°F (240°C)

Elevated Temperature Materials Carrier
Cargo tank truck or cargo truck carrying large metal pots that transport elevated-temperature materials.

Emergency Response Guidebook (ERG)
Manual provided by the U.S. Department of Transportation that aids emergency response personnel in identifying hazardous materials placards. It also offers guidelines for initial actions to be taken at haz mat incidents.

Emergency Valve
Self-closing tank outlet valve.

Emergency Valve Operator
Device used to open and close emergency valves.

Emergency Valve Remote Control
Secondary means, remote from tank discharge openings, for operation in event of fire or other accident.

Emulsion
An insoluble liquid suspended in another liquid.

Encapsulating
Completely enclosed or surrounded as in a capsule.

Engulf
In the General Emergency Behavior Model (GEBMO), the dispersion of material. An engulfing event is when matter and/or energy disperses and forms a danger zone.

EPA
Abbreviation for the U.S. Environmental Protection Agency. *See* Appendix A.

ERG
Abbreviation for *Emergency Response Guidebook.*

Etiological Agents
Living microorganisms, like germs, that can cause human disease; a biologically hazardous material.

Evaporation
Process of a solid or a liquid turning into gas.

Exothermal
Characterized by or formed with the evolution of heat.

Explosive Limit
See Flammable Limit.

Explosive Range
Range between the upper and lower flammable limits of a substance.

Explosives
Materials capable of burning or bursting suddenly and violently.

Exposure
(1) Structure or separate part of the fireground to which the fire could spread. (2) People, properties, systems, or portions of the environment that are or may be exposed to the harmful effects of a haz mat emergency.

Extremely Hazardous Substance
Chemicals determined by the Environmental Protection Agency (EPA) to be extremely hazardous to a community during an emergency spill or release as a result of their toxicities and physical/chemical properties. There are 402 chemicals listed under this category.

F

Fifth Wheel
Device used to connect a truck tractor or converter dolly to a semitrailer in order to permit articulation between the units. It is generally composed of a lower part consisting of a trunnion, plate, and latching mechanism mounted on the truck tractor (or dolly) and a kingpin assembly mounted on the semitrailer.

Fifth-Wheel Pickup Ramp
Steel plate designed to lift the front end of a semitrailer to facilitate the engagement of the kingpin onto the fifth wheel.

Fill Opening
Opening on top of a tank used for filling the tank; usually incorporated in a manhole cover.

Fire Tube
See Heating Tube.

Fissionable
Capable of splitting the atomic nucleus and releasing large amounts of energy.

Flame Impingement
Points at which flames contact the surface of a container or other structure.

Flammable Gas
Any material (except aerosols) that is a gas at 68°F (20°C) or less and that:

- Is ignitable at 14.7 psi (101.3 kPa) when in a mixture of 13 percent or less by volume with air

OR

- Has a flammable range at 14.7 psi (101.3 kPa) by volume with air at least 12 percent regardless of the lower limit

Flammable Limit
Percentage of a substance in air that will burn once it is ignited. Most substances have an upper (too rich) and lower (too lean) flammable limit. Also called Explosive Limit and Flammable Range.

Flammable Liquids
Any liquid having a flash point below 100°F (37.8°C) and having a vapor pressure not exceeding 40 psi absolute (276 kPa).

Flammable Materials
Substances that ignite easily and burn rapidly.

Flammable Range
See Flammable Limit.

Flammable Solid
Solid materials (other than explosives) that are liable to cause fires through friction or retained heat from manufacturing or processing or that ignite readily and then burn vigorously and persistently, creating a serious transportation hazard.

Flashing
Liquid-tight rail on top of a tank that contains water and spillage and directs it to suitable drains; may be combined with DOT overturn protection.

Flashing Drain
Metal or plastic tube that drains water and spillage from flashing to the ground.

Flash Point
Minimum temperature at which a liquid gives off enough vapors to form an ignitable mixture with air near the surface of the liquid.

Floating Roof Storage Tank
Atmospheric storage tank that stands vertically. It is wider than it is tall. The roof floats on the surface of the liquid to eliminate the vapor space.

Foam
Extinguishing agent formed by mixing a foam concentrate with water and aerating the solution for expansion; for use on Class A and Class B fires. Foam may be protein, synthetic, aqueous film forming, high expansion, or alcohol types.

Foam Blanket
Covering of foam applied over a burning surface to produce a smothering effect; can be used on nonburning surfaces to prevent ignition.

Foam Concentrate
Raw chemical compound solution that is mixed with water and air to produce foam.

Foam Eductors
Type of foam proportioner used for mixing foam concentrate in proper proportions with a stream of water to produce foam solution.

Foam Proportioner
Device that injects the correct amount of foam concentrate into the water stream to make the foam solution.

Foam Solution
Mixture of foam concentrate and water after it leaves the proportioner but before it is discharged from the nozzle and air is added to it.

Fork Pockets
Transverse structural apertures in the base of the container that permit entry of forklift devices.

Frangible
Breakable, fragile, or brittle.

Freight Container
See Container.

Frostbite
Local freezing and tissue damage caused by prolonged exposure to extreme cold.

Full Structural Protective Clothing
Protective clothing including helmets, self-contained breathing apparatus, coats and pants customarily worn by firefighters (turnout or bunker coats and pants), rubber boots, and gloves. It also includes covering for the neck, ears, and other parts of the head not protected by the helmet or breathing apparatus. When working with hazardous materials, bands or tape are added around the legs, arms, and waist.

Full Trailer
Truck trailer constructed so that all of its own weight and that of its load rests upon its own wheels; that is, it does not depend upon a truck tractor to support it. A semitrailer equipped with a dolly is considered a full trailer.

Gamma Radiation
Electromagnetic wave with no electrical charge. This type of radiation is extremely penetrating; very high energy X rays.

Gas
Compressible substance, with no specific volume, that tends to assume the shape of a container. Molecules move about most rapidly in this state.

Gas Chromatogram
Chart from a gas chromatograph tracing the results of analysis of volatile compounds by display in recorded peaks.

Gas Chromatograph
Device to detect and separate small quantities of volatile liquids or gases through instrument analysis.

Gas Chromatography
Characterizing volatilities and chemical properties of compounds that evaporate enough at low temperatures (about 120°F or 50°C) to provide detectable quantities in the air through the use of instrument analysis in a gas chromatograph.

Genetic Effect
Mutations or other changes that are produced by irradiation of the germ plasma; changes produced in future generations.

Gladhands
Fittings for connection of air brake lines between vehicles. Also called Hose Couplings, Hand Shakes, and Polarized Couplings.

Gross Weight
Weight of a vehicle or trailer together with the weight of its entire contents.

Half-life
Time required for half of the atoms of a radioactive substance to become disintegrated.

Halogenated Agents
Chemical compounds (halogenated hydrocarbons) that contain carbon plus one or more elements from the halogen series. Halon 1301 and Halon 1211 are most commonly used as extinguishing agents for Class B and Class C fires. Also called Halogenated Hydrocarbons.

Halogenated Hydrocarbons
See Halogenated Agents.

Hand Shakes
See Gladhands.

Hazard Area
Established area from which bystanders and unneeded rescue workers are prohibited.

Hazard Class
Group of materials designated by the Department of Transportation (DOT) that share a major hazardous property such as radioactivity or flammability.

Hazardous Atmosphere
Any atmosphere that is not conducive to the support of human life; includes atmospheres that contain toxic gases or vapors and atmospheres that are oxygen deficient or heated.

Hazardous Chemical
Defined by the Occupational Safety and Health Administration (OSHA) as any chemical that is a physical hazard or a health hazard to employees.

Hazardous Material
Any material that possesses an unreasonable risk to the health and safety of persons and/or the environment if it is not properly controlled during handling, storage, manufacture, processing, packaging, use, disposal, or transportation.

Hazardous Substance
Any substance designated under the Clean Water Act and the Comprehensive Environmental Response, Compensation and Liability Act (CERCLA) as posing a threat to waterways and the environment when released. (U.S. Environmental Protection Agency.)

Hazardous Wastes
Discarded materials regulated by the Environmental Protection Agency because of public health and safety concerns. Regulatory authority is granted under the Resource Conservation and Recovery Act. (U.S. Environmental Protection Agency.)

Head
Front and rear closure of a tank shell.

Heat Cramps
Heat illness resulting from prolonged exposure to high temperatures; characterized by excessive sweating, muscle cramps in the abdomen and legs, faintness, dizziness, and exhaustion.

Heat Exhaustion
Heat illness caused by exposure to excessive heat; symptoms include weakness, cold and clammy skin, heavy perspiration, rapid and shallow breathing, weak pulse, dizziness, and sometimes unconsciousness.

Heating Tube
Tube installed inside a tank to heat the contents. Also called Fire Tube.

Heat Rash
Condition that develops from continuous exposure to heat and humid air; aggravated by clothing that rubs the skin; reduces the individual's tolerance to heat.

Heat Stress
Combination of environmental and physical work factors that make up the heat load imposed on the body. The environmental factors that contribute to heat stress include air, temperature, radiant heat exchange, air movement, and water vapor pressure. Physical work contributes to heat stress by the metabolic heat in the body. Clothing also has an affect on heat stress.

Heat Stroke
Heat illness caused by heat exposure, resulting in failure of the body's heat regulating mechanism; symptoms include high fever of 105° to 106°F (40.5°C to 41.1°C); dry, red, hot skin; rapid, strong pulse; and deep breaths or convulsions. May result in coma or possibly death. Also called Sunstroke.

Heavy Metal
Generic term referring to lead, cadmium, mercury, and other elements that are toxic in nature. The term may also be applied to compounds containing these elements. Also called Toxic Element.

Hematotoxic Agent
Chemicals that affect the blood.

Hemispherical
Half a sphere in shape.

Hepatotoxic Agent
Chemicals that affect the liver.

Hitch
Connecting device at the rear of a vehicle used to pull a full trailer with provision for easy coupling.

HMR
Abbreviation for Hazardous Materials Regulations; developed and enforced by the U.S. Department of Transportation.

HMTA
Abbreviation for Hazardous Materials Transportation Act.

Hopper
(1) Any of various receptacles used for temporary storage of material. (2) Tank holding a liquid and

having a device for releasing its contents through a pipe. (3) Freight car with a floor sloping to one or more hinged doors for discharging bulk contents.

Horizontal Pressure Vessel
Pressurized tanks characterized by rounded ends. Capacities may range from 500 to 40,000 gallons (2 000 L to 160 000 L). Propane, butane, ethane, and hydrogen chloride are examples of materials stored in these tanks.

Horizontal Storage Tank
Atmospheric storage tank that is laid horizontally and constructed of steel.

Hose Couplings
See Gladhands.

Hose Troughs
See Hose Tube.

Hose Tube
Housing used on tank and bulk commodity trailers for the storage of cargo handling hoses. Also called Hose Troughs.

Hygroscopic
Ability of a substance to absorb moisture from the air.

Hypergolic
Chemical reaction between a fuel and an oxidizer that causes immediate ignition on contact without the presence of air. An example is the contact of fuming nitric acid and unsymmetrical dimethyl hydrazine (UDMH).

Hypergolic Materials
Materials that ignite when they come into contact with each other. The chemical reactions of hypergolic substances vary from slow reactions that may barely be visible to reactions that occur with explosive force.

Hypothermia
Abnormally low body temperature. Also called Systemic Hypothermia.

I

IDLH
See Immediately Dangerous to Life and Health.

IERG
Canadian *Initial Emergency Response Guide. See* Dangerous Goods Guide To Initial Emergency Response.

Ignition Temperature
Minimum temperature to which a fuel in air must be heated in order to start self-sustained combustion independent of the heating source.

Immediately Dangerous To Life And Health (IDLH)
Any atmosphere that poses an immediate hazard to life or produces immediate irreversible, debilitating effects on health. A companion measurement to the PEL, IDLH concentrations represent concentrations above which respiratory protection should be required. IDLH is expressed in ppm or mg/m^3.

IMO Type 5
See Pressure Intermodal Container.

Impingement
Come into sharp contact with.

IM Portable Tank
See Nonpressure Intermodal Tank Container.

Incident Commander (IC)
Person in charge of the incident management system during an emergency.

Individual Container
Product container used to transport materials in small quantities; includes bags, boxes, and drums.

Inert Gas
Gas that does not normally react chemically with a base or filler metal.

Infrared Analyzer
Instrument used to monitor gas and vapor exposures by measuring the infrared energy absorbed by the contaminant.

Ingestion
Taking in food or other substances through the mouth.

Inhalation
Taking in materials by breathing through the nose or mouth.

Intermodal Tank Container
Freight containers designed and constructed to be used interchangeably in two or more modes of transport.

Irritant
Liquid or solid that upon contact with fire or exposure to air gives off dangerous or intensely irritating fumes.

J

Jacket
Metal cover that protects the tank insulation.

Jackknife
Condition of truck tractor/semitrailer combination when their relative positions to each other form an angle of 90 degrees or less about the trailer kingpin.

Junction Box
See Connection Box.

K

Kingpin
Attaching pin on a semitrailer that connects with pivots within the lower coupler of a truck tractor or converter dolly while coupling the two units together.

L

Labels
Four-inch-square diamond markers required on individual shipping containers smaller than 640 cubic feet (18 m^3).

Lading
Freight or cargo that makes up a shipment.

Landing Gears
See Supports.

Leach
To pass out or through by percolation (gradual seepage).

Legs
See Supports.

LEL
See Lower Explosive Limit.

Lethal Concentration (LC_{50})
Concentration of an inhaled substance that results in the death of 50 percent of the test population. (The lower the value the more toxic the substance.) LC_{50} is an inhalation exposure expressed in parts per million (ppm), mg/liter, or mg/m^3.

Lethal Dose, 50 Percent (LD_{50})
Concentration of an ingested or injected substance that results in the death of 50 percent of the test population. (The lower the dose the more toxic the substance.) LD_{50} is an oral or dermal exposure expressed in mg/kg.

Lifter Roof Storage Tank
Atmospheric storage tank designed so that the roof floats on a slight cushion of vapor pressure. The liquid-sealed roof floats up and down with the vapor pressure. When the vapor pressure exceeds a designated limit, the roof lifts to relieve the excess pressure.

Light Box
See Connection Box.

Liquefied Compressed Gas
Gas that under the charging pressure is partially liquid at 70°F (21°C). Also called Liquefied Gas.

Liquefied Gas
See Liquefied Compressed Gas.

Liquefied Petroleum Gas (LPG)
Any of several petroleum products, such as propane or butane, stored under pressure as a liquid.

Liquid Oxygen (LOX)
Oxygen that is stored under pressure as a liquid.

Lower Explosive Limit (LEL)
Lowest percentage of fuel/oxygen mixture required to support combustion. Any mixture with a lower percentage would be considered too lean to burn.

Low-Pressure Storage Tank
Class of fixed-facility storage tanks that are designed to have an operating pressure ranging from 0.5 to 15 psig (4 to 103 kPa).

LOX
Acronym for Liquid Oxygen.

LPG
Abbreviation for Liquefied Petroleum Gas.

Manhole
(1) Hole through which a person may go to gain access to an underground or enclosed structure. (2) Openings usually equipped with removable, lockable covers large enough to admit a person into a tank trailer or dry bulk trailer. Also called Manway.

Manifold
Used to join a number of discharge pipelines to a common outlet.

Manway
See Manhole.

Material
Generic term used by first responders for substance involved in an incident. *Also see* Product.

Material Safety Data Sheet (MSDS)
Manufacturer's document containing information on a hazardous material. Property occupants should have these for all chemicals found on their site.

Mechanical Trauma
Injury, such as an abrasion, puncture, or laceration, resulting from direct contact with a fragment or a whole container.

Metabolism
Conversion of food into energy and waste products.

Miscibility
Two or more liquids' capability to mix together.

Mixture
Substance containing two or more materials not chemically united.

MSDS
Abbreviation for Material Safety Data Sheet.

Mutagen
Materials that cause changes in the genetic system of a cell in ways that can be transmitted during cell division. The effects of a mutagen may be hereditary.

N

NACA
Abbreviation for the National Agricultural Chemical Association.

National Fire Protection Association (NFPA)
Nonprofit educational and technical association devoted to protecting life and property from fire by developing fire protection standards and educating the public.

Nephrotoxic Agent
Chemicals that affect the kidneys.

Neurotoxic Agent
Chemicals that affect the central nervous system.

Neutron
(1) Part of the nucleus of an atom that has a neutral electrical charge. (2) Highly penetrating type of radiation that has no electrical charge.

NFPA
Abbreviation for National Fire Protection Association.

NFPA 704 Labeling System
Identifies hazardous materials in fixed facilities. The placard is divided into sections that identify the degree of hazard according to health, flammability, and reactivity as well as special hazards.

NFPA 704 Placard
Color-coded, symbol-specific placard affixed to a structure to inform of fire hazards, life hazards, special hazards, and reactivity potential.

NIOSH
Acronym for National Institute of Occupational Safety and Health.

Nonbulk Packaging
Package that has the following characteristics:

- Maximum capacity of 119 gallons (450 L) or less as a receptacle for a liquid
- Maximum net mass of 882 pounds (400 kg) or less and a maximum capacity of 119 gallons (450 L) or less as a receptacle for a solid
- Water capacity of 1,000 pounds (454 kg) or less as a receptacle for a gas

Nonflammable Gases
Compressed gases not classified as flammable.

Nonliquefied Gases
Gas, other than a gas in a solution, that under the charging pressure is entirely gaseous at 70°F (21°C).

Nonpressure Intermodal Tank Container
Portable tank that transports liquids or solids at a maximum pressure of 100 psig (700 kPa). Also called IM Portable Tank.

N.O.S.
Abbreviation for Not Otherwise Specified.

Noxious
Unwanted or troublesome; physical harm or destruction.

NRC
Abbreviation for National Response Center.

Objective
Purpose to be achieved by tactical units at an emergency.

Olfactory
Related to the sense of smell.

Organic Peroxide
Organic derivative of the inorganic compound hydrogen peroxide.

ORM-D
Abbreviation for Other Regulated Material; material, such as a consumer commodity, that presents limited hazard during transportation because of its form, quantity, and packaging.

OSHA
Acronym for U.S. Occupational Safety and Health Administration.

Outage
Difference between the full or rated capacity of a tank or tank car as compared to actual content.

Outlet Valve
In a tank piping system, the valve farthest downstream to which the discharge hose is attached.

Overturn Protection
Protection for fittings on top of a tank in case of rollover; may be combined with flashing rail or flashing box.

Oxidizer
Substance that yields oxygen readily and may stimulate the combustion of organic and inorganic matter.

Package Markings
Descriptive name, instructions, cautions, weight, and specification marks required on the outside of haz mat containers.

Packaging
(1) Broad term the Department of Transportation uses to describe shipping containers and their markings, labels, and/or placards. (2) Readying a victim for transport.

PCB
Abbreviation for Polychlorinated Biphenyls.

Permissible Exposure Limit (PEL)
Maximum time-weighted concentration at which 95 percent of exposed, healthy adults suffer no adverse effects over a 40-hour work week; an 8-hour time-weighted average unless otherwise noted. PELs are expressed in either ppm or mg/m^3. They are commonly used by OSHA and are found in the NIOSH *Pocket Guide to Chemical Hazards.*

Physical Properties
Properties that do not involve a change in the chemical identity of the substance. However, they affect the physical behavior of the material inside and outside the container, which involves the change of the state of the material; for example, the Boiling Point, the Specific Gravity, the Vapor Density, and Water Solubility.

Pick-Up Plate
Sloped plate and structure of a trailer, which is located forward of the kingpin and designed to facilitate engagement of the fifth wheel to the kingpin.

Pictogram
Drawing or symbol that indicates information.

Piggyback Transport
Rail flatcar used to transport highway trailers. Also called a Trailer-On-Flatcar (TOFC).

PIN
Acronym for Product Identification Number.

Placard
Diamond-shaped sign that is affixed to each side of a vehicle transporting hazardous materials. The placard indicates the primary class of the material and, in some cases, the exact material being transported; required on containers that are 640 cubic feet (18 m^3) or larger.

Poison
Any material that when taken into the body is injurious to health.

Polarized Couplings
See Gladhands.

Polar Solvent
Flammable liquids that have an attraction for water, much like a positive magnetic pole attracts a negative pole. Examples include alcohols, ketones, and lacquers.

Polymerization
Reactions in which two or more molecules chemically combine to form larger molecules. This reaction can often be violent.

Pressure Intermodal Container
Liquefied gas container designed for working pressures of 100 to 500 psig (700 kPa to 3 500 kPa). Also known as Spec 51 or internationally as IMO Type 5.

Pressure Tank Railcar
Tank railcars that carry flammable and nonflammable liquefied gases, poisons, and other hazardous materials. They are recognizable by the valve enclosure at the top of the car and the lack of bottom unloading piping.

Pressure Vessel
Fixed-facility storage tanks with operating pressures above 15 psig (103 kPa).

Prilled
Converted into spherical pellets.

Primary Label
Label placed on the container of a hazardous material to indicate the primary hazard.

Product
Generic term used in industry to describe a substance that is used or produced in an industrial process. *Also see* Material.

Product Identification Number (PIN)
Assigned by the United Nations and used in the Canadian *Dangerous Goods Initial Emergency Response Guide* and the DOT *Emergency Response Guidebook* to identify specific product names.

Proportioning Valve
Valve used to balance or divide the air supply between the aeration system and the discharge manifold of a foam system.

Props
See Supports.

Pump-Off Line
Pipeline that usually runs from the tank discharge openings to the front of the trailer.

Pyrophoric
Material that ignites spontaneously when exposed to air. Also called Air-Reactive Materials.

Radiated Heat
See Radiation.

Radiation
Transfer of heat energy through light by electromagnetic waves. Also called Radiated Heat.

Radiation, Nuclear
Product of a process known as radioactivity; the emission of alpha, beta, and gamma radiation.

Radioactive Material (RAM)
Material whose atomic nucleus spontaneously decays or disintegrates, emitting radiation.

Radiography
Process of making a picture on a sensitive surface by a form of radiation other than light.

Radiopharmaceutical
A radioactive drug used for diagnostic or therapeutic purposes.

Reaction, Chemical
Any change in the composition of matter that involves a conversion of one substance into another.

Reactive Materials
Substances capable of or tending to react chemically with other substances.

Reactivity
Ability of two or more chemicals to react and release energy and the ease with which this reaction takes place.

Refrigerated Liquid
See Cryogens.

Refrigeration Unit
Cargo space cooling equipment.

Relay Emergency Valve
Combination valve in an air brake system that controls brake application and also provides for

automatic emergency brake application should the trailer become disconnected from the towing vehicle.

Resources
All of the immediate or supportive assistance available, such as personnel, equipment, control agents, agencies, and printed emergency guides, to help control an incident.

Ring Stiffener
Circumferential tank shell stiffener that helps to maintain the tank cross section.

Roadside
Side of the trailer farthest from the curb when trailer is traveling in a normal forward direction (left-hand side); opposite to "curbside."

Rotary Gauge
Gauge for determining the liquid level in a pressurized tank.

Rupture Disk
Safety device that fails at a predetermined pressure and thus protects a pressure vessel from being overpressurized.

S

Safety Can
Flammable liquid container that has been approved by a suitable testing agency.

Safety Chain
Chain connecting two vehicles to prevent separation in the event the primary towing connection breaks.

Safety Relief Valve
Device on cargo tanks with an operating part held in place by a spring. The valve opens at preset pressures to relieve excess pressure and prevent failure of the vessel.

Sandshoe
Flat, steel plate that serves as ground contact on the supports of a trailer; used instead of wheels, particularly where the ground surface is expected to be soft.

Semitrailer
(1) Freight trailer that when attached is supported at its forward end by the fifth wheel device of the truck tractor. (2) Trucking rig made up of a tractor and a semitrailer.

Sensitizer
See Allergen.

Shipping Papers
Shipping order, bill of lading, manifest, waybill, or other shipping document issued by the carrier.

Side Rails
Upper Side Rails: Main longitudinal frame members of a tank used to connect the upper corner fittings. *Lower Side Rails:* Main longitudinal frame members of a tank used to connect the lower corner fittings.

Sliding Fifth Wheel
Fifth-wheel assembly capable of being moved forward or backward on the truck tractor to vary load distribution on the tractor and to adjust the overall length of combination.

Slurry
A watery mixture of insoluble matter (such as mud, lime, or plaster of paris).

Somatic
Pertaining to all tissues other than reproductive cells.

Sorbent
Granular, porous filtering material used in vapor- or gas-removing respirators.

Sorption
Method of removing contaminants; used in vapor- and gas-removing respirators.

SPEC 51
See Pressure Intermodal Container.

Special Protective Clothing
(1) Chemical protective clothing specially designed to protect against a specific hazard or corrosive substance. (2) High-temperature protective clothing including approach, proximity, and fire entry suits.

Specific Gravity
Weight of a substance compared to the weight of an equal volume of water at a given temperature.

Specific Heat
Ratio between the amount of heat required to raise the temperature of a specified quantity of a material and the amount of heat necessary to raise the temperature of an identical amount of water by the same number of degrees.

Splash Guard
Deflecting shield sometimes installed on tank trailers to protect meters, valves, etc.

Splitter Valve
Valve installed to divide the pipeline manifold.

Spontaneous Combustion
See Spontaneous Ignition.

Spontaneous Heating
Heating resulting from chemical or bacterial action in combustible materials that may lead to spontaneous ignition.

Spontaneous Ignition
Combustion of a material initiated by an internal chemical or biological reaction producing enough heat to cause the material to ignite. Also called Spontaneous Combustion.

Stabilization
Stage of an incident when the immediate problem or emergency has been controlled, contained, or extinguished.

Standard Transportation Commodity Code (STCC Number)
Numerical code used by the rail industry on the waybill to identify the commodity.

STCC Number
See Standard Transportation Commodity Code.

Stress
State of tension put on a shipping container by internal or external chemical, mechanical, or thermal change.

Suffocate
To die from being unable to breathe; to be deprived of air or to stop respiration as by strangulation or asphyxiation.

Sump
Low point of a tank at which the emergency valve or outlet valve is attached.

Sunstroke
See Heat Stroke.

Supports
Devices generally adjustable in height that are used to support the front end of a semitrailer in an approximately level position when disconnected from the towing vehicle. Formerly called Landing Gears, Props, Dollies, and Legs.

Switch List
List of cars on a track and instructions as to where those cars go within the yard.

Systemic Hypothermia
See Hypothermia.

T

Tactics
Methods of employing equipment and personnel to accomplish specific tactical objectives in order to achieve established strategic goals.

Tandem
Two-axle suspension.

Tare
The weight of an empty vehicle or container; subtracted from gross weight to ascertain net weight.

Technical Assistance
Personnel, agencies, or printed materials that provide technical information on handling hazardous materials or other special problems.

Teratogen
Chemicals that interfere with the normal growth of an embryo, causing malformations in the developing fetus.

Threshold Limit Value (TLV)
Concentration of a given material that may be tolerated for an 8-hour exposure during a regular workweek without ill effects.

Threshold Limit Value/Ceiling (TLV/C)
Maximum concentration that should not be exceeded, even instantaneously.

Threshold Limit Value/Short-Term Exposure Limit (TLV/STEL)
Fifteen-minute time-weighted average exposure. It should not be exceeded at any time nor repeated more than four times daily, with a 60-minute rest period required between each STEL exposure. These short-term exposures can be tolerated without suffering from irritation, chronic or irreversible tissue damage, or narcosis of a sufficient degree to increase the likelihood of accidental injury, impair self-rescue, or materially reduce worker efficiency. TLV/STELs are expressed in ppm and mg/m^3.

Threshold Limit Value/Time-Weighted Average (TLV/TWA)
Maximum airborne concentration of a material to which an average, healthy person may be exposed repeatedly for 8 hours each day, 40 hours per week without suffering adverse effects. They are based upon current available data and are adjusted on an annual basis.

TOFC
Abbreviation for Trailer-On-Flatcar. Also referred to as Piggyback Transport.

Tow Bar
Beam structure used to maintain the distance between a towed vehicle and the towing vehicle.

Toxic Atmosphere
Any area, inside or outside a structure, where the air contains substances harmful to human life or health when inhaled.

Toxic Element
See Heavy Metal.

Toxic Gas
(1) Product of combustion that is poisonous; a gas given off from toxic materials by exposure to an intense heat environment. (2) Any gas that contains poisons or toxins that are hazardous to life.

Toxicity
Ability of a substance to do harm within the body.

Toxic Material
Substances that can be poisonous if inhaled, swallowed, absorbed, or introduced into the body through cuts or breaks in the skin.

Toxin
Substance that has the property of being poisonous.

Trailer-On-Flatcar (TOFC)
See Piggyback Transport.

Train Consist
See Consist.

Transition
(1) Passage from one state, stage, subject, or place to another. (2) Section of a tank that joins two unequal cross sections.

Truck
Self-propelled vehicle carrying its load on its wheels; primarily designed for transportation of property rather than passengers.

Truck Tractor
Powered motor vehicle designed to pull a truck trailer.

Truck Trailer
Vehicle without motor power; primarily designed for transportation of property rather than passengers. It is drawn by a truck or truck tractor.

Tube Trailer
See Compressed Gas Trailer.

Twist Lock
Mechanically operated device located on the corners of a container chassis and on automatic lifting spreaders; used for restraining a container during transport or transfer.

U

UEL
See Upper Explosive Limit

Unstable Material
Material that is capable of undergoing chemical changes or decomposition with or without a catalyst.

Upper Coupler Assembly
Consists of upper coupler plate, reinforcement framing, and fifth-wheel kingpin mounted on a semitrailer. Formerly called Upper Fifth-Wheel Assembly.

Upper Explosive Limit (UEL)
Maximum concentration of vapor or gas in air that will allow combustion to occur. Concentrations above this are called "too rich" to burn.

Upper Fifth-Wheel Assembly
See Upper Coupler Assembly.

Utilidor
An insulated, heated conduit built below the ground surface or supported above the ground surface to protect the contained water, steam, sewage, and fire lines from freezing.

Vapor Density
Weight of a given volume of pure vapor or gas compared to the weight of an equal volume of dry

air at the same temperature and pressure. A vapor density less than 1 indicates a vapor lighter than air; a vapor density greater than 1 indicates a vapor heavier than air.

Vaporization
Passage from a liquid to a gaseous state. Rate of vaporization depends on the substance involved, heat, and pressure.

Vapor Pressure
Measure of the tendency of a substance to evaporate.

Venturi Principle
When a fluid is forced under pressure through a restricted orifice, there is a decrease in the pressure exerted against the side of the constriction and a corresponding increase in the velocity of the fluid. Because the surrounding air is under greater pressure, it rushes into the area of lower pressure.

Volatile
(1) Changing into vapor quite readily at a fairly low temperature. (2) Tending to erupt into violence; explosive.

W

Water Gel
Chemical solution that is gelled or partially solidified to make it easier to use or handle, for example gelatin dynamite (gelignite).

Water-Reactive Materials
Substances, generally flammable solids, that react in varying degrees when mixed with water or exposed to humid air.

Water Solubility
Ability of a liquid or solid to mix with or dissolve in water.

Waybill
Shipping paper used by a railroad to indicate origin, destination, route, and product. There is a waybill for each car, and it is carried by the conductor.

Introduction

Just as the technology explosion has changed society, so has it changed the jobs of emergency response personnel. No longer is there such a thing as a "routine incident." When exposed to fire conditions, synthetic materials in consumer products and industrial processes create smoke that is exceptionally toxic and dangerous to firefighters who are not properly equipped with protective breathing apparatus. The increased numbers of people infected with diseases, such as hepatitis and AIDS, create contamination concerns for emergency medical service workers. All emergency first responders will come into contact with hazardous materials at some point in their careers.

Today's first responder must be aware of the haz mat problem and must be able to cope with it. Hazardous materials are found in every jurisdiction, community, and workplace. Because of this, plans and standard operating procedures for safely handling haz mat incidents must be developed. First responders must be trained in accordance with these plans and procedures. Each person must clearly understand what his or her role will be when faced with an incident involving haz mat. First responders must know their limitations; they must understand when they cannot proceed any farther in an incident.

Emergency response organizations are obligated to prepare their personnel for dealing with hazardous materials when they are encountered during the course of normal activities. This obligation is not only dictated by governmental regulations (such as OSHA) or by consensus standards (such as the NFPA) but also by the bounds of the moral obligation of an employer to provide a safe working atmosphere for his or her employees.

PURPOSE AND SCOPE

This manual is written for individuals who are mandated by law or called upon by necessity to prepare for and respond to emergency incidents that involve hazardous materials. First responders include firefighters, police officers, emergency medical service workers, industrial emergency response personnel, utility workers, and other members of private industry. This manual provides first responders with the information they need to take appropriate initial defensive actions when hazardous materials are encountered.

This manual covers the requirements of NFPA 472, *Standard for Professional Competence of Responders to Hazardous Materials Incidents*, for awareness and operational levels. Also included are the requirements of 29 CFR 1910.120, *Hazardous Waste Site Operations and Emergency Response* (HAZWOPER), section (q), for first responders at the awareness and operational levels.

This manual covers the extent and nature of today's haz mat problem and the responsibility of the first responder. The manual also addresses the following topics:

- Properties of hazardous materials
- Recognition and identification
- Command and control
- Personal protective equipment
- Hazard analysis and risk assessment
- Container assessment and handling
- Decontamination procedures

Chapter 1

Photo Courtesy of Scott D. Christiansen, Minot, N.D..

HAZ MAT

Introduction To Hazardous Materials

LEARNING OBJECTIVES

This chapter provides information that will assist the reader in meeting the objectives from NFPA 472, *Standard for Professional Competence of Responders to Hazardous Materials Incidents* and 29 CFR 1910.120 that are listed below. The objective numbers are also noted directly in the text in the sections where they are addressed. Objectives in the list below that are denoted with an asterisk (*) are global in nature and are covered by reading the chapter in its entirety.

AWARENESS LEVEL

- Identify the definition of hazardous materials, or dangerous goods (in Canada). [29 CFR 1910.120(q)(6)(i)(A)]
- Define a hazardous substance. [29 CFR 1910.120(q)(6)(i)(A)]
- Define the role of the First Responder Awareness individual. [29 CFR 1910.120(q)(6)(i)(E)]

NFPA 472: 2-2.1.1 Identify the definition of hazardous materials (or dangerous goods, in Canada).

NFPA 472: 2-2.1.4 Identify the difference between hazardous materials incidents and other emergencies.

NFPA 472: 2-2.1.5 Identify typical occupancies and locations in the community where hazardous materials are manufactured, transported, stored, used, or disposed of.

OPERATIONAL LEVEL

- Define basic hazardous materials terms. [29 CFR 1910.120(q)(6)(ii)(C)]
  Hazardous Materials (DOT)
  Hazardous Substance (CERCLA)
  Hazardous Chemicals (OSHA)
  Extremely Hazardous Substance (EPA)
  Dangerous Goods (Canada)
  Hazardous Waste (RCRA)
- Define the role of the First Responder Operational individual. [29 CFR 1910.120(q)(6)(ii)]

Chapter 1

Introduction To Hazardous Materials

Every year in the United States and Canada, billions of tons of chemicals and manufactured substances are produced, stored, used, and transported. Many consumer needs could not be met without using these chemicals and substances. Nonetheless, in addition to their necessary and beneficial uses, many of these materials pose considerable risks to the public and to the environment. Those substances that possess harmful characteristics are considered to be hazardous materials (or dangerous goods in Canada). The harmful characteristics of these substances are of concern to all emergency first responders.

For the purpose of this manual, the term *hazardous materials* (or *haz mat*) is used throughout the text. However, different terms are used by other agencies in various jurisdictions. In some cases, these agencies or jurisdictions have slightly different definitions for the specific terms that they use. The terms and their definitions are listed here for reference.

[NFPA 472: 2-2.1.1]; [29 CFR 1910.120(q)(6)(i)(A)]; [29 CFR 1910.120(q)(6)(ii)(C)]

Hazardous Material — (As defined by the U.S. Department of Transportation [DOT] in 49 CFR 171.8) A substance or material, including a hazardous substance, that has been determined by the Secretary of Transportation to be capable of posing an unreasonable risk to health, safety, and property when transported in commerce, and which has been so designated.

Hazardous Waste — (As defined by the Department of Transportation in 49 CFR 171.8) Any material that is subject to the Hazardous Waste Manifest Requirements of the U.S. Environmental Protection Agency specified in 40 CFR, part 262.

[29 CFR 1910.120(q)(6)(i)(A)]

Hazardous Substance — (As defined by 42 USC 9601, Comprehensive Environmental Response, Compensation and Liability Act [CERCLA], section 101[14]):

- Any substance designated via the Federal Water Pollution Control Act.
- Any element, compound, mixture, solution, or substance designated pursuant to CERCLA.
- Any hazardous waste having the characteristics identified under or listed pursuant to the Solid Waste Disposal Act (SWDA). (**NOTE:** This does not include those SWDA wastes whose regulation has been suspended by an act of Congress.)
- Any toxic pollutant listed under section 307(a) of the Federal Water Pollution Control Act.
- Any hazardous air pollutant listed under section 112 of the Clean Air Act.
- Any imminently hazardous chemical substance or mixture with respect to which the administrator has taken action pursuant to section 7 of the Toxic Substances Control Act.

NOTE: Section 101(14) of CERCLA does not include petroleum or any petroleum by-product, unless it is specifically listed or designated as a hazardous substance.

Hazardous Chemical — (As defined by the Occupational Safety and Health Administration [OSHA] in 29 CFR 1910.1200) Any chemical that is a physical hazard or a health hazard.

Extremely Hazardous Substance — (As defined by the Environmental Protection Agency in 40 CFR 355.20) Any chemical that must be reported to the appropriate authorities if released above the threshold reporting quantity. These hazardous substances are listed and identified in Title III of Superfund Amendments and Reauthorization Act (SARA) of 1986.

Dangerous Good — (As defined by the Canadian Transportation Commission) Any product, substance, or organism included by its nature or by the regulation in any of the classes listed in the schedule. (**NOTE:** The schedule is the nine United Nations Classes of Hazardous Materials. See "United Nations Classifications System" in Chapter 3.) 5/95

[NFPA 472: 2-2.1.5]

Hazardous materials are found in every community and workplace — there are no exceptions. When responding to an incident, first responders will encounter hazardous materials either in transportation or in fixed facilities. There are a number of transportation modes that are used to move hazardous materials (Figures 1.1 a through e). These include the following:

- Roadways
- Railways
- Waterways
- Airways
- Pipelines

Most jurisdictions will have at least two of these transportation modes to deal with.

Obviously, when considering hazardous materials in fixed facilities, first responders tend to draw immediate concern to large manufacturing and storage facilities. These types of occupancies often have large quantities of hazardous materials on site. However, jurisdictions that do not have large manufacturing and storage facilities are not com-

Figure 1.1a Tanker trucks are found in every jurisdiction.

Figure 1.1b Railroads service most jurisdictions.

Figure 1.1c Ships can carry large quantities of hazardous materials.

Figure 1.1d Cargo planes may carry limited quantities of hazardous materials.

Figure 1.1e Liquid and gaseous products are transported through pipelines.

pletely void of hazardous materials in fixed locations. Hazardous materials will be encountered in seemingly ordinary locations such as the following (Figures 1.2 a through e):

- Service stations
- Hardware stores
- Doctors' offices
- School laboratories
- Agricultural stores or co-ops
- Farms
- Residences

Regardless of the level of potential emergency in a given jurisdiction, haz mat first responders must be knowledgeable of, and prepared to take, the actions necessary to protect people, the environment, and property from harm. To accomplish these responsibilities, the first responder must learn the following:

Figure 1.2a Hazardous materials are found in school laboratories.

Figure 1.2b Hardware stores contain many consumer products, such as paints, thinners, and garden chemicals, that fall into the category of hazardous materials.

Figure 1.2c Medical buildings, such as doctors' offices and hospitals, contain hazardous chemicals, etiological hazards, and radiation hazards.

Figure 1.2d Agricultural stores contain large volumes of hazardous chemicals.

Figure 1.2e Various types of poisons and other hazardous chemicals may be found on a typical farm.

- Methods of pre-incident planning, recognition, and incident control
- Types, properties, and characteristics of hazardous materials
- Methods of transporting and storing hazardous materials
- Proper handling of hazardous materials
- Appropriate defensive actions to take in emergencies involving hazardous materials

- Local, state, and federal regulations governing hazardous materials

The Occupational Safety and Health Administration (OSHA), the Environmental Protection Agency (EPA), and the Canadian Workplace Hazardous Materials Information System (WHMIS) require that responders to haz mat incidents meet specific training standards. The OSHA versions of these legislative mandates are outlined in Section (q) of 29 CFR 1910.120, *Hazardous Waste Operations and Emergency Response* (HAZWOPER). The EPA version is in 40 CFR Part 311, which is identical to the OSHA regulation.

In addition to government regulations, the National Fire Protection Association (NFPA) has several standards that set requirements for personnel who respond to haz mat emergencies. These requirements are detailed in the following:

NFPA 471, *Recommended Practice for Responding to Hazardous Materials Incidents*

NFPA 472, *Standard for Professional Competence of Responders to Hazardous Materials Incidents*

NFPA 473, *Standard for Competencies for EMS Personnel Responding to Hazardous Materials Incidents*

[NFPA 472 2: 2.1.4]

A haz mat incident is one that involves a substance that has been released or is on fire. Because of this, the material poses an unreasonable risk to people, the environment, and property. It is almost certain that haz mat incidents will be more complex than the "standard" emergency incident faced by any first responder (Figure 1.3). However, by knowing some basic concepts, the haz mat first responder may be able to reduce injury, loss of life, and environmental/property losses caused by these incidents.

Figure 1.3 Haz mat incidents require coordination among many agencies. *Courtesy of Martin Grube.*

The first responder must determine whether the incident fits the definition of a haz mat incident as soon as possible upon arrival at the emergency scene. For example, first responders are commonly dispatched to motor vehicle accidents. The initial dispatch may indicate that a passenger vehicle and a tractor-trailer truck have collided. First responders arriving at the scene should be able to determine fairly quickly whether or not the truck is carrying hazardous materials. If the truck is carrying a load of lumber, the incident most likely will not be considered a major haz mat incident (although leaking diesel fuel tanks may pose a minor threat). However, if the truck is a tanker that contains anhydrous ammonia, a major haz mat emergency may be declared.

Some haz mat incidents may evolve out of nonemergency situations. An example of this would be a situation that involves police serving a routine search warrant. Upon entering the structure, they discover large quantities of chemicals and equipment, indicating the presence of a clandestine drug lab. This incident has just changed from a simple warrant service to a major haz mat incident. Special equipment and procedures will be required to stabilize and dismantle the evidence at the scene. All first responders must understand how certain conditions will change a seemingly routine situation into a complicated haz mat incident.

As the use, transportation, storage, and disposal of hazardous materials increase, there will be an inevitable increase in the number of haz mat incidents. The fact that these materials are being transported, stored, and disposed of more safely than ever will not eliminate these incidents. The causes of haz mat incidents, such as human error, package failure, and vehicle accidents/derailments, cannot be regulated by law. Government regulations do, however, cover the following issues related to hazardous materials:

- Packaging
- Labeling
- Placarding

- Use
- Training of personnel
- Inspection and operation of fixed facilities
- Transportation vehicles and methods

The haz mat first responder must be prepared to deal with a breakdown in the regulating of these issues.

As in any emergency situation, several operations may have to be performed simultaneously at a haz mat incident. First responders may be involved concurrently with things such as:

- Rescue
- Evacuation or sheltering-in-place
- Exposure protection
- Confinement or containment
- Emergency notification

This manual provides the necessary foundation for continuing education in relation to hazardous materials and their properties. This knowledge is essential for the safety of first responders and the communities they serve.

WHO MUST BE TRAINED?

There are two levels of first responders: first responder awareness and first responder operational. It is important to know and understand the first responder's role at each of these levels.

First Responder Awareness Level

[29 CFR 1910.120(q)(6)(i)(E)]

A first responder at the awareness level is an individual who in the course of his or her normal duties may be the first to discover a haz mat incident. First responders may be employed as the following:

- Firefighter (Figure 1.4)
- Private industry employee
- Law enforcement officer
- Member of an emergency medical service
- Municipal employee
- Transportation employee
- Utility worker
- Member of the military

Figure 1.4 Firefighters respond to all types of haz mat incidents.

Individuals trained to the awareness level are expected to assume the following responsibilities when faced with an incident involving hazardous materials:

- Suspect or recognize the presence of a hazardous material.
- Protect themselves.
- Call for appropriate assistance.
- Secure the area (Figure 1.5).

Figure 1.5 Securing the area around the incident is one of the basic measures that the first responder must take.

First Responder Operational Level

[29 CFR 1910.120(q)(6)(ii)]

A first responder at the operational level is an individual who responds as part of his or her normal duties in a defensive manner to releases, or potential releases, of hazardous materials. This responder is expected to protect individuals, the environment, and property from the effects of the release. Responsibilities of the first responder at the operational level include the four awareness level responsibilities PLUS confining the release in a defensive fashion from a safe distance.

WARNING:

First responders must know the limitations of their training and act only within those parameters. For safety reasons, they must not exceed the responsibilities of their roles.

POTENTIAL FOR HAZARDOUS MATERIALS INCIDENTS

The magnitude of a haz mat incident can vary from the relatively isolated effects of breaking a 1-ounce (28 g) container filled with an etiological agent to the widespread destruction caused by the failure of a liquid propane railroad tank car. In any size of incident, the risks and hazards associated with the hazardous material will be present. All haz mat incidents can be viewed as learning experiences, regardless of their magnitude. Unfortunately, history shows many of these lessons have been costly in terms of lives lost and of environment and property damaged or destroyed. The documentation of every response should be examined, analyzed, and critiqued by all responders. By studying this documentation, the first responder can avoid mistakes and use successful tactics and techniques on the next response.

The potential for a haz mat incident can exist any time during the life of the material. Some materials are hazardous as a raw material while others become dangerous only when mixed or refined (Figure 1.6). Materials may also be found as hazardous materials in storage. Materials that are ready to be used or consumed in any physical state may be stored as a gas or a liquid. For example, chlorine may be stored and used as a gas in industry, but it is used in the liquid state in household bleach. Quantity also plays a role in the potential for an incident as materials may be stored in varying quantities from small packages to large, bulk containers. For example, an incident involving a propane tank attached to a backyard barbecue grill would not have the same magnitude as an overturned tanker of propane on a city street.

Figure 1.6 Some materials become more dangerous after they have been refined.

Because millions of tons of hazardous waste is produced annually in North America, many areas of the country experience problems with hazardous waste disposal. Some waste is treated to remove or destroy its dangerous properties while some is sold to other companies for raw material. Most wastes must be transported to, and stored in, specially classified and permitted hazardous waste treatment, storage, and disposal facilities.

For many years, hazardous waste has been improperly stored or disposed of. Authorities have found sites that contain large quantities of mixed materials, including flammable liquids, heavy metals, and various carcinogens, mutagens, and teratogens. These wastes have been found in places such as lagoons, pits, septic tanks, storage vessels, and drums (Figure 1.7).

Wastes have been spread or dumped indiscriminately into streams and along roads. Over time, haz mat containers begin to leak and release their materials into the surrounding earth, water, and air. Many of these substances are magnified in the food chain; they are not filtered by the earth and

Figure 1.7 Areas where there are large quantities of hazardous wastes pose a serious threat to the environment and to first responders.

do not degrade for years. The long-range effects of these materials are not yet fully understood, but cases of groundwater and human contamination are well documented.

Although dangers are still present, properly regulated, permitted repositories allow the first responder to know, through pre-incident planning, the associated risks involved. To properly dispose of hazardous wastes, personnel must deposit the materials in a special container specified by law and must secure the container in a facility permitted as a hazardous waste site. The cost of properly (and legally) disposing of these wastes is high, but improper disposal can be exceedingly expensive to the involved community in terms of cleanup and human distress.

INCIDENT ASSESSMENT

First responders are expected to take reasonable actions to protect lives, the environment, and property. It is unreasonable for rescuers to risk their lives or health when only the environment or property is threatened. When human lives are at risk, however, the first responder must be prepared to weigh the situation according to the best facts known and then decide whether the benefits outweigh the risks. Adopting a policy of cautious assessment before taking action is vital. The first responder should assess the following variables:

- Risk to rescuers
- Probability of victim survival
- Difficulty of rescue
- Capabilities and resources of on-scene forces
- Possibility of explosions or sudden material releases
- Available escape routes and safe havens
- Constraints of time and distance

More detailed information on making these decisions is contained throughout the remainder of this manual.

WHO HANDLES HAZARDOUS MATERIALS INCIDENTS?

Many incidents, in order to be controlled, require the coordinated effort of several agencies. These agencies include the following (Figure 1.8):

- Fire service
- Law enforcement
- Emergency medical service
- Private concerns such as the material's manufacturer and shipper
- Government agencies with mandated interests such as health and environmental resources departments

Figure 1.8 Many different agencies may respond to haz mat incidents.

- Various technical support groups such as CHEMTREC/CANUTEC
- Specialized emergency response groups

To avoid jurisdictional and command disputes, the specific agency responsible for handling and coordinating response activities should be identified before an incident happens. The responsible or "lead" agency can then begin documenting the identities and capabilities of nearby support sources. Two goals can be achieved when agencies formulate their haz mat pre-incident plans together:

- Vital resource information can be easily shared.
- Rapport between the participating agencies can be developed.

The occurrence of a serious haz mat incident is not the occasion to find out that a neighboring department or industry cannot provide desperately needed equipment, personnel, or technical expertise. Proper planning and preparation should lead to a safe and successful response to a haz mat incident.

Chapter 2

Photo Courtesy of Scott D. Christiansen, Minot, N.D.

HAZ MAT

Properties Of Hazardous Materials

LEARNING OBJECTIVES

This chapter provides information that will assist the reader in meeting the objectives from NFPA 472, *Standard for Professional Competence of Responders to Hazardous Materials Incidents* that are listed below. The objective numbers are also noted directly in the text in the sections where they are addressed. Objectives in the list below that are denoted with an asterisk (*) are global in nature and are covered by reading the chapter in its entirety.

AWARENESS LEVEL

NFPA 472: 2-2.3.1 Identify the ways hazardous materials are harmful to people, the environment, and property at hazardous materials incidents.

NFPA 472: 2-2.3.2 Identify the general routes of entry for human exposure to hazardous materials.

OPERATIONAL LEVEL

NFPA 472: 3-2.3.1.1 Match the following chemical and physical properties with their significance and impact on the behavior of the container and/or its contents:

(a) Corrosivity (pH);
(b) Flammable (explosive) range;
(c) Flash point;
(d) Form (solid, liquid, gas);
(e) Ignition (autoignition) temperature;
(f) Reactivity;
(g) Specific gravity;
(h) Toxic products of combustion;
(i) Vapor density; and
(j) Water solubility

NFPA 472: 3-2.3.7 Identify the health and physical hazards that could cause harm.

NFPA 472: 3-2.3.7.1 Identify the health hazards associated with the following terms:

(a) Asphyxiant;
(b) Irritant/corrosive;
(c) Sensitizer/allergen;
(d) Convulsant; and
(e) Chronic health hazard

NFPA 472: 3-3.3.2.1 Identify skin contact hazards encountered at hazardous materials incidents.

NFPA 472: 3-4.3.4 Identify the symptoms of heat and cold stress.

Chapter 2

Properties Of Hazardous Materials

Hazardous materials may be elements, compounds, or mixtures found in gaseous, liquid, or solid states or in combinations of these states. The hazards they present may range from insignificant to catastrophic, depending on the material involved. The flammability of hazardous materials ranges from negligible (for materials that will not burn) to extremely flammable. Materials range in reactivity from those that are usually stable to those that detonate. The radioactive and biochemical effects of hazardous materials may be short lived or may last for generations.

Very few hazardous materials are without some detrimental effects on the human body. An exposure may be *acute* (single occurrence) or *chronic* (long-term, reoccurring) and may have health effects that are immediate or delayed. Many hazardous materials have a major impact on human health. Therefore, working with these materials requires that personnel use special personal protective equipment. This equipment offers protection that is far superior to that provided by the clothing worn for standard fire fighting.

This chapter looks at many of the various properties that make materials hazardous to humans. Included is information on how people may be exposed to various substances and how those exposed to hazardous materials are affected.

HEALTH AND SAFETY

[NFPA 472: 2-2.3.1]; [NFPA 472: 3-2.3.7]

Obviously, the health and safety of first responders and civilians are the primary considerations when dealing with hazardous materials. The following sections will discuss the many threats and harmful effects associated with hazardous materials (Table 2.1):

- Thermal
- Mechanical

TABLE 2.1
Effects Of Hazardous Materials Exposures On Humans

Exposure	Related Energy	Effects
Thermal	Temperature extremes	Burn, frostbite
Mechanical	Direct contact/fragments	Blisters, bruises, lacerations
Poisonous	Poisons or toxins	Damage to internal organs or body systems
Corrosive	Chemical	Burns or tissue damage
Asphyxiation	Oxygen deficiency	Affect the respiratory system
Radiation	Radiation	Injury to individual or future generations
Etiological	Living microorganism	Diseases (e.g., hepatitis, tuberculosis)

- Poisonous
- Corrosive
- Asphyxiation
- Radiation
- Etiological

Thermal Effects

[NFPA 472: 3-4.3.4]

Thermal effects are related to temperature extremes. Temperature extremes affect the way an individual works and the amount of work an individual is capable of performing. Excessive heat exposure can cause heat stress, heat cramps, heat exhaustion, heat rashes, and heatstroke. Excessive cold exposure from cryogenic substances (such as refrigerated liquids) and liquefied gases presents the potential for serious tissue damage.

HEAT EXPOSURE

Wearing personal protective equipment is intended to protect the individual from a hostile environment. Most types of personal protective equipment inhibit the body's ability to disperse heat, which is magnified because the first responder is usually performing strenuous work while wearing the equipment (Figure 2.1). With these facts established, it is important for the first responder to understand the impact of heat stress (Table 2.2). The body functions efficiently in a narrow temperature range. The core temperature of an individual is the deep temperature of the body — not the skin or extremities temperature. The body can sustain a small fluctuation of 2 degrees below and 3 degrees above the normal core temperature of 99.7°F or 38°C (98.6°F or 37°C oral). If this range is exceeded, a health threat exists.

Figure 2.1 Haz mat protective equipment places considerable physical stress on the person. *Courtesy of Joe Marino.*

Heat Rash

A heat rash may occur on the skin as a result of continuous exposure to heat and humid air. This condition is aggravated by clothing rubbing the skin. Heat rashes reduce the individual's tolerance to heat. Heat rashes can also become annoying.

Heat Cramps

Heat cramps can occur after heavy exertion and exposure to high temperatures. Cramps develop in the body's extremities as a result of dehydration and excessive loss of salt. Symptoms of heat cramps include the following:

TABLE 2.2
Condition And Symptoms Of Heat Exposure

Condition	Symptoms
Heat Stroke	Lack of perspiration Shallow breathing Rapid pulse Headache Weakness Body temperature of 105°F (41°C) or higher Hot, dry, red skin Confusion Convulsion Loss of consciousness
Heat Cramps	Muscle cramps Heavy perspiration Physical weakness Moist skin
Heat Exhaustion	Mildly elevated temperature Weak pulse Dizziness Profuse sweating Cool, moist, pale skin
Heat Rash	Intolerance to heat Annoyance

- Muscle cramps
- Heavy perspiration
- Physical weakness
- Moist skin

Heat Exhaustion

Heat exhaustion can result from prolonged physical work in a hot environment (Figure 2.2). In these cases, the body is not capable of releasing excessive heat, causing a mild form of traumatic shock. Symptoms of heat exhaustion include the following:

- Mildly elevated temperature
- Weak pulse
- Dizziness
- Profuse sweating
- Cool, moist, pale skin

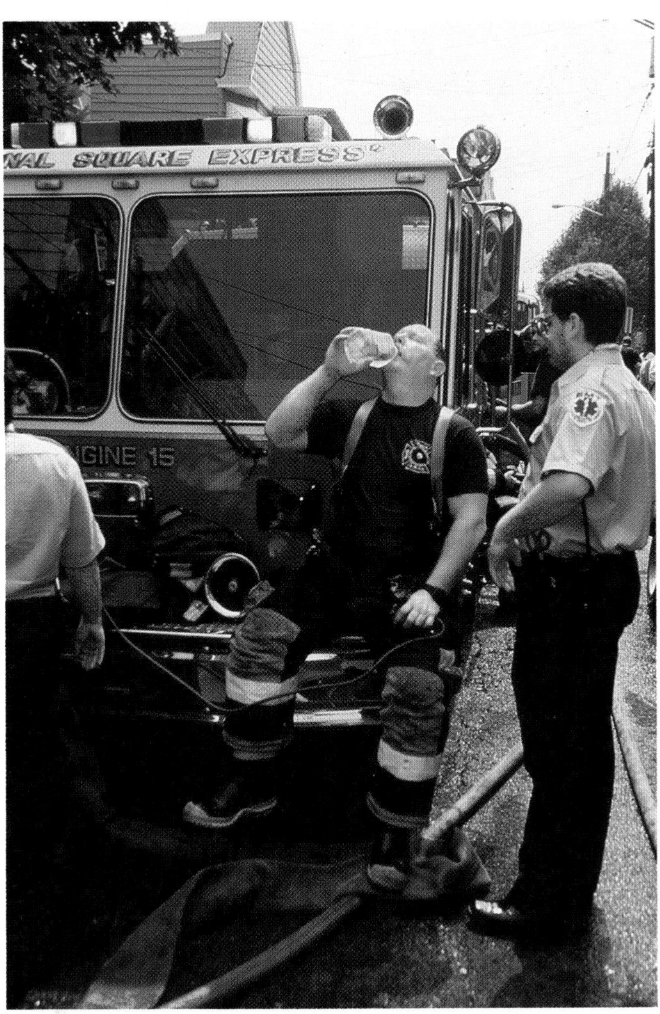

Figure 2.2 First responders who work in protective clothing are susceptible to heat exhaustion. *Courtesy of Ron Jeffers.*

Heatstroke

If preventive measures are not taken when the body passes the heat exhaustion stage, heatstroke occurs. This is a serious medical emergency, and if action is not taken, heatstroke can become a life-threatening situation. Symptoms of heatstroke include the following:

- Lack of perspiration
- Shallow breathing
- Rapid pulse
- Headache
- Weakness
- Body temperature of 105°F (41°C) or higher
- Hot, dry, red skin
- Confusion
- Convulsion
- Loss of consciousness

EFFECTS OF HEAT EXPOSURE

Responders wearing protective clothing need to be monitored for effects of heat exposure. The following list contains numerous ways to prevent and/or reduce the effects of heat exposure.

- Plenty of fluids should be available. Water or commercial body-fluid-replenishment drink mixes may be used to prevent dehydration. Even if the first responder is not thirsty, he or she should be encouraged to drink generous amounts of fluids both before and during operations. Carbonated beverages should be avoided. Drinking 7 ounces (200 ml) of fluid every 15 to 20 minutes is better than drinking large quantities once an hour. Balanced diets will normally provide enough salts to avoid cramping problems.

  CAUTION: Do not allow individuals to drink or eat within a contaminated atmosphere.

- First responders should be given long cotton undergarments or similar types of clothing to provide natural body ventilation.
- Mobile showers and misting facilities should be provided. They reduce the body temperature and cool protective clothing.

- Rest areas, such as shaded and air-conditioned areas, should be provided (Figure 2.3).
- Responders exposed to extreme temperatures or those performing difficult tasks should be rotated frequently.
- Liquids such as alcohol, coffee, and caffeinated drinks should be avoided — or their intake should be minimized — before working. These beverages can contribute to dehydration and heat stress.
- Responders should be encouraged to maintain their physical fitness.

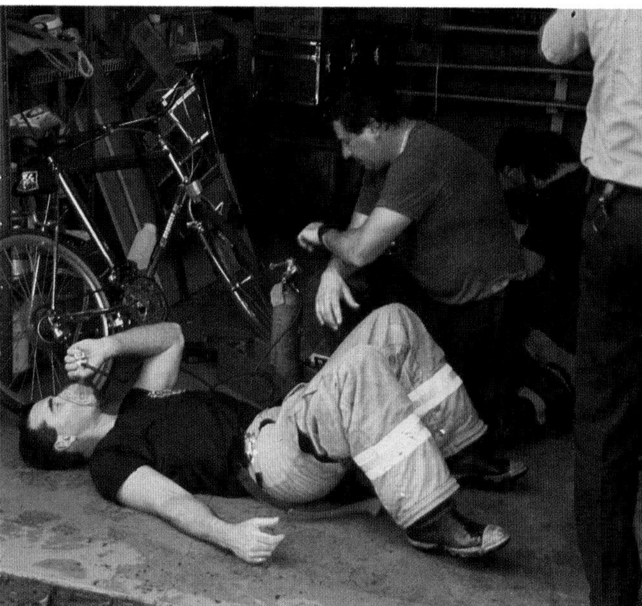

Figure 2.3 First responders should rest in a shaded or in an air-conditioned area. *Courtesy of Ron Jeffers.*

COLD EXPOSURE

Cold exposure is a concern when dealing with cryogenic and liquefied gases. A *liquefied gas*, such as propane or carbon dioxide, is one that at the charging pressure is partially liquid at 70°F (21°C).

A *cryogen* is a gas that turns into a liquid at or below -130°F (-90°C). Cryogens are sometimes called refrigerated liquids. These substances are commonly stored and transported in their liquid state. Examples of cryogenic materials include liquid oxygen (LOX), nitrogen, helium, hydrogen, and liquid natural gas (LNG). At these extremely cold temperatures, cryogens have the ability to instantly freeze materials, including human tissue, on contact. Some cryogens have hazardous properties in addition to the cold hazard. An example of this would be fluorine, which is also a corrosive, an oxidizer, and a poison.

Cryogenic and liquefied gases vaporize rapidly when released from their containers. A liquid spill or leak will boil into a much larger vapor cloud. These vapor clouds can be extremely dangerous if the vapors are flammable. Both types of liquids cause freeze burns, which are treated as cold injuries according to their severity (Table 2.3).

TABLE 2.3
Condition And Symptoms Of Cold Exposure

Condition	Symptoms
Frost Nip/Incipient Frostbite	Whitening or blanching of skin
Superficial Frostbite	Waxy or white skin Firm touch to outer layers of skin Underneath tissue is resilient (flexible)
Deep Frostbite	Cold skin Pale skin Solid, hard skin
Systemic Hypothermia	Shivering Sleepiness, apathy, listlessness Core Temperature of 95°F (35°C) or less Slow pulse Slow breathing Glassy eyes Unconsciousness Freezing of extremities Death

Any clothing saturated with the cryogenic material must be removed immediately. This is particularly important if the vapors are flammable or if they are oxidizers. The first responder could not escape flames from clothing-trapped vapors if the vapors were to ignite.

Mechanical Effects

Mechanical trauma refers to damage that occurs as a result of direct contact with an object. The two most common types are striking and friction exposures. The trauma can be mild, moderate, or severe and can occur in a single event. In haz mat situations, a striking injury would most likely be as a result of the failure of a pressurized container (Figure 2.4). Catastrophic failure of the container can result in bruises, punctures, or even avulsions (part of the body being torn off) when the person is struck by the container or pieces of the container. Striking injuries from sharp objects can result in lacerations and punctures.

Friction injuries are less common in haz mat operations. These injuries occur as a result of portions of the body rubbing against an abrasive or otherwise irritating surface, causing raw skin, blisters, and brush burns. In haz mat situations, friction injuries are most commonly caused by contact between protective clothing and the skin.

Poisonous Effects

[NFPA 472: 3-2.3.7.1] #15 Exam

Exposures to poisons cause damage to organs or other parts of the body and may even cause death. There are numerous types of poisons. For example, some poisons, such as halogenated hydrocarbons (nephrotoxic agents), affect the kidneys. Hematotoxic agents, such as benzene, nitrites, naphthalene, and arsine, affect the blood. #17 Exam The organophosphate pesticides, such as parathion, are neurotoxic agents that affect the central nervous system. Hepatotoxic agents, such as ammonia, carbon tetrachloride, and phenols, are examples of materials that affect the liver. Other poisons are known as carcinogens, mutagens, and teratogens, which are covered later in this chapter.

The method by which poisons attack the body varies depending on the type of poison. Irritants and asphyxiants interfere with the oxygen flow to the lungs and the blood. Nerve poisons act on the body's central nervous system by disrupting the brain control center by blocking the nerve impulses that control the circulatory and respiratory systems. Poisons, and the measurements of their toxicity, are discussed in more detail later in this chapter.

Corrosive Effects

[NFPA 472: 3-2.3.7.1 (c)]; [NFPA 472 :3-2.3.1.1 (a)]

Chemical exposures that destroy or burn living tissues and have destructive effects on other materials are called *corrosives*. Corrosives in contact with combustibles can result in a fire or an explosion. Corrosives are divided into two groups: acids and bases.

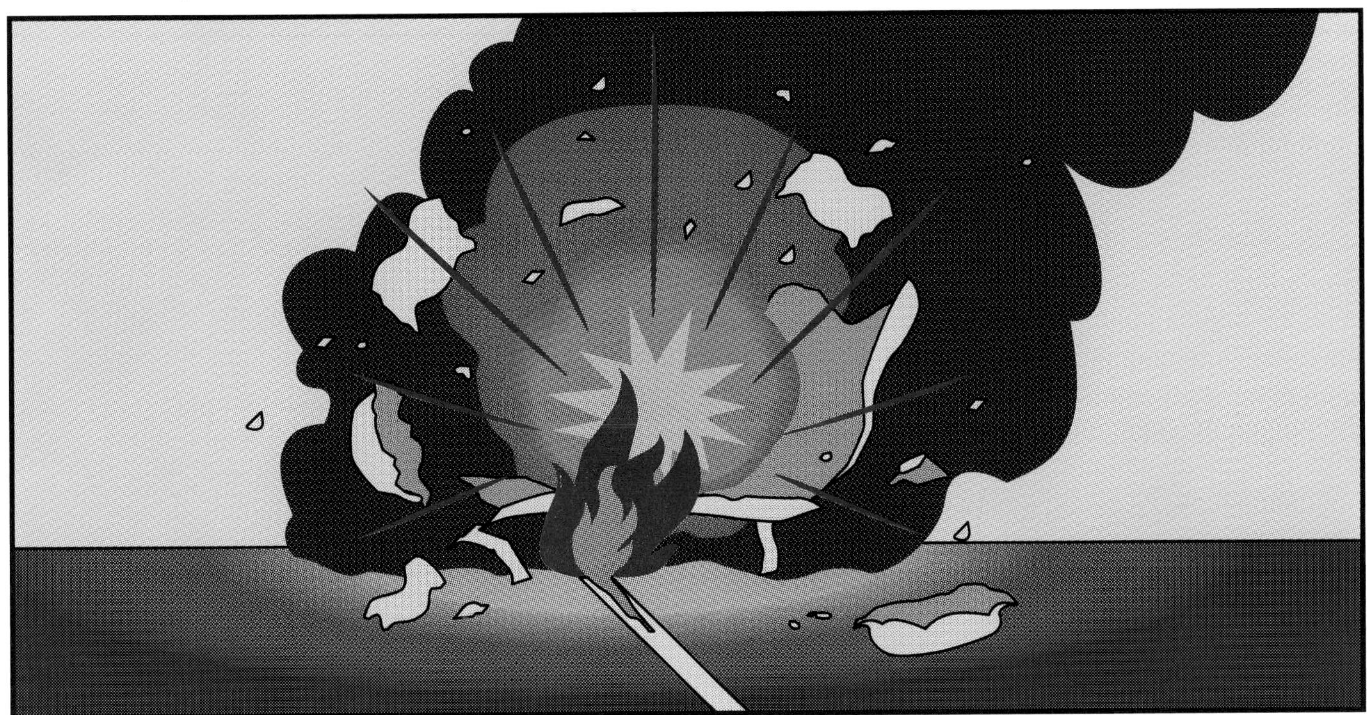

Figure 2.4 Pressurized containers may fail in a dramatic fashion.

Hydrochloric acid, nitric acid, and sulfuric acid are all common acids. An acid may cause severe chemical burns to flesh and permanent eye damage. Typically there will be pain associated with an acid on contact.

Bases break down fatty skin tissues and can penetrate deeply into the body. Caustic soda, potassium hydroxide, and other alkaline materials are bases. Contact with a base does not normally cause immediate pain. A sign of exposure to a base is a greasy or slick feeling of the skin, which is caused by the breakdown of the fatty tissues.

General symptoms of external corrosive exposures include the following:

- Burning around the eyes, nose, and mouth
- Nausea and vomiting
- Difficulty breathing, swallowing, or coughing
- Localized burning or skin irritation

Any hint of exposure should alert the first responder to withdraw to safety and assess the cause. Any sudden deterioration, melting, or discoloration of equipment should cause the first responder to have serious safety concerns.

Asphyxiation Effects

[NFPA 472: 3-2.3.7.1 (a)]

Asphyxiants affect the oxygenation of the body and generally lead to suffocation. Asphyxiants can be divided into two classes: simple and chemical.

Simple asphyxiants are generally inert gases that displace the oxygen necessary for breathing. These inert gases dilute the oxygen concentration below the level required by the human body. These gases may also displace the oxygen normally present — just as a gas that is heavier than air will fill a basement. Examples of simple asphyxiants are acetylene, carbon dioxide, helium, hydrogen, nitrogen, methane, and ethane (Figure 2.5).

Chemical asphyxiants, also called blood poisons, are substances that prohibit the body from using oxygen. There are three ways that this can occur:

- Compounds such as carbon monoxide (CO) react more readily with the blood than does oxygen. In this reaction, hemoglobin bonds to the CO instead of the oxygen, forming carbon dioxide (CO_2). The CO_2 is then transported to the cells, which die of oxygen starvation.
- Compounds such as hydrazine liberate hemoglobin from the red blood cells, leaving a lack of transport for the oxygen.
- Compounds such as benzene and toluene cause a malfunction in the oxygen-carrying ability of red blood cells.

In all three ways, the cells of the body are starved for oxygen, even though oxygen is available. Other examples of chemical asphyxiants are hydrogen cyanide, aniline, acetonitrile, and hydrogen sulfide.

Figure 2.5 Acetylene is an asphyxiant.

Radiation Effects

Radiation events can cause somatic effects (injury to individuals) and genetic effects (changes to future generations). Radiation can have internal and external effects on the human body. External radiation effects come from radioactive sources outside the body. Internal radiation effects occur when radioactive materials enter the body through respiration, ingestion, or skin penetration. The severity of the injury will depend on the type of radiation, the dose rate, the body part exposed, and the total dose received.

Radiation effects include radiation sickness, radiation injury, or radiation poisoning. Radiation sickness is caused by exposure to large amounts of radiation. The initial symptoms of radiation sickness include nausea, vomiting, and malaise. Radiation injuries generally occur from high amounts of the less-penetrating types of radiation like beta particles. These injuries are most commonly in the form of burns. Radiation injury is usually confined to the hands, because such large amounts of exposure generally occur during improper handling of radiation sources. Radiation poisoning is caused by dangerous amounts of internal radiation. The internal radiation can cause medical problems such as anemia or cancer. Internal exposure from alpha particles is the most common cause of radiation poisoning.

Radiation sickness, injury, and poisoning are NOT contagious or infectious. Treating or helping a victim who has been exposed to radiation will not expose emergency response personnel to radiation. First responders may be exposed to the same source as the victim, but the victim will not cause harm because the victim cannot become radioactive. However, if the victim is covered with a radioactive material, such as dust, first responders may become contaminated if they come into contact with the material. Radiation sickness may be a result of this contact.

The potential for radiation exposure exists when first responders respond to some facilities such as medical centers, industrial operations, power plants, and research facilities. Alpha, beta, x-ray, gamma, and neutron are the five basic types of radiation. The following sections examine in detail each type of radiation and cover the basics of how these agents affect the body.

ALPHA RADIATION

Alpha radiation consists of particles having large mass and a positive electrical charge. This radiation form is least dangerous as a threat to external portions of the body but is very dangerous when ingested, inhaled, or otherwise allowed to enter the body. The alpha particles may initially attach themselves to bone structure. However, in time they disintegrate and settle into organs such as the liver, kidneys, spleen, and lungs. Alpha radiation is easily guarded against. A sheet of paper can effectively block an external alpha particle. Personal protective equipment including self-contained breathing apparatus will effectively prevent injury by alpha radiation.

BETA RADIATION

Beta radiation particles are very small. They are about 1/7000 the size of alpha particles but have more penetrating power. The beta particle has a negative electrical charge. Large quantities can seriously damage skin tissue. Beta particles entering the body through damaged skin, ingestion, or inhalation cause organ damage similar to that caused by alpha particles. Shielding from beta particles can be achieved by using heavy plastic, wood, or thin metal; the particles will not penetrate any of these. Personal protective equipment including self-contained breathing apparatus is required to protect against beta particle hazards (Figure 2.6).

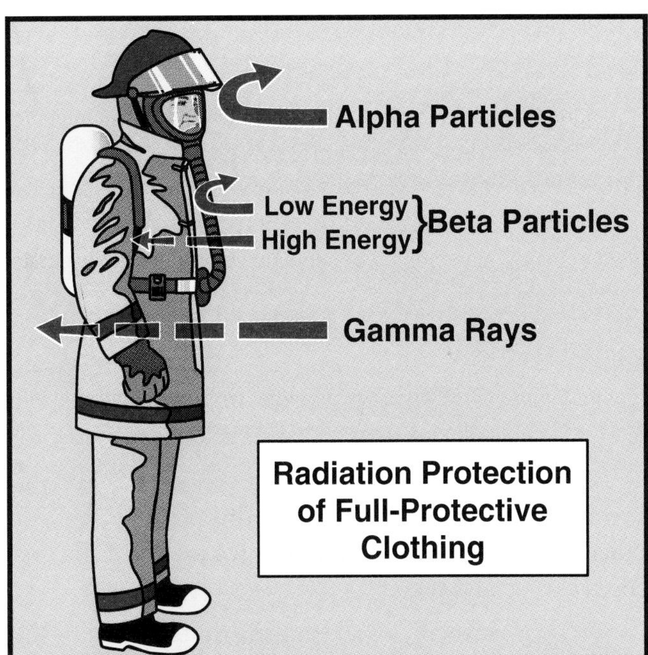

Figure 2.6 Standard firefighter protective equipment shields only against alpha radiation and low energy beta radiation.

X-RAY AND GAMMA RADIATION

X-ray and gamma radiation are electromagnetic forms of radiation bearing no particular electrical charge. The two forms are very similar, differing mainly in their origin. X-ray radiation rays arise from a complete atom, while gamma rays arise solely from the nucleus of the atom. The rays from both move at the speed of light (186,000 miles/second [299,300 km/second]). When controlled, these types of radiation have useful purposes such as medical applications. However, when uncontrolled, these two are the most dangerous forms of penetrating radiation. They may cause somatic (injury) and genetic effects (changes passed on to future generations) to those who are exposed to them. Controlled gamma radiation is used for x-rays and many other purposes. Materials that are very dense, such as lead, effectively shield against gamma and x-ray radiation. Standard fire fighting clothing provides no protection against the penetrating power of gamma rays.

NEUTRONS

This type of radiation has a physical mass like alpha radiation but has no electrical charge. Neutrons are highly penetrating. Uncontrolled radioactive reactions produce neutrons along with gamma radiation. Neutron radiation is difficult to measure in the field and is usually estimated based on gamma measurements. Neutron radiation is not commonly used in commercial or industrial operations. It is most likely to be encountered in research laboratories. The health hazard that neutrons present arises from the fact that they cause the release of secondary radiation.

RADIATION PROTECTION STRATEGIES

Time, distance, and shielding are three ways to provide protection from external radiation during an emergency:

- Time — The shorter the exposure time, the smaller the total radiation dose.
- Distance — The farther the distance from the source, the smaller the dose. As the distance doubles, the amount of radiation decreases by the square of the distance. In other words, doubling the distance from the radioactive source gives one-fourth the exposure.
- Shielding — Certain materials, such as lead, earth, concrete, and water, will prevent penetration of some of the radioactive particles. The thickness of the shielding should depend on the type of material, the type of radiation, and the distance from the source (Figure 2.7). Wearing personal protective equipment, including self-contained breathing apparatus, will generally protect emergency response personnel from internal radiation caused by ingestion, inhalation, or skin absorption. It will not protect against the effects of the more powerful penetrating forms of radiation.

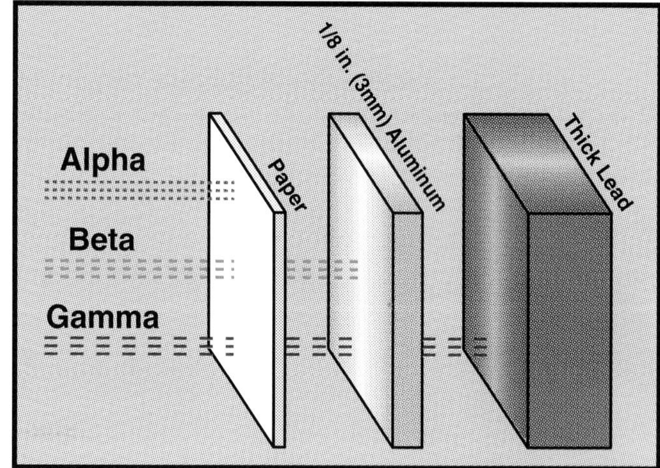

Figure 2.7 Distance and shielding are good protection against radiation.

Etiological Effects

An *etiological effect* is the exposure to a microorganism, or its toxin, that may result in a severe, disabling disease. Examples of diseases associated with etiological events are hepatitis, AIDS, tuberculosis, and typhoid. These hazards are present in biological and medical laboratories or when dealing with people who are carriers of such diseases. Most of these diseases are carried in the body fluids and are transmitted by contact with the fluids. In most cases, simple protective garments provide an effective barrier against these fluids (Figure 2.8).

Other Hazards

There are other hazards associated with haz mat incidents. In addition to those previously described, there are irritants, sensitizers/allergens, convulsants, and chronic health hazards (carcinogens, mutagens, teratogens).

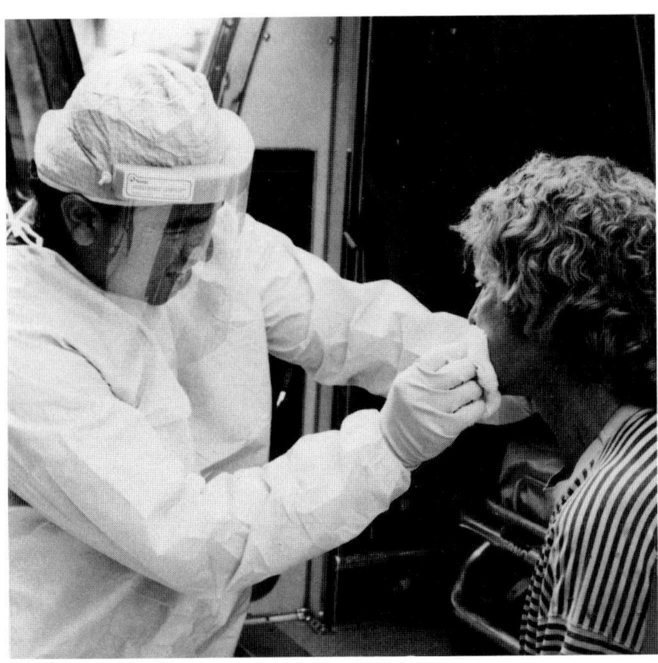

Figure 2.8 Proper protective equipment provides effective protection against the transmission of communicable diseases. *Courtesy of Phoenix (AZ) Fire Department.*

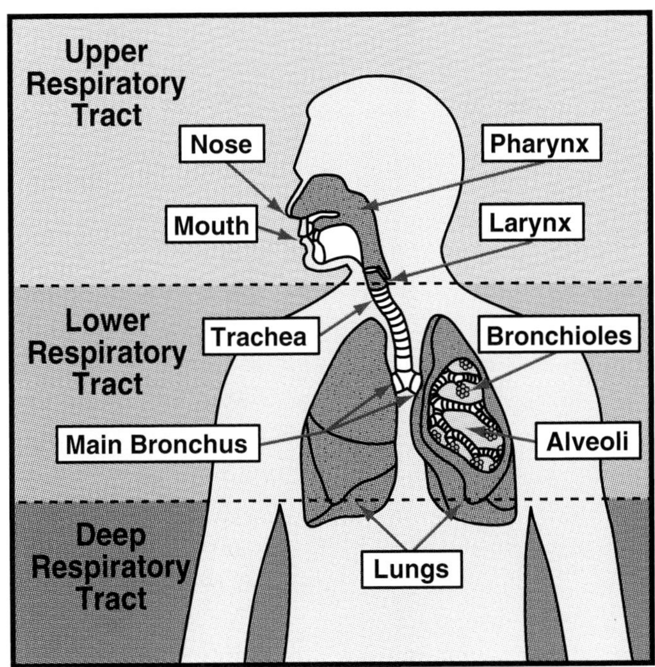

Figure 2.9 The respiratory tract is easily affected by inhaling hazardous chemicals.

IRRITANTS

[NFPA 472: 3-2.3.7.1 (b)]

Irritants primarily affect the respiratory system. They are toxins that cause temporary but sometimes severe inflammation to the eyes, skin, or respiratory tract. An irritant gives off vapors that attack the mucous membranes of the body such as the surfaces of the eyes, nose, mouth, throat, and lungs. There are three types of respiratory irritants: upper, lower, and deep (terminal) respiratory tract (Figure 2.9).

SENSITIZERS/ALLERGENS

[NFPA 472: 3-2.3.7.1 (c)]

Sensitizers/allergens will cause an allergic reaction after repeated exposure. Some individuals exposed to a material may not be abnormally affected at first but may experience significant and dangerous effects in the presence of the material if ever exposed again. An example is an individual's severe reaction to a subsequent bee sting.

CONVULSANTS

[NFPA 472: 3-2.3.7.1 (d)]

Convulsants are poisons that will cause an exposed individual to have seizures. A sense of suffocation, dyspnea, and muscular rigidity develop. The spasms that occur can be very painful to the individual. Muscle spasms may begin soon after the individual is exposed and may occur at varying intervals from 3 to 30 minutes. Death can result from asphyxiation or exhaustion. Some materials that are considered convulsants are strychnine, organophosphate, carbamates, and infrequently used drugs such as picrotoxin.

CHRONIC HEALTH HAZARDS

[NFPA 472: 3-2.3.7.1 (e)]

Chronic health hazards, such as carcinogens, mutagens, and teratogens, are permanent and irreversible conditions.

Carcinogens

Carcinogens are cancer-causing agents. Exact data is not available on the amounts of exposure needed for individual chemicals to cause cancer. However, exposure to only small amounts of some substances for a short time can cause long-term consequences. Examples of known or suspected carcinogenic hazardous materials are polyvinyl chloride, asbestos, some chlorinated hydrocarbons, arsenic, nickel, some pesticides, and many plastics. Some carcinogens may be contained within the smoke from fires to which first responders are dispatched (Figure 2.10).

Figure 2.10 In addition to dangerous levels of heat, smoke contains many hazardous particles and gases.

The ultimate problem with carcinogenic materials and other poisons is the unknown long-term effects of exposure. Disease and complications can occur as long as 20 years after the exposure. Therefore, it is extremely important that appropriate personal protective clothing including self-contained breathing apparatus be worn at all fires and haz mat emergencies.

Mutagens

Mutagens are materials that cause changes in the genetic system of a cell in ways that can be transmitted during cell division. Exposed individuals may transmit undesirable mutations to a later generation. In simple terms, the individual who was exposed to the mutagen (chemical) may not be hurt, but his or her offspring can be. Radiation is one mutagen where the exposure and the effect have been correlated. Other mutagens include benzene and ethylene oxide.

Teratogens

Teratogens cause congenital malformation. In simpler terms, a teratogen will interfere with the normal growth of an embryo, causing malformations in the developing fetus. If these chemicals are administered to or absorbed by pregnant women, they produce a deformed offspring. Some materials classified as teratogens are ionizing radiation, ethyl alcohol, methyl mercury, thalidomide, and dioxins. Some infections, such as rubella, are also classified as teratogens. The major difference between a teratogen and a mutagen is that a mutagen affects the genetic system and a teratogen does not. Therefore, a teratogen is not hereditary, whereas a mutagen's effects may be hereditary.

ROUTES OF ENTRY

[NFPA 472: 2-2.3.2]

A hazardous material may enter the body through inhalation, ingestion, injection, or absorption through the skin or eyes (Figure 2.11). The following sections highlight each of these methods.

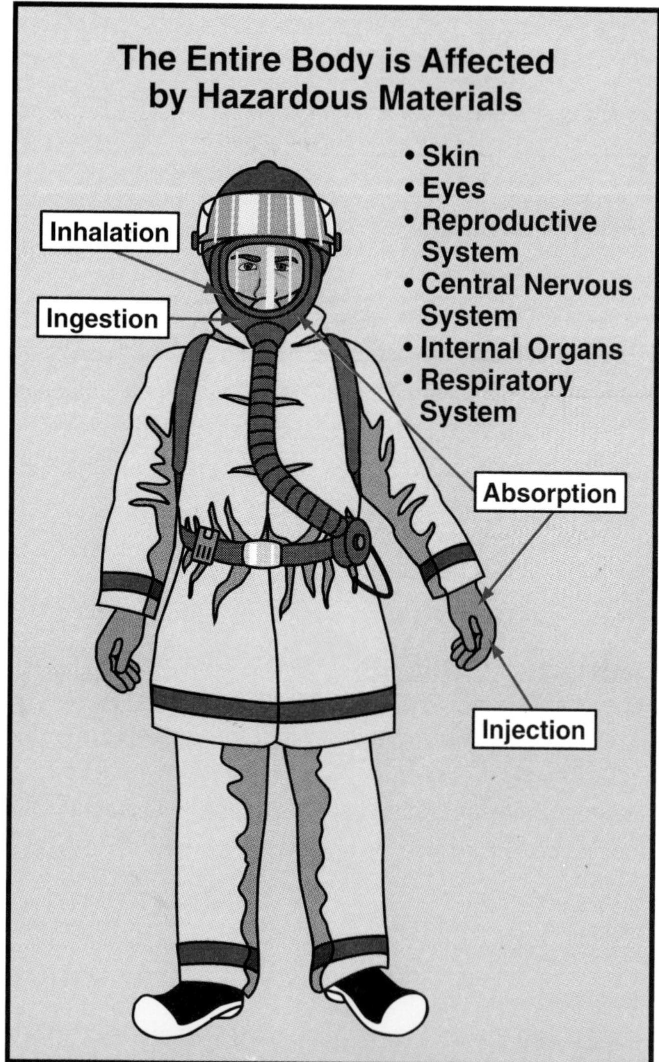

Figure 2.11 All avenues through which hazardous materials enter the body must be protected.

Inhalation

Inhalation is the process of taking in materials by breathing through the nose or mouth. Hazardous vapors, smoke, gases, liquid aerosols, fumes, and suspended dusts may be inhaled if self-contained breathing apparatus is not worn.

Ingestion

Ingestion is the process of taking in materials through the mouth by means other than simple inhalation. Poor hygiene after handling a hazardous material can lead to ingestion. Eating, drinking, and smoking contaminated products allow the hazardous materials to enter the body. Ingestion can also occur when particles of insoluble materials become trapped in the mucous membranes and are swallowed after being cleared from the respiratory tract. Tobacco, food, and drinks should be prohibited in the haz mat area. Use extreme caution in obtaining and dispensing drinking water. Water should be drawn from a known clean source and dispensed in disposable cups to reduce the risk of internal contamination (Figure 2.12). Make sure that all first responders are completely decontaminated and have clean hands before eating.

Figure 2.12 Use disposable cups when dispensing fluids at the emergency scene. *Courtesy of Phoenix (AZ) Fire Department.*

Injection

Injection is the process of taking in materials through a puncture or stick with a needle. The hazardous material can be injected into the bloodstream, skin, muscle, or any place a needle can be inserted. This hazard is especially realistic for emergency medical personnel who handle syringes and IV needles (Figure 2.13).

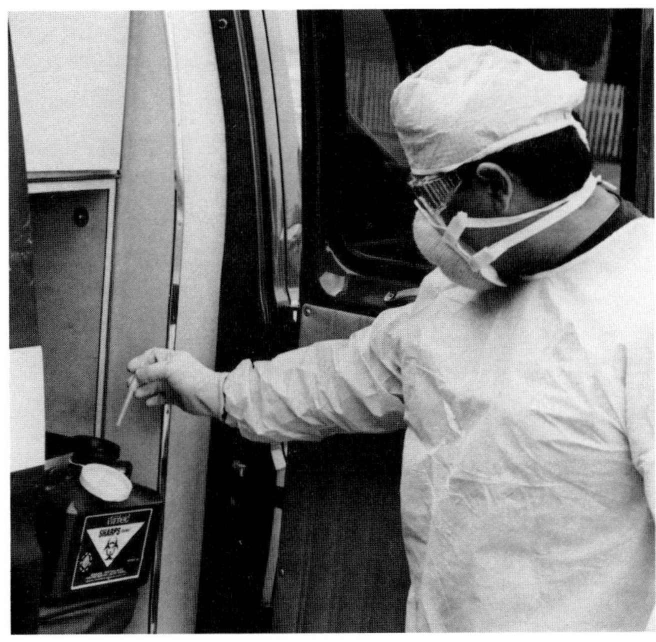

Figure 2.13 Place medical needles in appropriate disposal containers to prevent accidental needle "sticks." *Courtesy of Phoenix (AZ) Fire Department.*

Absorption

[NFPA 472: 3-3.3.2.1]

Absorption is the process of taking in materials through the skin or eyes. Some materials pass easily through areas of the body where the skin is the thinnest, allowing the least resistance to penetration. The eyes, wrists, neck, hands, groin, underarms, and breaks in the skin are areas of concern. Many poisons are easily absorbed into the body system in this manner. Others can enter the system easily through the unknowing act of touching a contaminated finger to one's eye (Figure 2.14).

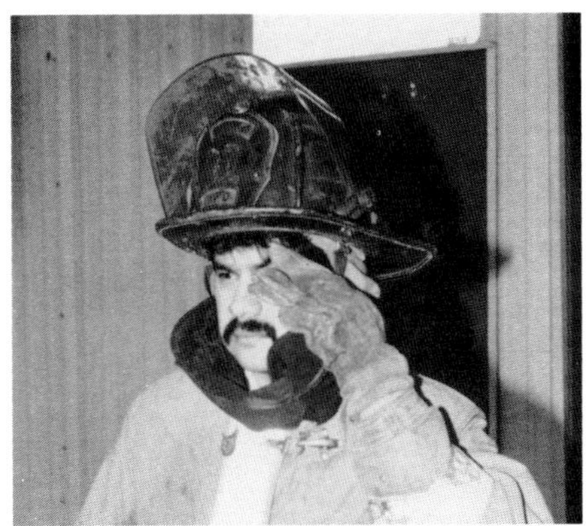

Figure 2.14 A first responder should avoid touching his or her face with dirty hands.

EFFECTS OF EXPOSURES

#15 Exam

Once they are in the system, hazardous materials have varying effects on the body. Many materials, such as asbestos, mercury, silica, and heavy metals, attack the body internally but have no external effect. Other hazardous materials, such as chlorine, sulfuric acid, anhydrous ammonia, and isopropyl alcohol, affect the body both internally and externally. A material may severely damage the exterior skin while at the same time attack internally as a poison. Many haz mat incidents will have materials in concentrations sufficient to be toxic.

Hazardous materials produce a wide range of physical symptoms. Symptoms may not be immediately apparent and can be masked by common illnesses like the flu or by smoke inhalation. Some general symptoms of exposure to hazardous materials include the following:

- Confusion, light-headedness, anxiety, and dizziness
- Blurred or double vision
- Changes in skin color or blushing
- Coughing or painful respiration
- Tingling or numbness of extremities
- Loss of coordination
- Nausea, vomiting, abdominal cramping, and diarrhea
- Changes in behavior or mannerisms
- Unconsciousness

At the first signs or symptoms of exposure, first responders and their partners should withdraw to the predetermined safe area. Upon withdrawal, they should report this condition immediately.

STATES OF MATTER

[NFPA 472: 3-2.3.1.1 (d)]

Everything in the world is made of substances called matter. Matter is found in three states: gas, liquid, and solid. Water, in its various forms, is used to explain the states of matter:

- Gases are fluids that have neither independent shape nor volume. Gases tend to expand indefinitely. Steam is the gaseous form of water (Figure 2.15).
- Liquids are fluids that have no independent shape but do have a specific volume. Liquids flow in accordance with the laws of gravity. Water, as it flows from a tap, is a liquid.
- Solids are substances that have both a specific shape (without a container) and volume. Ice is the solid form of water.

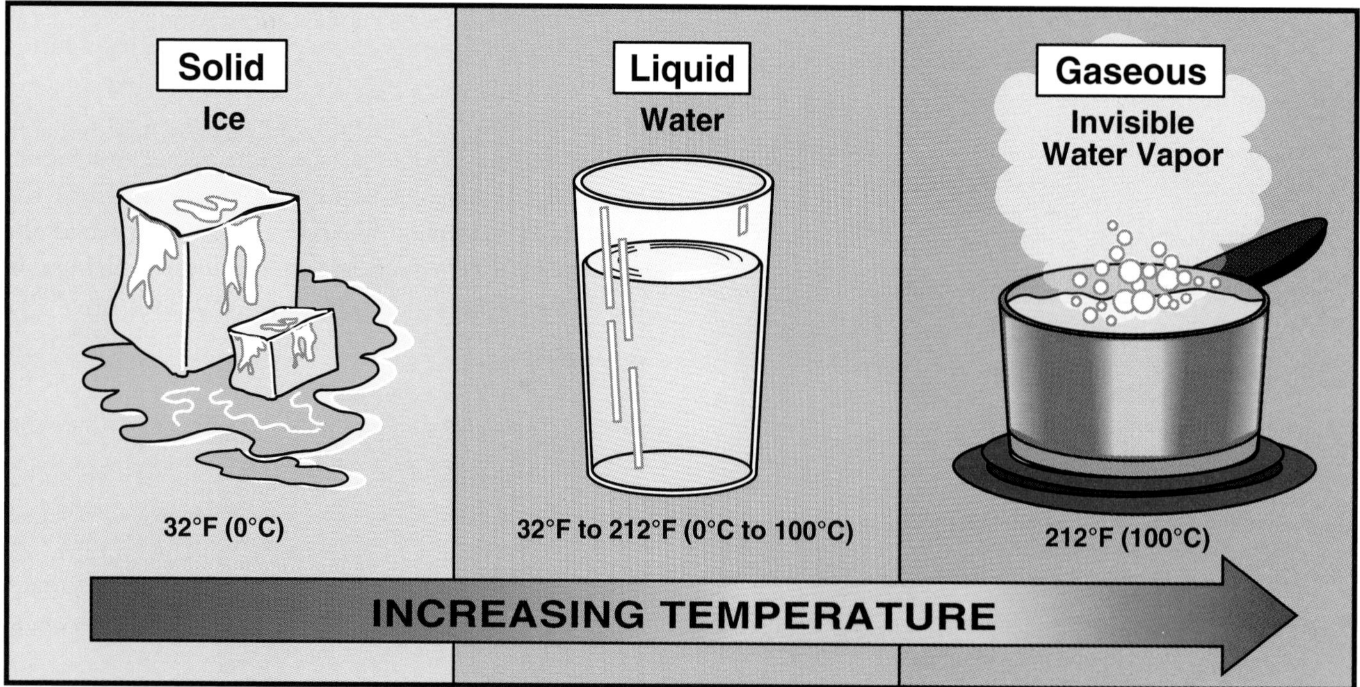

Figure 2.15 Water is a good example of the states of matter. Below 32°F (0°C), it is in a solid state (ice). Above 212°F (100°C), it is in a gaseous state (steam).

In general, gases pose more of a potential danger to first responders than do liquids. Gases are more difficult to contain, and they ignite more readily. Solid substances typically pose the least amount of danger.

CHARACTERISTICS OF HAZARDOUS MATERIALS

An uncontrolled release of hazardous materials from a container can create a variety of problems. The first responder needs to know the symptoms and effects of an exposure to hazardous materials. How are harmful exposures limited to a safe level? How do hazardous materials react when they contact other chemicals, humans, and the environment?

The principal dangers of hazardous materials are the health risks they present to humans and their flammability and reactivity characteristics. However, many hazardous materials have more than one dangerous characteristic. The most prudent action to take when dealing with an unknown chemical or a chemical with more than one hazard is to assume the worst hazard and provide the largest safety margin possible. These materials can produce extremely pronounced effects on the unprotected or inadequately protected human body.

CAUTION: Appropriate personal protective equipment, including self-contained breathing apparatus, must be used by all personnel working at haz mat incidents (Figure 2.16).

Figure 2.16 First responders must wear appropriate protective equipment when working at an incident.

The following sections explore some of the more important concepts and terms associated with health, flammability, and reactivity hazards.

Factors Related To Health Hazards

Many hazardous materials exhibit characteristic signs when spilled or released. For example, nitric acid is only one of many chemicals that produce a colored vapor cloud. Other materials have distinctive and irritating or pungent odors. Use extreme caution when approaching and working at a haz mat incident. Many vapor clouds that have a particular color in heavy concentrations may still be deadly when diluted enough to be colorless. Some substances, such as the cyanides, cannot be detected by the senses because they are essentially odorless and colorless. Further, some chemicals, such as narcotics and anesthetics, deaden the olfactory nerves, thus affecting the first responder's ability to sense them. The first responder must NEVER rely solely on sight or smell and must absolutely NEVER rely on taste or touch to detect the presence of leaking or spilled hazardous materials.

The damage that a hazardous material will inflict on a human being depends on the material's concentration and the dose received. Various types of safety limits have been developed to provide guidance for exposure to various substances. The following sections highlight these limits.

THRESHOLD LIMIT VALUES (TLV)

Threshold limit values (TLV) are established by the American Conference of Governmental Industrial Hygienists (ACGIH). They are published in *Threshold Limit Value and Biological Exposure Indices.* Always use the most current edition of the *TLV Indices* because they are constantly being revised. Several different types of TLVs are time-weighted averages, short-term exposure limits, and ceiling levels.

Threshold Limit Value/Time-Weighted Average (TLV-TWA)

The *TLV-TWA* of a material is the maximum airborne concentration to which an average, healthy person may be exposed repeatedly for 8 hours each day, 40 hours per week without suffering adverse effects. These TLVs are based upon current available data and are adjusted on an annual basis.

TLV-TWAs are expressed in parts per million (ppm) and milligrams per cubic meter (mg/m^3). The lower the TLV the more toxic the substance. At a haz mat incident, emergency response personnel should never assume that the concentration is below the TLV-TWA. TLV-TWAs are designed primarily to be used for exposure in the workplace and are not applicable for use for exposure in haz mat emergencies. However, they are useful during pre-incident planning.

Threshold Limit Value/Short-Term Exposure Limit (TLV-STEL)

The *TLV-STEL* value is the 15-minute, time-weighted average exposure that should not be exceeded at any time or repeated more than four times a day. A 60-minute rest period is required between each TLV-STEL exposure. This short-term exposure can be tolerated without suffering from irritation or chronic or irreversible tissue damage. Also, during this short time, none of the following conditions should occur: narcosis of a sufficient degree to increase the likelihood of accidental injury, impairment of self-rescue, or reduction of worker efficiency. TLV-STELs are also expressed in ppm and mg/m^3. Under no circumstances is an employee allowed to work in an area with the TLV-STEL value longer than 15 minutes without proper personal protective equipment.

Threshold Limit Value/Ceiling Level (TLV-C)

This value is a maximum concentration that should never be exceeded. When working in conditions where the hazardous material concentration is equal to or greater than the TLV-C, appropriate personal protective equipment including respiratory protection must always be used. TLV-C is reported in ppm or mg/m^3.

PERMISSIBLE EXPOSURE LIMIT (PEL)

Permissible exposure limits are very similar to TLV-TWAs. The difference is that a TLV-TWA is determined by the American Conference of Governmental Industrial Hygienists (ACGIH), and PELs are adopted by the Occupational Safety and Health Administration (OSHA) upon recommendation by the ACGIH or the National Institute of Occupational Safety and Health (NIOSH). *PELs* are the maximum allowable amount of exposure for an 8-hour day. PELs are expressed in terms of ppm and mg/m^3.

LETHAL DOSE (LD_{50})

The *lethal dose* of a substance is the minimum amount of solid or liquid that when ingested, absorbed, or injected through the skin will be fatal to 50 percent of all subjects exposed to that dosage. LD_{50} is an oral or dermal exposure expressed in mg/kg. The lower the number the more toxic the material. Note that this does not mean that the other half of the subjects will necessarily be all right. They may be very sick or almost dead, but only half will actually die.

LETHAL CONCENTRATION (LC_{50})

The *lethal concentration* of a substance is the minimum concentration of an inhaled substance in the gaseous state that will be fatal to 50 percent of the test group. These values are generally established by testing the effects of exposure on animals under laboratory conditions. Lethal concentration is given in LC_{50} values. As the value decreases, it becomes more toxic. LC_{50} is expressed in ppm, mg/m^3, and milligrams per liter (mg/L). First responders exposed to hazardous materials should know that exertion, stress, and their individual metabolism or chemical sensitivities (allergies) may make them more vulnerable to the harmful effects of hazardous materials. The 50 percent of the population not killed may suffer effects ranging from no response to severe injury.

IMMEDIATELY DANGEROUS TO LIFE AND HEALTH (IDLH)

Immediately Dangerous to Life and Health (IDLH) is an atmospheric concentration of any toxic, corrosive, or asphyxiating substance that poses an immediate threat to life. It can cause irreversible or delayed adverse health effects and can interfere with the individual's ability to escape from a dangerous atmosphere. IDLHs are expressed in ppm or mg/m^3. At IDLH levels, respiratory protection is required.

Factors Related To Flammability Hazards

In addition to the health hazards previously listed, many materials are flammable. The obvious hazard of flammable materials is the damage caused to life and property when they burn. The flammability of a material depends on such properties as its flash point, autoignition temperature, and flammable (explosive) range. Other important considerations when dealing with a flammable liquid include specific gravity, vapor density, boiling point, and water solubility. It is vital to have this information for fireground tactics.

FLASH POINT 

[NFPA 472: 3-2.3.1.1 (c)]

The *flash point* is the minimum temperature at which a liquid fuel gives off sufficient vapors to form an ignitable mixture with air near its surface. At this temperature, the vapors will flash (in the presence of an ignition source) but will not continue to burn (Figure 2.17). It is important to know that flammable liquids themselves do not burn but that the vapors they produce can ignite. As the temperature of the liquid increases, more vapors are emitted. Vapors are given off below the flash point but not in sufficient quantities to ignite. Flammable liquids, such as gasoline and acetone, have flash points well below 100°F (38°C). Combustible liquids, such as fuel and lubricating oils, have flash points above 100°F (38°C). Flammable gases have no flash points because they are already in the gaseous state. With the exception of a few solids, such as naphthalene with a flash point of about 174°F (79°C), solids are not considered to have flash points.

The flash point should not be confused with the fire point. The *fire point* is the temperature at which enough vapors are given off to support continuous burning. This temperature is usually only slightly higher than the flash point.

AUTOIGNITION TEMPERATURE 

[NFPA 472: 3-2.3.1.1 (e)]

The *autoignition temperature* of a substance is the minimum temperature to which the fuel in air must be heated to initiate self-sustained combustion without initiation from an independent ignition source. All flammable materials have autoignition temperatures. The autoignition temperature is considerably higher than the flash and fire points. The temperature at which gasoline can be ignited by an independent ignition source is -45°F (-43°C). The autoignition temperature of gasoline is about 536°F (280°C). This is the temperature at which gasoline will ignite independent of a heat source.

FLAMMABLE (EXPLOSIVE) RANGE 

[NFPA 472: 3-2.3.1.1 (b)]

The *flammable (explosive) range* is the percentage of the gas or vapor concentration in air that will burn if ignited. Below the flammable range or the lower explosive limit (LEL), a gas or vapor concentration is too lean to burn (too little fuel and too

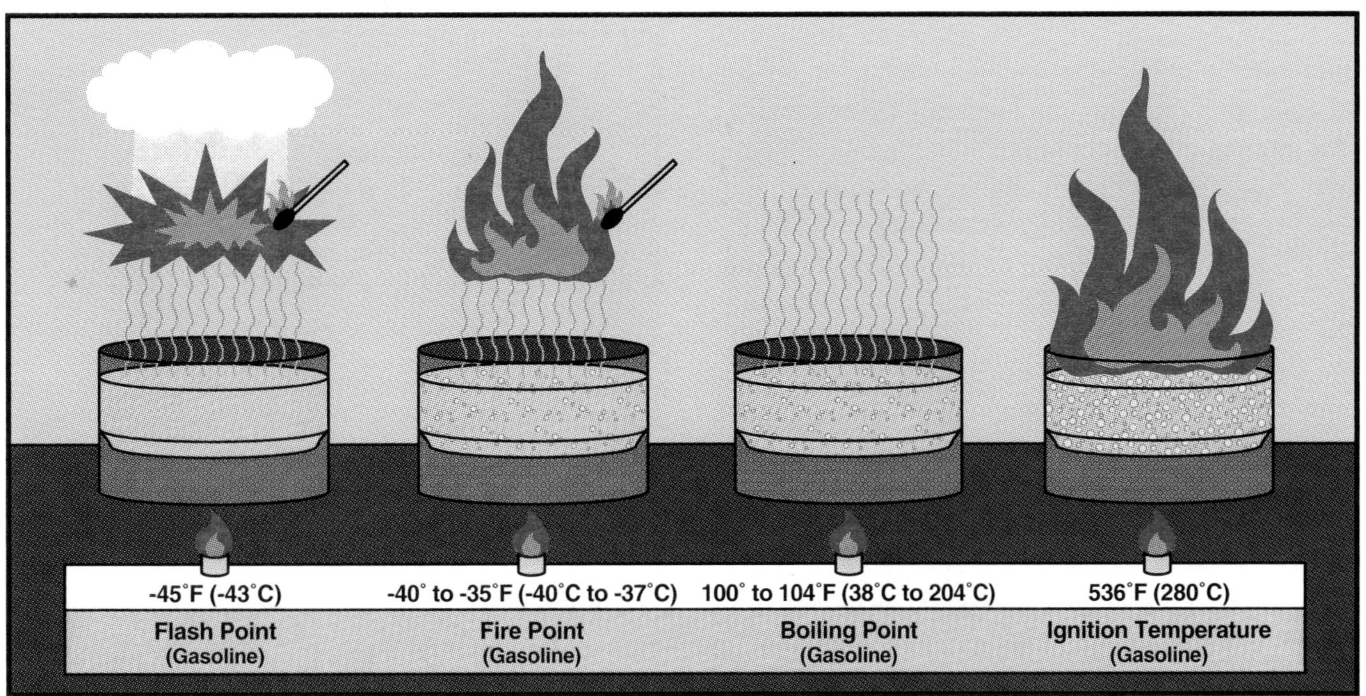

Figure 2.17 Flammable liquids must vaporize to burn. Vaporization is initiated at the flash point.

much oxygen) (Figure 2.18). Above the flammable range or upper explosive limit (UEL), the gas or vapor concentration is too rich to burn (too much fuel and too little oxygen). Within the upper and lower limits, the gas or vapor concentration will burn rapidly if ignited.

CAUTION: Any concentration above 10 percent of the LEL must be considered as a serious ignition potential and require special considerations. Ventilating concentrations above the UEL will result in an explosive mixture.

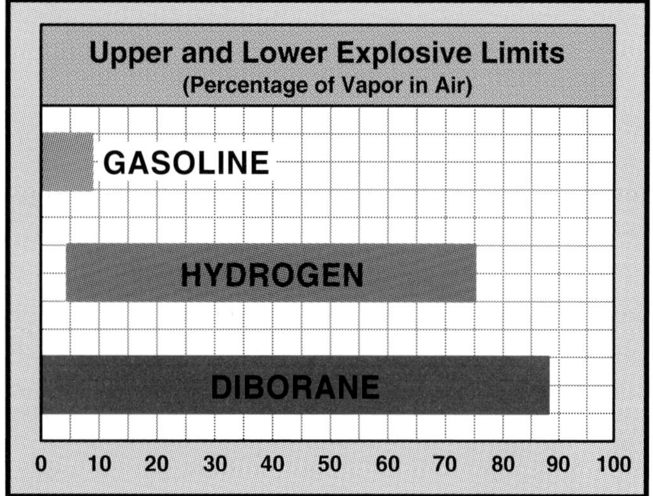

Figure 2.18 The flammable range defines the percentage of vapor in air that burns.

SPECIFIC GRAVITY

[NFPA 472: 3-2.3.1.1 (g)]

Specific gravity is the weight of a substance compared to the weight of an equal volume of water at a given temperature. Water is assigned the specific gravity of 1.0. If a substance has a specific gravity lower than 1.0, it will float on water. If the specific gravity is greater than 1.0, the substance will sink in water (Figure 2.19). Most flammable liquids have specific gravities less than 1.0 and float on the surface of water.

VAPOR DENSITY

[NFPA 472: 3-2.3.1.1 (i)]

Vapor density is similar to specific gravity except that it compares the density of gas to the density of air. Air is given a vapor density of 1.0. A vapor density less than 1.0 indicates that the substance is lighter than air. If the vapor density is

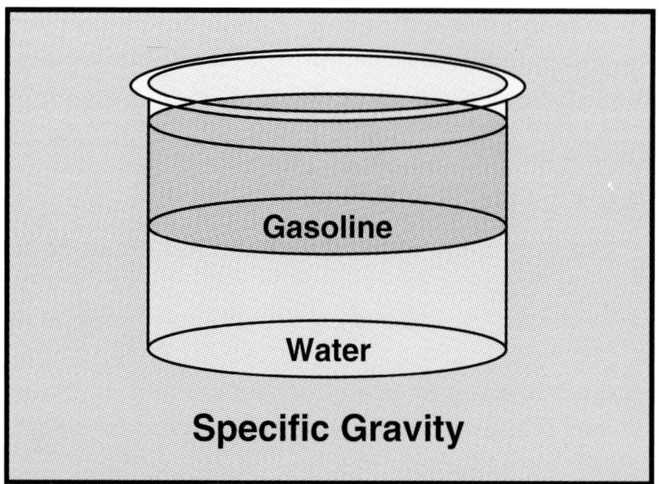

Figure 2.19 Specific gravity is a comparison of the weight of a substance to the weight of an equal volume of water.

greater than 1.0, the substance is heavier than air. It is useful to know the vapor density of a gas at an incident involving a leak or spill. If the gas or vapor is lighter than air, it will rise to the upper levels of a confined area or will rapidly dissipate if outdoors. If the gas is heavier than air, it will hang close to the ground and possibly contact an ignition source (Figure 2.20). The spread of the vapors cannot be predicted exactly from the vapor density, which is easily affected by topography, weather conditions, and the vapor mixture with air. However, knowing vapor density will give a general idea of what to expect from a specific gas.

Figure 2.20 Vapor density is a comparison of the weight of a gas to the weight of an equal volume of air.

BOILING POINT

[NFPA 472: 3-2.3.1.1]

The *boiling point* of a substance is the temperature at which it most rapidly changes from a liquid to a gas or when the rate of evaporation exceeds the rate of condensation. At 212°F (100°C), water rapidly changes to steam, which is water's gaseous form. At temperatures below the boiling point, liquids change to gas but at a slower rate. This slow change is called *evaporation*. As the gas accumulates in a closed container, pressure eventually builds to a point at which the gas begins to condense at the same rate as it evaporates. This equilibrium point is called the *vapor pressure*. As the temperature of the liquid increases, the vapor pressure in the container also increases.

WATER SOLUBILITY

[NFPA 472: 3-2.3.1.1 (j)]

The *water solubility* (miscibility) of a flammable liquid refers to the liquid's ability to mix with water. When a nonwater-soluble liquid is combined with water, the two liquids remain separated. When a water-soluble liquid is combined with water, the two liquids mix easily. Because diluting a water-soluble flammable liquid, such as acetone, with water increases the flammable liquid's flash point, it can be a method of control. However, it will also increase the volume of runoff. Knowledge of a flammable liquid's water solubility is also useful in determining the proper extinguishing agent to be used on the liquid. Water-soluble flammable liquids are known as *polar solvents*. Nonwater-soluble flammable liquids are known as *hydrocarbons*.

TOXIC PRODUCTS OF COMBUSTION

[NFPA 472: 3-2.3.1.1 (h)]

It is important to understand what products of combustion are given off by a burning material. Many materials give off a toxic product when they burn, and this could be life threatening. Examples would include hydrogen cyanide, produced from burning nitrogen-containing materials; hydrogen chloride, from burning polyvinyl chloride; and acrolein, which is a potent irritant produced from burning polyethylene. Even seemingly routine structure fires can give off many deadly gases (Figure 2.21). This is why the use of full personal protective clothing and SCBA is crucial.

Factors Related To Reactivity Hazards

[NFPA 472: 3-2.3.1.1 (f)]

The *reactivity* of a substance is its ability to undergo a chemical reaction with another substance. An unstable substance is one capable of undergoing spontaneous change. With advances in

Figure 2.21 A common structure fire gives off large volumes of toxic products of combustion. *Courtesy of Ron Jeffers.*

modern technology, more and more reactive and unstable materials are being used for various processes. Under emergency conditions, however, they can be extremely destructive to life and property.

Unstable materials can violently decompose with little or no outside stimulus. Materials that crystalize or deteriorate, such as picric acid, ether, dynamite, and organic peroxides, are examples of unstable materials. Explosives, such as nitroglycerin, are also examples of unstable materials. These materials may explode if jarred, heated, or contaminated by other substances. A highly reactive substance may be hypergolic, pyrophoric, or water reactive.

HYPERGOLIC MATERIALS

Hypergolic materials are those that ignite when coming into contact with each other. Hypergolic reactions involve the mixing of two types of chemicals: a fuel and an oxidizer. The chemical reactions of hypergolic substances vary from slow reactions that may be barely visible to reactions that occur with explosive force. Spills of liquid oxygen (LOX) onto asphalt, macadam, or blacktop have been shown to cause this type of violent reaction. Hypergolic materials have found extensive use as rocket fuels in space exploration and in the military. One common rocket fuel is a mixture of nitric acid or nitrogen tetroxide with hydrazine.

PYROPHORIC MATERIALS

Pyrophoric materials are elements that react and ignite on contact with air. Under normal storage and transportation conditions, these materials are packed with inert substances or stored under pressure in sealed containers to prevent any contact with air. Examples of pyrophoric materials are white phosphorus, molten sodium, cesium, potassium, aluminum alkyls, rubidium, powdered titanium, and powdered uranium. These materials and their derivatives are found in a wide variety of industrial and commercial uses.

WATER-REACTIVE MATERIALS

Water-reactive materials are substances, generally flammable solids, that react in varying degrees when either mixed with water or exposed to humid air. Identifying water-reactive materials is extremely important in fighting fires because water is the most commonly used extinguishing agent (Figure 2.22).

The chemical reactions of a substance will vary. Water-reactive materials, such as lithium and finely divided magnesium, burn with such intensity that they decompose water into separate hydrogen and oxygen molecules. The hydrogen then burns while the oxygen supports the combustion. Adding water to these chemical reactions intensifies the fire rather than suppresses it. Nonburning magnesium powder, potassium, and rubidium will decompose water, and the reaction may create enough heat to ignite the hydrogen. Sodium and cesium react explosively on contact with water. Other materials produce highly flammable gases when in contact with water. For instance, acetylene gas is formed when water comes into contact with calcium carbide.

Knowledge and extreme caution are vital in handling emergencies involving reactive materials. When in doubt, keep people and equipment upwind, uphill, and back a safe distance or in protected locations until pertinent facts are established and definite plans can be formulated.

Figure 2.22 Using water on combustible metal fires, such as the magnesium engine in this car, poses a threat to first responders. *Courtesy of Linda Gheen.*

Chapter 3

Photo Courtesy of Scott D. Christiansen, Minot, N.D.

Recognizing And Identifying Hazardous Materials

LEARNING OBJECTIVES

This chapter provides information that will assist the reader in meeting the objectives from NFPA 472, *Standard for Professional Competence of Responders to Hazardous Materials Incidents* and 29 CFR 1910.120 that are listed below. The objective numbers are also noted directly in the text in the sections where they are addressed. Objectives in the list below that are denoted with an asterisk (*) are global in nature and are covered by reading the chapter in its entirety.

AWARENESS LEVEL

- Recognize the presence of a hazardous substance in an emergency. [29 CFR 1910.120(q)(6)(i)(c)] [29 CFR 1910.120(q)(6)(i)(d)]

NFPA 472: 2-2.1 The first responder at the awareness level shall, given various facility and/or transportation situations, with and without hazardous materials present, identify those situations where hazardous materials are present.

NFPA 472: 2-2.1.2 Identify the DOT hazard classes and divisions of hazardous materials and identify common examples of materials in each hazard class or division.

NFPA 472: 2-2.1.3 Identify the primary hazards associated with each of the DOT hazard classes and divisions of hazardous materials by hazard class or division.

NFPA 472: 2-2.1.6 Identify typical container shapes that may indicate hazardous materials.

NFPA 472: 2-2.1.7 Identify facility and transportation markings and colors that indicate hazardous materials, including:
(a) UN/NA identification numbers;
(b) NFPA 704 markings;
(c) Military hazardous materials markings;
(d) Special hazard communication markings;
(e) Pipeline marker; and
(f) Container markings.

NFPA 472: 2-2.1.7.1 Given an NFPA 704 marking, identify the significance of the colors, numbers, and special symbols.

NFPA 472: 2-2.1.8 Identify U.S. and Canadian placards and labels that indicate hazardous materials.

NFPA 472: 2-2.1.9 Identify the basic information on material safety data sheets (MSDS) and shipping papers that indicates hazardous materials.

NFPA 472: 2-2.1.9.1 Identify where to find material safety data sheets (MSDS).

NFPA 472: 2-2.1.9.2 Identify entries on a material safety data sheet that indicate the presence of hazardous materials.

NFPA 472: 2-2.1.9.3 Identify the entries on shipping papers that indicate the presence of hazardous materials.

NFPA 472: 2-2.1.9.4 Match the name of the shipping papers found in transportation (air, highway, rail, and water) with the mode of transportation.

NFPA 472: 2-2.1.9.5 Identify the person responsible for having the shipping papers in each mode of transportation.

NFPA 472: 2-2.1.9.6 Identify where the shipping papers are found in each mode of transportation.

NFPA 472: 2-2.1.9.7 Identify where the papers may be found in an emergency in each mode of transportation.

NFPA 472: 2-2.1.10 Identify examples of clues (other than occupancy/location, container shape, markings/color, placards/labels, and shipping papers) that use the senses of sight, sound, and odor to indicate hazardous materials.

NFPA 472: 2-2.1.11 Describe the limitations of using the senses in determining the presence or absence of hazardous materials.

NFPA 472: 2-2.2.1 Identify difficulties encountered in determining the specific names of hazardous materials in both facilities and transportation.

NFPA 472: 2-2.2.2 Identify sources for obtaining the names of, UN/NA identification numbers for, or types of placard associated with hazardous materials in transportation.

NFPA 472: 2-2.2.3 Identify sources for obtaining the names of hazardous materials in a facility.

NFPA 472: 2-2.3 The first responder at the awareness level shall, given the identity of various hazardous materials (name, UN/NA identification number, or type placard), identify the fire, explosion, and health hazard information for each material using the current edition of the *Emergency Response Guidebook.*

NFPA 472: 2-2.3.3 Given the current edition of the *Emergency Response Guidebook*, identify the three methods for determining the appropriate guide page for a specific hazardous material.

NFPA 472: 2-2.3.4 Given the current edition of the *Emergency Response Guidebook*, identify the two general types of hazards found on each guide page.

NFPA 472: 2-4.1.5 Given the identity of various hazardous materials (name, UN/NA identification number, or type placard), identify the following response information using the current edition of the *Emergency Response Guidebook:*

(a) Emergency action (fire, spill, or leak and first aid);
(b) Personal protective equipment necessary; and
(c) Initial isolation and protective action distances.

OPERATIONAL LEVEL

NFPA 472: 3-2.1.1 Given examples of various hazardous materials containers, identify the general shapes of containers for liquids, gases, and solids.

NFPA 472: 3-2.1.1.1 Given examples of the following tank cars, identify each tank car by type:

(a) Nonpressure tank cars with and without expansion domes;
(b) Pressure tank cars; and
(c) Cryogenic liquid tank cars.

NFPA 472: 3-2.1.1.2 Given examples of the following intermodal tank containers, identify each intermodal tank container by type:

(a) Nonpressure intermodal tank containers; and
(b) Pressure intermodal tank containers.

NFPA 472: 3-2.1.1.3 Given examples of the following cargo tanks, identify each cargo tank by type:

(a) MC-306/DOT-406 cargo tanks;
(b) MC-307/DOT-407 cargo tanks;
(c) MC-312/DOT-412 cargo tanks;
(d) MC-331 cargo tanks;
(e) MC-338 cargo tanks; and
(f) Dry bulk cargo tanks.

NFPA 472: 3-2.1.1.4 Given examples of the following facility tanks, identify each fixed facility tank by type:

(a) Nonpressure facility tanks; and
(b) Pressure facility tanks.

NFPA 472: 3-2.1.2 Given examples of facility and transportation containers, identify the markings that differentiate one container from another.

NFPA 472: 3-2.1.2.1 Given examples of the following transport vehicles and their corresponding shipping papers, identify the vehicle or tank identification marking in all applicable locations:

(a) Rail transport vehicles, including tank cars;
(b) Intermodal equipment including tank containers; and
(c) Highway transport vehicles, including cargo tanks.

NFPA 472: 3-2.1.2.2 Given examples of facility containers, identify the markings indicating container size, product contained, and/or site identification numbers.

NFPA 472: 3-2.1.3 Given examples of facility and transportation situations involving hazardous materials, identify the name(s) of the hazardous material(s) in each situation.

NFPA 472: 3-2.1.3.1 Identify the following information on a pipeline marker:

(a) Product;
(b) Owner; and
(c) Emergency telephone number.

NFPA 472: 3-2.1.3.2 Given a pesticide label, identify each of the following pieces of information; then match the piece of information to its significance in surveying the hazardous materials incident:

(a) Name of pesticide;
(b) Signal word;
(c) Pest control product (PCP) number (in Canada);
(d) Precautionary statement;
(e) Hazard statement; and
(f) Active ingredient.

NFPA 472: 3-2.1.5 Give examples of ways to verify information obtained from the survey of a hazardous materials incident.

NFPA 472: 3-2.2 The first responder at the operational level shall, given known hazardous materials, collect hazard and response information using material safety data sheets (MSDS), CHEMTREC/CANUTEC, and contacts with the shipper/manufacturer.

NFPA 472: 3-2.2.2 Identify two ways to obtain a material safety data sheet (MSDS) in an emergency.

NFPA 472: 3-2.2.3 Using a material safety data sheet (MSDS) for a specified material, identify the following hazard and response information:

(a) Physical and chemical characteristics;
(b) Physical hazards of the material;
(c) Health hazards of the material;

(d) Signs and symptoms of exposure;
(e) Routes of entry;
(f) Permissible exposure limits;
(g) Responsible party contact;
(h) Precautions for safe handling (including hygiene practices, protective measures, procedures for cleanup of spills or leaks);
(i) Applicable control measures including personal protective equipment; and
(j) Emergency and first aid procedures.

NFPA 472: 3-2.2.4 Identify the following:
(a) The type of assistance provided by CHEMTREC/CANUTEC;
(b) How to contact CHEMTREC/CANUTEC; and
(c) The information to be furnished to CHEMTREC/CANUTEC.

NFPA 472: 3-2.2.5 Identify two methods of contacting the manufacturer or shipper to obtain hazard and response information.

NFPA 472: 3-2.3.1 Given situations involving known hazardous materials, interpret the hazard and response information obtained from the current edition of the *Emergency Response Guidebook*, material safety data sheets (MSDS), CHEMTREC/CANUTEC, and shipper/manufacturer contacts.

NFPA 472: 3-2.4.1 Identify a resource for determining the size of an endangered area of a hazardous materials incident.

NFPA 472: 3-2.4.2 *Given the dimensions of the endangered area and the surrounding conditions at a hazardous materials incident, estimate the number and type of exposures within that endangered area.

NFPA 472: 3-2.4.3 Identify resources available for determining the concentrations of a released hazardous material within an endangered area.

NFPA 472: 3-2.4.4 Identify the factors for determining the extent of physical, health, and safety hazards within the endangered area of a hazardous materials incident given the concentrations of the released material.

NFPA 472: 3-4.4.3 Identify the location and use of the mechanical, hydraulic, and air emergency remote shutoff devices as found on MC-306/DOT 406 and MC-331 cargo tanks.

Chapter 3

Recognizing And Identifying Hazardous Materials

The complexity and potential harm from haz mat incidents require that all first responders make a firm commitment to preparation. It is important to use the talents of every member of the responsible department. To form a cohesive response, everyone must participate in planning, preparing, and training for haz mat incidents.

The very nature of haz mat incidents places extreme pressure on first responders to make decisions quickly and accurately. First responders can reduce the number of on-site decisions by the following: preparing pre-incident plans, establishing procedures, and tentatively preassigning actions before the incident occurs. With the groundwork laid, first-arriving companies can concentrate on the situation and operate more safely and efficiently. Planning reduces oversights, confusion, and duplication of effort, and it results in a desirable outcome. Furthermore, planning identifies the following information:

- Location and quantities of hazardous materials in the area
- Dangers of the hazardous materials
- Possible difficulties of property access
- Inherent limitations of the department to control certain types of haz mat emergencies (Figure 3.1)

Figure 3.1 Check around the outside of the building for signs of hazardous materials.

The local emergency planning committee may provide technical assistance to first responders in preparing pre-incident plans. More information on pre-incident planning may be obtained from the *Hazardous Materials Emergency Planning Guide* (NRT-1) from the National Response Team.

Historically, the failure of first responders to recognize the presence and potential harm of hazardous materials at accidents, fires, spills, and other emergencies has caused unnecessary casualties. Haz mat incidents can be controlled only when the personnel involved have sufficient information to make informed decisions. Part of this data must include a size-up of all materials that may be hazardous. The time and effort devoted to a positive identification of the contents of buildings, vehicles, and containers result in greater safety for first responders and the community.

Once the presence of a hazardous material is detected, first responders can use a number of resources to identify the material and its hazards accurately and to specify recommended protective measures. First responders who know the properties of substances can perform tasks more confidently and can evaluate changing conditions more accurately. First responders must be diligent and observant of the hazardous materials present at every emergency. This chapter explains how to identify hazardous materials properly during pre-incident planning and during emergencies; it also describes formal and informal methods that are used to detect the presence of hazardous materials.

INFORMAL METHODS OF IDENTIFICATION

[NFPA 472: 2-2.1.10]

There are a number of informal ways to recognize the presence of hazardous materials. The first responder may be alerted to the involvement of hazardous materials by the "circumstantial evidence" of informal identification. These telltale signs include, but are not limited to, the following indicators:

- Verbal reports of bystanders or responsible persons
- Occupancy type
- Incident location
- Visual physical/chemical indicators
- Trade and common names

The first responder must never rely solely on informal methods of identification to arrive at decisions that may put human life at risk. Nonetheless, informal clues to identification can be useful before formal identification is completed.

Verbal Reports

The first evidence that hazardous materials are present at an emergency may come from a knowledgeable or responsible person at the site. This person may have vital information about the events leading up to the emergency, the materials involved, and the human or property exposures. Whether this person is questioned by a dispatcher over the telephone or by the first responders at the scene, emergency personnel must be prepared to make maximum use of this resource (Figure 3.2). Officials at the scene who receive verbal reports of the involvement of hazardous materials should determine and note the answers to the following questions:

- Who are the reporting parties, and how did they get the information?
- What materials are involved, and how did the party identify them?
- How much material is involved, exposed, spilled, or leaking?
- Is the material leaking from a vessel or escaping under pressure? What is the estimated flow rate? Is it a static or flowing spill? Are sealed containers subject to the physical damage of fire exposure?
- What is the number, location, and condition of personnel needing rescue?
- Is any other information pertinent or peculiar to the situation?

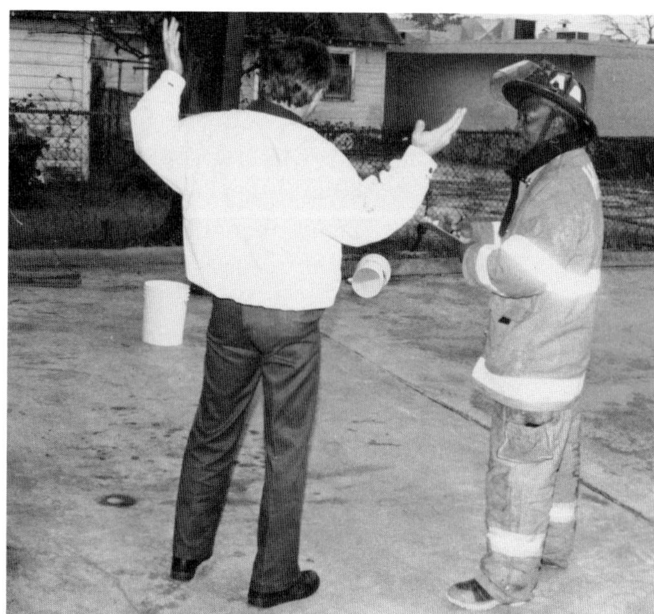

Figure 3.2 Civilians on the scene may provide valuable information to first responders.

Occupancy Type

The occupancy use for a particular structure may warn the first responders that hazardous materials may be involved. The following are all highly probable locations for using and storing hazardous materials:

- Fuel storage facilities
- Paint supply stores
- Plant nurseries
- Doctors' and dentists' offices

- Photo processing laboratories
- Dry cleaners
- Plastic and high-technology factories
- Metal-plating businesses
- Mercantile concerns

There should be a pre-incident plan prepared and available for use for these locations. Be wary of using informal methods of identification in these cases. Chemicals may be stored or used unsafely. Materials may have been transferred from their original, labeled containers, or dangerous substances may have been hidden to avoid detection by chance inspection.

Location

The location of emergency incidents may provide informal evidence of haz mat involvement. Docks or piers, railroad sidings, airplane hangars, truck terminals, and other places of material transfer are likely locations for haz mat accidents (Figure 3.3). Without seeing the material itself, responders can begin to deduce the potential of an incident by observing the presence of the following equipment:

- Off-loading hose
- Forklifts
- Dollies and hand trucks
- Booms
- A-frames
- Ramps
- Assorted rigging

Figure 3.3 Airport hangars have large quantities of hazardous materials.

Local experience with transportation accidents also indicates where to expect haz mat problems. Each of the following modes of transportation, such as highways, rail, water, air, and pipeline, have particular locations where frequent accidents occur.

HIGHWAYS

The potential for frequent accidents are likely to occur at the following locations:

- Designated truck routes
- Blind intersections
- Poorly marked or poorly engineered interchanges
- Areas frequently congested by traffic
- Heavily traveled roads
- Sharp turns (Figure 3.4)
- Steep grades
- Highway interchanges and ramps (Figure 3.5)

Figure 3.4 Accidents frequently occur at sharp bends in the road.

Figure 3.5 Accidents frequently occur on highway entrance/exit ramps.

RAIL

Accidents are likely to occur at the following places:

- Depots, terminals, and switch or classification yards
- Sections of poorly laid or poorly maintained track
- Steep grades and severe curves
- Shunts and sidings
- Uncontrolled crossings (Figure 3.6)

WATER

Accidents are likely to occur at the following places:

- Difficult passages at bends or other threats to navigation
- Bridges and other crossings (Figure 3.7)
- Piers and docks
- Shallows
- Locks
- Loading stations

AIR

Accidents are likely to occur at the following places:

- Fueling ramps (Figure 3.8)
- Repair and maintenance hangars
- Freight terminals

Figure 3.6 Pay particular attention to railroad crossings that are not equipped with warning signals.

PIPELINE

Accidents are likely to occur at the following places:

- Exposed crossings over waterways or roads (Figure 3.9)
- Pumping stations (Figure 3.10)
- Construction and demolition sites
- Intermediate or final storage facilities (Figure 3.11)

Figure 3.7 Ships carry a variety of hazardous materials.

Figure 3.8 Fuel spills are common at airports.

Figure 3.9 Pipes running over roadways can be struck accidentally by vehicles.

Figure 3.10 Pipeline pumping stations may pose problems, even in rural areas.

Figure 3.11 Tank farms contain millions of gallons of product.

IDENTIFYING LOCATION-SPECIFIC FACTORS

As with hazardous occupancies, problem locations should be identified and evaluated during pre-incident planning. Some remote pre-entry observation/assessment stops should be included in the plan so that responding companies can stop briefly to assess the situation for unusual conditions.

Binoculars, spotting scopes, camera lenses, or sight scopes can be invaluable equipment for the observer. It allows the observer to identify scene conditions from a distant location that is safe (Figure 3.12). The assessment stop is also the perfect place for all members of the crew to complete donning their appropriate protective clothing and respiratory protection (Figure 3.13). It is at the assessment stop that first responders should report any unusual conditions to the communications center and reevaluate the situation. Also, the location should be used as a temporary staging area if reconnaissance teams must approach on foot.

Figure 3.12 View the scene from a distance before getting too close.

Figure 3.13 Firefighters should don full personal protective equipment while the size-up is being conducted.

Visual Physical/Chemical Indications

[NFPA 472: 2-2.1.11]

Tangible evidence of hazardous materials, such as a spreading vapor cloud, a pair of gloves melting, fish dying, container deterioration, etc., are signs that physical and/or chemical actions and reactions are taking place.

CAUTION: The first responder must remember, however, that using the human senses indiscriminately to detect the presence of hazardous materials is both unreliable and unsafe (Figure 3.14).

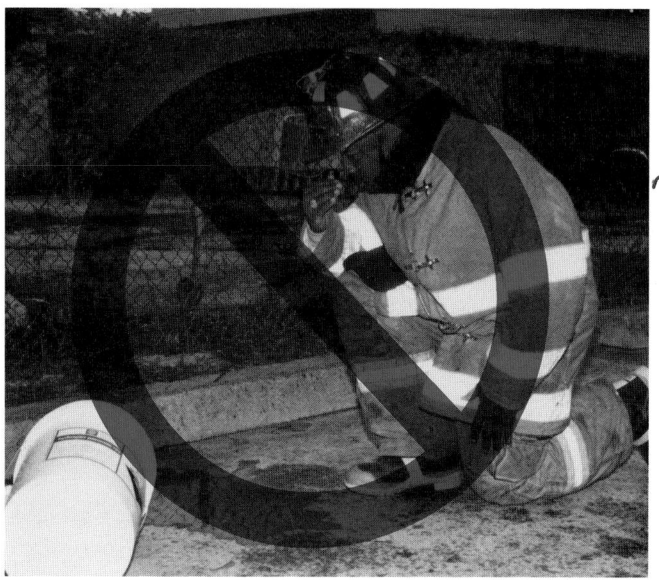

Figure 3.14 Never try to identify a substance by taste, touch, or smell.

Physical actions are processes that do not change the elemental composition of the materials involved. One example is a liquefied, compressed material changing to a gas as it escapes from a vessel. The resulting white vapor cloud is another physical change, which is the condensation of moisture in the air by the expanding material. Indications of a physical action include the following:

- Rainbow sheen on water surfaces
- Wavy vapors over a volatile liquid
- Frost near a leak
- Containers deformed by the force of an accident (Figure 3.15)
- Operation of pressure-relief devices
- Pinging or popping of heat-exposed vessels

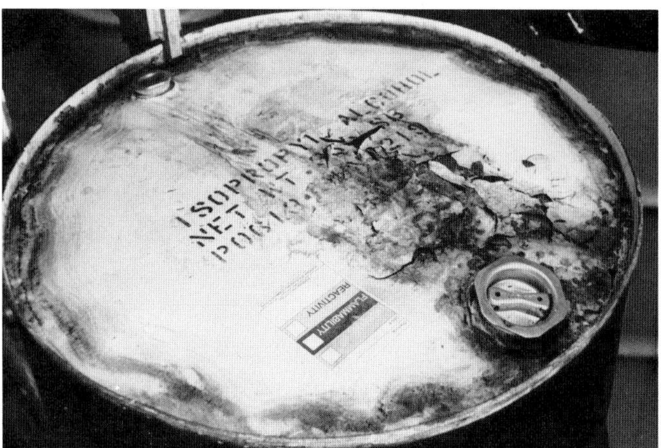

Figure 3.15 Damaged containers may signal that a problem is present.

Chemical reactions convert one substance into another. Visual and sensory evidence of chemical reactions include the following:

- Extraordinary fire conditions
- Peeling or discoloration of a container's finish
- Spattering or boiling of unheated materials
- Distinctively colored vapor clouds
- Smoking or self-igniting materials
- Unexpected deterioration of equipment
- Peculiar smells
- Unexplained changes in ordinary materials
- Symptoms of chemical exposure

All responders should be especially watchful for the onset of any symptoms of chemical exposure that can occur separately or in clusters, depending on the chemical. First responders should be alert for the following symptoms of chemical exposure:

- *Changes in respiration:* difficult breathing, increase or decrease in respiration rate, tightness of the chest, irritation of the nose and throat, and/or respiratory arrest
- *Changes in consciousness:* dizziness, lightheadedness, drowsiness, confusion, fainting, and/or unconsciousness
- *Abdominal distress:* nausea, vomiting, and/or cramping
- *Change in activity level:* fatigue, weakness, stupor, hyperactivity, restlessness, anxiety, giddiness, and/or faulty judgment
- *Visual disturbances:* double vision, blurred vision, cloudy vision, burning of the eyes, and/or dilated or constricted pupils
- *Skin changes:* burning sensations, reddening, paleness, fever, and/or chills
- *Changes in excretion or thirst:* uncontrolled tears, profuse sweating, mucus flowing from the nose, diarrhea, frequent urination, bloody stool, and/or intense thirst
- *Pain:* headache, muscle ache, stomachache, chest pain, and/or localized pain at sites of substance contact

At the onset of physical symptoms, those people exposed should withdraw immediately to a safe location for decontamination, treatment, and observation or medical treatment (Figure 3.16).

CAUTION: Anyone in comparable protective equipment might have the same symptoms and should also withdraw.

Promptly reassess the situation when workers become ill. Remember, not all chemical exposures result in immediate symptoms. In fact, symptoms may not appear until hours after the exposure.

Figure 3.16 Seek treatment as soon as symptoms of exposure occur. *Courtesy of Ron Jeffers.*

FORMAL METHODS OF IDENTIFICATION

[NFPA 472: 3-2.1.5]; [NFPA 472: 3-2.2.5]; [NFPA 472: 3-2.3.1]

Although informal methods of identification are helpful, it is only through positive identification of the materials involved that first responders may fully develop a sound defense strategy. The information gathered then allows more technically qualified responders to mitigate the incident. The following sections highlight some of the more common, formal methods of haz mat identification.

DOT Emergency Response Guidebook (ERG)

[NFPA 472: 2-2.3]; [NFPA 472: 3-2.4.1]; [NFPA 472: 3-2.4.4]

The *Emergency Response Guidebook (ERG)* was developed by the U.S. Department of Transportation (DOT) for use by first responders such as firefighters, police, and emergency medical services personnel (Figure 3.17). The *ERG* is a basic guide for *initial* actions to be taken by first responders to protect themselves and the general public when responding to incidents involving hazardous materials. The *ERG* is intended to be carried in every public safety vehicle in the country. It will also be useful in handling incidents in other modes of transportation and at transportation facilities such as terminals and warehouses (Table 3.1). (**NOTE:** The information provided in this section is based on the 1993 edition of the *Emergency Response Guidebook*. Current plans are for the *ERG* and the Canadian *Dangerous Goods Guide to Initial Emergency Response (IERG)* to be combined into a common guide for all of North America in 1995.) Copies of the *ERG/IERG* may be available through local or state/province emergency management offices. (**NOTE:** For purposes of consistency and clarity, the *Dangerous Goods Guide to Initial Emergency Response* will be referred to by its more common names, *IERG* or *Initial Emergency Response Guide*, throughout the remainder of this manual.)

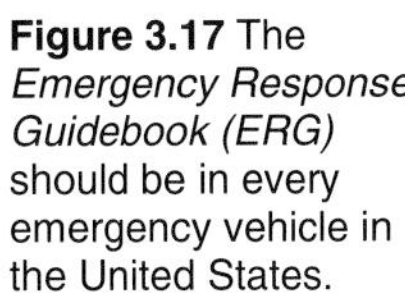

Figure 3.17 The *Emergency Response Guidebook (ERG)* should be in every emergency vehicle in the United States.

TABLE 3.1
DOT *Emergency Response Guidebook* Color Codes

Page Color	Identifying Factor	Information Contained
Yellow	ID Number	Product name Guide number highlighted— Poisonous effects
Blue	Product Name	ID number Guide number highlighted— Poisonous effects
Orange	Guide Number	Potential hazards— Fire and explosion Health hazards Emergency actions— Scene control Fire Spill or leak First aid Special
Green	ID Number	ID name Initial isolation zone Protective action distance
No Color	Plain Paper	Instructions, definitions, and explanations

The primary objective of the *ERG* is to direct first responders to the appropriate guide page (orange-bordered pages) as quickly as possible upon their arrival at the scene of a haz mat incident. The document is designed to get first responders to this point by providing several different means by which to identify the material(s) involved in the incident. In the event the material cannot be specifically identified by the methods provided, the *ERG* provides guidance for initiating safety actions that are applicable to all materials until a specific decision can be made.

The sequential steps to follow when identifying a specific material and then arriving at the appropriate guide page are outlined on the first page of the *ERG* under the heading, "How to Use this Guidebook During an Incident Involving Hazardous Materials." These "how to" instructions are actually guidance for identifying the material through several cross-referenced indexes.

UN/NA IDENTIFICATION NUMBER INDEX

[NFPA 472: 2-2.3.3]

This index (yellow-bordered pages) lists the UN/NA four-digit identification number in numerical order with its proper material shipping name and the guide number assigned to it (Figure 3.18). Therefore, if the identification number is known, first responders can turn to this index and locate the following information:

- Identification number (first column)
- Appropriate guide page (second column) that outlines the initial actions to be taken
- Name of the material to which the number is assigned (third column)

ID No.	Guide No.	Name of Material
2330	**27**	HENDECANE
2330	**27**	UNDECANE
2331	**60**	ZINC CHLORIDE, anhydrous
2332	**26**	ACETALDEHYDE OXIME
2333	**28**	ALLYL ACETATE
2334	**28**	ALLYLAMINE
2335	**28**	ALLYL ETHYL ETHER
2336	**28**	ALLYL FORMATE
2337	**57**	PHENYL MERCAPTAN
2338	**28**	BENZOTRIFLUORIDE
2339	**27**	2-BROMOBUTANE
2340	**27**	BROMOETHYL ETHYL ETHER
2341	**27**	BROMOMETHYLBUTANE
2342	**27**	BROMOMETHYLPROPANE
2343	**27**	BROMOPENTANE
2344	**29**	BROMOPROPANE
2345	**29**	BROMOPROPYNE
2346	**26**	BUTANEDIONE
2346	**26**	DIACETYL
2347	**27**	BUTANETHIOL
2347	**27**	BUTYL MERCAPTAN
2348	**26**	BUTYL ACRYLATE
2350	**26**	BUTYL METHYL ETHER
2351	**26**	BUTYL NITRITE
2352	**26**	BUTYL VINYL ETHER
2353	**29**	BUTYRYL CHLORIDE
2354	**28**	CHLOROMETHYL ETHYL ETHER
2356	**26**	CHLOROPROPANE
2357	**68**	CYCLOHEXYLAMINE
2358	**27**	CYCLOOCTATETRAENE
2359	**29**	DIALLYLAMINE
2360	**28**	DIALLYETHER
2361	**68**	DIISOBUTYLAMINE
2362	**27**	1,1-DICHLOROETHANE
2363	**27**	ETHYL MERCAPTAN
2364	**26**	PROPYL BENZENE
2366	**26**	DIETHYL CARBONATE
2367	**27**	METHYL VALERALDEHYDE
2368	**26**	PINENE
2369	**26**	ETHYLENE GLYCOL MONO-BUTYL ETHER
2370	**27**	HEXENE
2371	**27**	ISOPENTENE
2372	**26**	BIS(DIMETHYLAMINO)ETHANE
2372	**26**	1,2-DI-(DIMETHYLAMINO) ETHANE
2373	**26**	DIETHOXYMETHANE
2374	**26**	DIETHOXYPROPENE
2375	**28**	DIETHYL SULFIDE
2376	**26**	DIHYDROPYRAN
2377	**27**	1,1-DIMETHOXYETHANE
2378	**28**	DIMETHYLAMINOACETONITRILE
2379	**27**	1,3-DIMETHYLBUTYLAMINE
2380	**26**	DIMETHYLDIETHOXYSILANE
2381	**27**	DIMETHYLDISULFIDE
2382	**57**	DIMETHYLHYDRAZINE, symmetrical
2383	**68**	DIPROPYLAMINE
2384	**26**	DIPROPYL ETHER
2385	**26**	ETHYL ISOBUTYRATE
2386	**26**	ETHYL PIPERIDINE
2387	**27**	FLUOROBENZENE
2388	**27**	FLUOROTOLUENE
2389	**26**	FURAN

Figure 3.18 The identification number index pages have yellow borders.

MATERIAL NAME INDEX

[NFPA 472: 2-2.3.3]

This index (blue-bordered pages) lists in alphabetical order the names of materials that should appear on shipping documents (Figure 3.19). Therefore, if first responders have the name of the material involved, they may go directly to this index and identify the guide page applicable for the material. (**NOTE:** First responders should always double-check the proper spelling of any material. Many materials have like-sounding pronunciations and little deviation in spelling, which may easily lead to an incorrect identification and guide page reference.) The first column has the proper shipping name, the second column has the guide page, and the third column has the identification number.

TABLE OF PLACARDS AND THE INITIAL RESPONSE GUIDES TO USE ON-SCENE

[NFPA 472: 2-2.3.3]

This table of illustrated placards should be used by first responders only when the material cannot be specifically identified by using shipping papers, numbered placards affixed to vehicles, or identification numbers. This index allows first responders to match the appearance of a placard affixed to a vehicle with a similarly appearing placard illustrated in the index. The assigned guide page number is entered next to the illustrated placard.

GUIDE NUMBER 11

If identification of a material cannot be made using any of the index methods, first responders should turn to Guide Number 11 (the first guide appearing in the orange-bordered pages) and take actions accordingly (Figure 3.20). Guide Number 11 is designed to address all incident contingencies in a basic and standardized fashion with a heavier-than-ordinary emphasis upon first responder and general public safety. This guide is to be used only until the material is fully identified, at which time the appropriate guide number assigned to the material should be used.

INITIAL ACTION GUIDES

[NFPA 472: 2-2.3.4]; [NFPA 472: 2-4.1.5 a through c]

Each numbered guide provides only the most essential information in a brief and practical form. The guides identify the most significant potential hazards and offer information and guidance to first responders on *initial actions* to take upon arrival at a haz mat scene. The guides offer information on the following situations:

- Potential hazards (health, fire, or explosion)
- Preliminary emergency actions for self-protection and protection of others
- Fire instructions
- Leak or spill guidance
- First aid procedures

Because many hazardous materials represent similar types of hazards that call for similar initial emergency response actions, there are only 77 guides. These guides accurately group not only the materials listed in the name and identification number indexes but also thousands of materials shipped under generic names such as "FLAMMABLE LIQUID N.O.S."

Name of Material	Guide No.	ID No.
MERCURY POTASSIUM IODIDE	**53**	1643
MERCURY SALICYLATE	**53**	1644
MERCURY SULFATE	**53**	1645
MERCURY THIOCYANATE	**53**	1646
MESITYLENE	**26**	2325
MESITYL OXIDE	**26**	1229
METAL ALKYL, n.o.s.	**40**	2003
METAL ALKYL HALIDE, n.o.s.	**40**	3049
METAL ALKYL HYDRIDE, n.o.s.	**40**	3050
METAL ALKYL SOLUTION, n.o.s.	**40**	9195
METAL CATALYST, dry	**37**	2881
METAL CATALYST, finely divided,	**32**	1378
activated or spent, wet with not less than 40% water or other suitable liquid	**32**	1332
METALDEHYDE	**32**	3089
METAL POWDER, flammable, n.o.s.	**28**	2396
METHACRYLALDEHYDE	**60**	2531
METHACRYLIC ACID, inhibited	**28**	3079
METHACRYLONITRILE, inhibited	**26**	2614
METHALLYL ALCOHOL	**17**	1971
METHANE, compressed	**22**	1972
METHANE, refrigerated liquid (cryogenic liquid)	**59**	9265
METHANESULFONYL CHLORIDE	**28**	1230
METHANOL	**57**	2605
METHOXYMETHYL ISOCYANATE	**27**	2293
METHOXYMETHYLPENTANONE	**26**	3092
1-METHOXY-2-PROPANOL	**26**	1231
METHYL ACETATE	**26**	1232
METHYL ACETONE	**17**	1060
METHYL ACETYLENE and PROPADIENE MIXTURE, stabilized		
METHYLACRYLATE, inhibited	**26**	1919
METHYLAL	**26**	1234
METHYL ALCOHOL	**28**	1230
METHYL ALLYL CHLORIDE	**26**	2554
METHYL ALUMINUM SESQUIBROMIDE	**40**	1926
METHYL ALUMINUM SESQUICHLORIDE	**40**	1927
METHYLAMINE, anhydrous	**19**	1061
METHYLAMINE, aqueous solution	**68**	1235
METHYL AMYL ACETATE	**26**	1233
METHYL AMYL ALCOHOL	**26**	2053
METHYL AMYL KETONE	**26**	1110
METHYLANILINE	**57**	2294
METHYL BENZOATE	**31**	2938
METHYLBENZYL ALCOHOL (alpha)	**55**	2937
METHYL BROMIDE	**55**	1062
METHYL BROMIDE and CHLOROPICRIN MIXTURE	**55**	1581
METHYL BROMIDE and ETHYLENE DIBROMIDE MIXTURE, liquid	**55**	1647
METHYL BROMIDE and NON-FLAMMABLE COMPRESSED GAS MIXTURE	**15**	1955
METHYL BROMOACETATE	**58**	2643
METHYL BUTANONE	**26**	2397
METHYLBUTENE	**26**	2460
2-METHYL-1-BUTENE	**26**	2459
2-METHYL-2-BUTENE	**26**	2460
3-METHYL-1-BUTENE	**26**	2561
METHYLBUTYLAMINE	**29**	2945
METHYL-tert-BUTYL ETHER	**26**	2398
METHYL BUTYRATE	**26**	1237
METHYL CHLORIDE	**18**	1063

Figure 3.19 The product name index pages have blue borders.

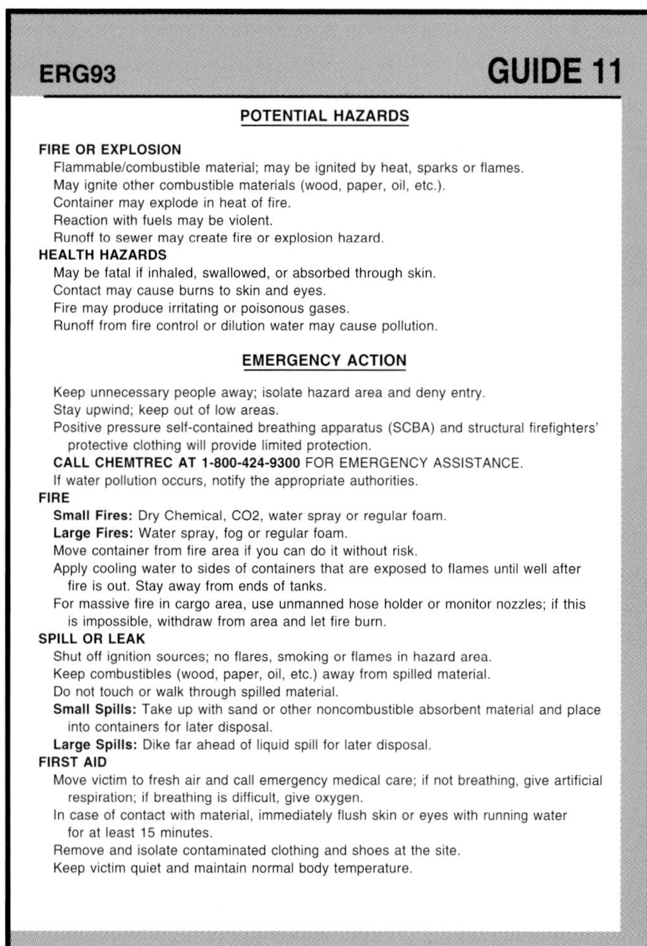

ERG93 **GUIDE 11**

POTENTIAL HAZARDS

FIRE OR EXPLOSION

Flammable/combustible material; may be ignited by heat, sparks or flames.
May ignite other combustible materials (wood, paper, oil, etc.).
Container may explode in heat of fire.
Reaction with fuels may be violent.
Runoff to sewer may create fire or explosion hazard.

HEALTH HAZARDS

May be fatal if inhaled, swallowed, or absorbed through skin.
Contact may cause burns to skin and eyes.
Fire may produce irritating or poisonous gases.
Runoff from fire control or dilution water may cause pollution.

EMERGENCY ACTION

Keep unnecessary people away; isolate hazard area and deny entry.
Stay upwind; keep out of low areas.
Positive pressure self-contained breathing apparatus (SCBA) and structural firefighters' protective clothing will provide limited protection.
CALL CHEMTREC AT 1-800-424-9300 FOR EMERGENCY ASSISTANCE.
If water pollution occurs, notify the appropriate authorities.

FIRE

Small Fires: Dry Chemical, CO2, water spray or regular foam.
Large Fires: Water spray, fog or regular foam.
Move container from fire area if you can do it without risk.
Apply cooling water to sides of containers that are exposed to flames until well after fire is out. Stay away from ends of tanks.
For massive fire in cargo area, use unmanned hose holder or monitor nozzles; if this is impossible, withdraw from area and let fire burn.

SPILL OR LEAK

Shut off ignition sources; no flares, smoking or flames in hazard area.
Keep combustibles (wood, paper, oil, etc.) away from spilled material.
Do not touch or walk through spilled material.
Small Spills: Take up with sand or other noncombustible absorbent material and place into containers for later disposal.
Large Spills: Dike far ahead of liquid spill for later disposal.

FIRST AID

Move victim to fresh air and call emergency medical care; if not breathing, give artificial respiration; if breathing is difficult, give oxygen.
In case of contact with material, immediately flush skin or eyes with running water for at least 15 minutes.
Remove and isolate contaminated clothing and shoes at the site.
Keep victim quiet and maintain normal body temperature.

Figure 3.20 The action guide pages have orange borders.

TABLE OF INITIAL ISOLATION AND PROTECTIVE ACTION DISTANCES

USE THIS TABLE WHEN THE MATERIAL IS NOT ON FIRE

ID No.	NAME OF MATERIAL	SMALL SPILLS (Leak or spill from a small package or small leak from a large package.) First, ISOLATE in all directions- (Feet)	SMALL SPILLS Then, PROTECT those persons in the DOWNWIND direction (Miles)	LARGE SPILLS (Leak or spill from a large package or spill from many small packages.) First, ISOLATE in all directions- (Feet)	LARGE SPILLS Then, PROTECT those persons in the DOWNWIND direction (Miles)
1143	CROTONALDEHYDE, inhibited	150	0.2	150	0.4
1163	DIMETHYLHYDRAZINE, unsymmetrical	1200	4	1500	5
1182	ETHYL CHLOROFORMATE	150	0.2	150	0.2
1185	ETHYLENEIMINE, inhibited	600	2	900	3
1238	METHYL CHLORO-CARBONATE	150	0.8	600	2
1238	METHYL CHLORO-FORMATE	150	0.8	600	2
1239	METHYL CHLOROMETHYL ETHER	150	0.2	150	0.4
1244	METHYLHYDRAZINE	1500	5	1500	5
1259	NICKEL CARBONYL	1500	5	1500	5
1380	PENTABORANE	1500	5	1500	5
1510	TETRANITROMETHANE	150	0.4	150	0.8
1541	ACETONE CYANOHYDRIN	150	0.2	150	0.2
1556	METHYLDICHLOROARSINE	150	0.2	150	0.2
1560	ARSENIC CHLORIDE	1200	4	1500	5
1560	ARSENIC TRICHLORIDE	1200	4	1500	5
1569	BROMOACETONE	150	0.2	150	0.2
1580	CHLOROPICRIN	600	2	900	3
1581	CHLOROPICRIN and METHYL BROMIDE MIXTURE	150	0.8	1200	4
1581	METHYL BROMIDE and CHLOROPICRIN MIXTURE	150	0.8	1200	4

Figure 3.21 The initial action and isolation table pages have green borders.

First responders should be aware that explosives are not listed individually in the *ERG* by identification number. If the shipping paper or placard identifies the material as Division 1.1, 1.2, 1.3, 1.5, or 1.6 explosive, the *ERG* directs first responders to Guide 46. If the shipping paper identifies the material as a Division 1.4 explosive, first responders are directed to Guide 50. The instructions relating to guides for explosives are highlighted in the "how to" guidance on the first page of the *ERG*.

TABLE OF INITIAL ISOLATION AND PROTECTIVE ACTION DISTANCES

Another feature of the *ERG* that is important to first responders is the highlighted entries in both the Identification Number Index and the Name Index. These highlighted materials represent poison and poison inhalation risks and are also listed in the Table of Initial Isolation and Protective Action Distances in the *ERG* (green-bordered pages) (Figure 3.21). Anytime an entry in the Identification Number Index and the Name Index is highlighted and there is no fire involved with the incident, first responders *must* refer to this table of isolation and protective action distances and be guided accordingly.

Transport Canada Dangerous Goods Initial Emergency Response Guide

The Canadian government maintains a Transport of Dangerous Goods Directorate as a part of Transport Canada. This organization publishes the *Dangerous Goods Guide to Initial Emergency Response (IERG)*, a publication that is similar to the DOT *ERG*. The placarding/labeling system is also similar to that used in the United States, and both are reciprocal. The main difference is that there are additional or different placards in the *IERG*:

- Corrosive gas
- Miscellaneous dangerous goods (different symbol)
- Dangerous wastes (Table 3.2)

The *IERG* begins with a chart of placards and labels. It provides brief explanations of words and terms used in the guide, the guide's purpose and use, the five steps to emergency response, and tips on contacting and using the resources of CANUTEC.

The next two sections are printed in contrasting colors. The first section is color-coded orange and lists materials by their names. The second section is color-coded green and lists materials by product identification numbers (PIN). Each list refers the user to 52 two-page numbered "guides." This dual-listing system lets the user refer to the appropriate guide from the shipping name or from the PIN. The guides are color-coded yellow and contain information on response for the initial phase of the operation. The book ends with a transportation identification chart.

The initial response guides can be used to develop basic strategies for first responders during the early stages of a haz mat incident. Each guide gives several suggestions for handling the possible fire, explosion, health, and pollution risks posed by the identified substance. There are obvious limitations to a system that has only a limited number of guides for the thousands of chemicals in existence. Once the guide has served its purpose, first responders must obtain more complete information about the materials involved. Any number of factors, many of which will be discussed later, can affect the direction and distance of spreading leaks or spills of liquids, vapors, and gases.

If only the material's general class can be identified from the pictorial placard, the responder can refer to the "Chart of Placards and Labels" at the front of the book. This table refers the user to an all-purpose action guide for each general classification.

If the material cannot be identified by the name or placard, the responder can refer to the "Rail Car and Road Trailer Identification Chart" at the back of the book and then determine a general guide page based on the shape of the container.

TABLE 3.2
Dangerous Goods Initial *Emergency Response Guidebook* Color Codes

Color	Identifying Factor	Information Contained
Orange	Product name	Guide number PIN
Green	PIN	Guide number Product number
Yellow	Guide number	Potential hazards— Fire or explosion Health Public safety— Protective clothing Evacuation Fire Spill or leak First aid
No Color	Plain paper	Instructions, definitions, and explanations

Finally, if the material involved cannot be identified by the identification number, name, pictorial placard, or shape of container, the responder should use Guide 01. However, because of its nonspecific nature, Guide 01 should be used only as a last resort.

The responder must continue to try to acquire more information about the material. Although the *IERG* does not contain a Table of Isolation and Protective Action Distances as the *ERG* does, initial evacuation distances are listed in each guide for large spills and fires.

Copies of the *IERG* are available by contacting the following: *Dangerous Goods Guide To Initial Emergency Response*, The Canadian Government Publishing Centre, Supply and Services Canada, Ottawa, Ontario, Canada K1A 0S9. As stated previously in the *ERG* section, the *IERG* and the *ERG* will be combined into a single guide in 1995.

Shipper's Emergency Response Center

Anyone who offers a hazardous material for transportation must enter on the shipping document a 24-hour emergency response telephone number. This emergency response telephone must be monitored at all times by a person who has knowledge of the material or has immediate access to a knowledgeable person. In the event a shipping paper is unavailable or the telephone number does not appear on the document, the first responder should call CHEMTREC/CANUTEC.

Emergency Information Centers — CHEMTREC/CANUTEC

[NFPA 472: 3-2.2.4 a through c]

Emergency information centers, such as CHEMTREC (U.S.) and CANUTEC (Canada), are principal providers of immediate technical assistance. Many companies maintain emergency phones and response teams, but most prefer to be contacted through CHEMTREC and CANUTEC. Both organizations operate 24 hours a day, 365 days a year (Figures 3.22 a and b).

When calling CHEMTREC or CANUTEC, the first responder should provide the following information:

- Name of the caller and a call-back number
- Location of the incident
- Names of the material, shipper, and manufacturer
- Type of container or vehicle
- Rail car or truck number
- Carrier's name
- Consignee (material destination)
- Local conditions
- Action already taken

When callers contact CHEMTREC or CANUTEC, they should be ready to record the information and suggestions offered. These organizations provide immediate information on the material's properties, its hazards, and suggested control techniques. CHEMTREC has over 1 million material safety data sheets on file. All of these are available to be faxed to the scene of an emergency. They can also serve as a communication link with other technical support.

These organizations are capable of accessing data bases, coordinating information sources, and notifying shippers of the incident. The shipper may decide to activate its own response team or to coordinate other mutual aid assistance. CHEMTREC facilitates telephone conference calls between responders and the shipper.

CHEMTREC (which is run by the Chemical Manufacturer's Association) works closely with DOT and the Coast Guard's National Response Center (NRC). (**NOTE:** The responsibility for reporting incidents to the NRC lies with the shipper, facility manager, or vehicle operator.) Emergency service contact with CHEMTREC *does not* fulfill the incident reporting requirements set forth in federal regulations for transporters. Responders must also be aware of the following limitations of these systems:

Figure 3.22a CHEMTREC provides technical information to on-scene first responders. *Courtesy of CHEMTREC.*

Figure 3.22b CANUTEC is based in Canada. *Courtesy of CANUTEC.*

- Not every chemical manufacturer provides data to these organizations.
- The relevance of information these sources provide can only be as good as the information that first responders relay to the system.
- Much of the tactical information is general; the system will not make decisions for the incident commander.
- These systems are not infallible.

CHEMTREC and CANUTEC may provide assistance in reaching other associations that can provide technical assistance:

- National Agricultural Chemical Association (NACA)
- Fertilizer Institute
- LP-Gas Association
- Chlorine Institute
- Compressed Gas Association

To obtain assistance from CHEMTREC during a *chemical emergency*, first responders should call them as soon as possible at 1-800-424-9300. For *nonemergency* information, CHEMTREC may be reached at 1-800-262-8200 on weekdays from 9:00 a.m. to 6:00 p.m. Eastern standard time. In Canada, to obtain assistance from CANUTEC, first responders should call them collect as soon as possible at 0-613-996-6666.

Material Safety Data Sheet (MSDS)

[NFPA 472: 2-2.1.9.1;]; [NFPA 472: 3-2.2.2]

The best source of information on a specific hazardous material is the manufacturer's data sheet known as a material safety data sheet (MSDS). State and federal legislation on hazard communication, right-to-know, and mandatory local notification on hazards make the MSDS a necessity. First responders can acquire an MSDS from the manufacturer of the material, the supplier, the facility Hazard Communication Plan, or the Local Emergency Planning Committee. Sometimes the MSDS will be attached to the shipping papers.

[NFPA 472: 2-2.1.9]; [NFPA 472: 3-2.2.3 a through h]

Minimal content of the MSDS sheet is mandated by the U.S. Department of Labor, Occupational Safety and Health Administration (OSHA). A sample of an MSDS is shown in Figure 3.23. There is no set format for the MSDS, but each sheet has eight sections that contain the following information:

Section I

- Manufacturer's name and address
- Emergency telephone number
- Information telephone number
- Signature and date

Section II — Hazardous Ingredients

- Common name
- Chemical name
- CAS number
- OSHA Permissible Exposure Limit (PEL)
- ACGIH Threshold Limit Value (TLV)
- Other Exposure Limits

Section III — Physical and Chemical Characteristics

- Boiling point
- Specific gravity
- Vapor pressure
- Melting point
- Vapor density
- Evaporation rate
- Solubility in water
- Appearance and odor

Section IV — Fire and Explosion Hazard Data

- Flash point
- Flammable limits (LEL, UEL)
- Extinguishing media
- Special fire fighting procedures
- Unusual fire and explosion hazards

Section V — Reactivity Data

- Stability (stable/unstable conditions to avoid)
- Incompatibility (materials to avoid)
- Hazardous decomposition or by-products
- Hazardous polymerization (may or may not occur, conditions to avoid)

Material Safety Data Sheet

Hoechst Celanese

Chemical Group
Hoechst Celanese Corporation
*P.O. Box 819005/Dallas, Texas 75381-9005
*Information phone: 214 277 4000
Emergency phone: 800 424 9300 (CHEMTREC)

Ethylene oxide

Issued December 31, 1992

#40

Identification

Product name: Ethylene oxide

Chemical name: Ethylene oxide

Chemical family: Epoxide

Formula: $(CH_2)_2O$

Molecular weight: 44

CAS number: 75-21-8

CAS name: Oxirane

Synonyms: Dihydrooxirene; dimethylene oxide; 1,2-epoxyethane; oxiran; oxirane; oxacyclopropane; oxane; oxidoethane; alpha, beta-oxidoethane; EO; EtO.

***Transportation information**
Shipping name: Ethylene Oxide
Hazard class: 2.3, Poisonous Gas
Subsidiary hazard: 2.1, Flammable Gas
United Nations no.: UN1040
Packing group: 1
Emergency Response Guide no.: 69
DOT Reportable Quantity: 10 lb/4.54 kg

Physical data

Boiling point (760 mm Hg): 10.7°C (51°F)

Freezing point: -112.5°C (-171°F)

Specific gravity (H_2O=1 @ 20/20°C): 0.8711

Vapor pressure (20°C): 1094 mm Hg

Vapor density (Air =1 @ 20°C): 1.5

Solubility in water (% by WT @ 20°C): Complete

Percent volatiles by volume: 100

Appearance and odor: Colorless gas with sweet ether-like odor. Odor threshold: 500 ppm.

Fire and explosion hazard data

Flammable limits in air, % by volume
Upper: 100
Lower: 3.0

Flash point (test method):
Tag open cup (ASTM D1310): <0°F (<–18°C)
Tag closed cup (ASTM D56): –4°F (–20°C)

Extinguishing media:
Use water (flood with water), CO_2, dry chemical or alcohol-type aqueous film-forming foam. Allow to burn if flow cannot be shut off immediately.

Special fire-fighting procedures:
*If potential for exposure to vapors or products of combustion exists, wear complete personal protective equipment, including self-contained breathing apparatus with full facepiece operated in pressure-demand or other positive-pressure mode.

Dilution of ethylene oxide with 23 volumes of water renders it non-flammable. A ratio of 100 parts water to one part ethylene oxide may be required to control build-up of flammable vapors in a closed system. Water spray can be used to reduce intensity of flames and to dilute spills to nonflammable mixture. Use water spray to cool fire-exposed structures and vessels. Ethylene oxide is an NFPA Class 1A flammable liquid with a 51°F boiling point. Locations classified as hazardous because of the presence of ethylene oxide are designated Class 1.

Unusual fire and explosion hazards:
Rapid, uncontrolled polymerization can cause explosion under fire conditions. Vapor is heavier than air and can travel considerable distance to a source of ignition and flashback. Will burn without the presence of air or other oxidizers.

Component information (See Glossary at end of MSDS for definitions)

Component, wt. % (CAS number)	Exposure levels: OSHA PEL TWA	Exposure levels: ACGIH TLV ®TWA	IDLH	Subject to SARA §313 reporting?
Ethylene oxide, 99.95% (75-21-8)	1ppm(2); 5 ppm excursion limit	1ppm(2)	800 ppm	Yes

(1) All components listed as required by federal, California, New Jersey and Pennsylvania regulations.
(2) Suspectd human carcinogen.

Special hazard designations

	HMIS	NFPA	Key
Health:	3	3	0 = Minimal
Flammability:	4	4	1 = Slight
Reactivity:	3	3	2 = Moderate
Personal protective equipment:	G	—	3 = Serious
			4 = Severe

SARA §311 hazard categories

Acute health:	Yes
Chronic health:	Yes
Fire:	Yes
Sudden release of pressure:	Yes
Reactive:	Yes

Reactivity data

Stability:
Potentially unstable

Hazardous polymerization:
Can occur.UNCONTROLLED POLYMERIZATION CAN CAUSE RAPID EVOLUTION OF HEAT AND INCREASED PRESSURE WHICH CAN RESULT IN VIOLENT RUPTURE OF STORAGE VESSELS OR CONTAINERS.

Conditions to avoid:
Heat, sparks, flame.

Materials to avoid:
Acetylide-forming metals (for example, copper, silver, mercury and their alloys): alcohols; amines; mercaptans; metallic chlorides; aqueous alkalis; mineral acids; oxides; strong oxidizing agents (for example, oxygen, hydrogen peroxide, or nitric acid).

Hazardous combustion or decomposition products:
Carbon monoxide.

Health data

Effects of exposure/toxicity data
Acute

Ingestion (swallowing): Can cause stomach irritation, also liver and kidney damage. Moderately toxic to animals (oral LD_{50}, rats: 0.1 g/kg).

Inhalation (breathing): Can cause irritation of nasal passages, throat and lungs; lung injury; nausea; vomiting; headache; diarrhea; shortness of breath; cyanosis (blue or purple coloring of the skin); and pulmonary edema (accumulation of fluid in the lungs) - signs and symptons can be delayed for several hours. Slightly toxic to animals (inhalation LC_{50}, rats, 4 hrs: 1460 ppm).

Skin contact: Can lead to severe reddening and swelling of the skin, with blisters.

*New or revised information; previous version dated October 1,1991.

5/95

Figure 3.23 A sample material safety data sheet (MSDS). *Courtesy of Hoechst Celanese.*

Ethylene oxide

page 2

#40

Issued December 31,1992

Sensitization (allergic reaction) possible. Large amounts evaporating from skin can cause frostbite.

Eye contact: Contact with liquid can cause severe injury to the cornea. High exposure to vapors is irritating.

Chronic

Mutagenicity: *In vitro*, mutagenic. *In vivo*, mutagenic. Human and animal inhalation studies show genetic material (DNA) damage, including hemoglobin alkylation, unscheduled DNA synthesis, sister chromatid chromosomal aberration, functional sperm abnormalities and dominant lethal effects.

Carcinogenicity: Carcinogenic to animals (inhaled, caused tumors and leukemia in rats at concentrations down to 10 ppm; caused tumors and lymphomas in mice at concentrations down to 50 ppm). Suspected of carcinogenic potential in humans (ACGIH). Listed as an experimental animal carcinogen and probable human carcinogen (IARC, OSHA, NTP).

Reproduction: No evidence of effect on female reproduction. Damages embryo and fetus. No evidence of malformed offspring. Causes dominant lethal effects (see "Mutagenicity" section).

Other: In animals, can damage the nervous system (for example, inhalation can cause paralysis of the hind legs of rats). Adversely affects blood and liver.

Medical conditions aggravated by exposure:

Significant exposure to this chemical may adversely affect people with chronic disease of the respiratory system, skin, blood, liver, central nervous system, kidneys and/or eyes.

Emergency and first aid procedures

Ingestion (swallowing): Patient should be made to drink large quantities of water. Then, induce vomiting by pressing finger down throat. Contact a physician immediately.

Inhalation (breathing): Remove patient from contaminated area. If breathing has stopped, give artificial respiration, then oxygen if needed. Contact a physician immediately.

Skin contact: Remove contaminated clothing. Wash contaminated skin with soap and water for 15 minutes. Contact a physician immediately.

Eye contact: Flush eyes with water for at least 15 minutes, lifting the upper and lower eyelids. Contact a physician immediately.

Spill or leak procedures

***Steps to be taken if material is released or spilled:**

Eliminate ignition sources. Ethylene oxide/air mixtures may detonate upon ignition. Avoid eye or skin contact; see "Special protection information" section for respirator information. Evacuate all personnel from the area except for those engaged in stopping the leak or in clean-up. Flood affected area with water. Use water spray to disperse vapors. Avoid runoff into storm sewers and ditches which lead to natural waterways. Call the National Response Center (800 424 8802) if the quantity spilled is equal to or greater than the reportable quantity (10 lb/day) under CERCLA "Superfund."

***Waste disposal method:**

All notification, clean-up and disposal should be carried out in accordance with federal, state and local regulations. Preferred methods of waste disposal are incineration or biological treatment in federal/state approved facility.

***Hazardous waste (40 CFR 261):**

Yes; hazardous waste codes U115, DOO1.

Special protection information

***Respiratory protection:**

Based on contamination level and working limits of the respirator, use a respirator approved by NIOSH/MSHA (the following are the minimum recommended equipment).

For ethylene oxide concentrations of:

$\geq$1 ppm and $\leq$50 ppm — Air-purifying respirator with full facepiece and ethylene oxide approved canister.

>50 ppm and <800 ppm — Positive-pressure full facepiece supplied-air respirator, or continuous-flow full facepiece supplied-air respirator.

$\geq$800 ppm or unknown concentration (such as in emergencies) — Positive-pressure self-contained breathing apparatus with full facepiece. Positive-pressure supplied-air respirator with full facepiece equipped with an auxiliary positive-pressure self-contained breathing apparatus escape system.

Ventilation

Local exhaust: Recommended as the sole means of controlling employee exposure.

Mechanical (general): Not recommended as the sole means of controlling employee exposure.

Protective gloves:

Butyl rubber.

Eye protection:

Chemical safety goggles. Contact lenses should not be worn if exposure to ethylene oxide is likely to occur.

***Additional protective equipment:**

For operations where spills or splashing can occur, use chemical protective clothing, including gloves and boots. A safety shower and eye bath should be readily available.

Special precautions

***Precautions to be taken in handling and storing:**

Storage vessels should be insulated and should have pressure relief valves. To avoid product contamination, install double-block valves on outlet of storage vessel. Electrical installations should be in accordance with Article 501 of the National Electrical Code. Do not incinerate ethylene oxide cartridges, tanks or other containers. Store in a cool, well-ventilated area. Replace or repair protective equipment that has been torn or otherwise damaged. Protective clothing wet with ethylene oxide should be immediately removed while under a safety shower. Decontaminate soiled clothing thoroughly before re-use. Contaminated leather articles should be destroyed. Do not expose to temperataures above 21°C (70°F). Keep away from heat, sparks and flame. Keep containers closed when not in use. Always open containers slowly to allow any excess pressure to vent. Use only DOT-approved containers. Use spark-resistant tools. Do not load into compartments adjacent to heated cargo. When transferring follow proper grounding procedures. Use with adequate ventilation. Do not store near combustible materials. Avoid breathing gas. Avoid contact with eyes, skin and clothing. Wash thoroughly with soap and water after handling. Do not enter storage area unless adequately ventilated.

*New or revised information; previous version dated October 1, 1991.

Glossary for Components information table

ACGIH - American Conference of Governmental Industrial Hygienists
CAS - Chemical Abstracts Service
Ceiling - The concentration that should not be exceeded during any part of the working day.
IDLH - Immediately Dangerous to Life or Health
OSHA - Occupational Safety and Health Administration
PEL - Permissible exposure limit
SARA - Superfund Amendments and Reauthorization Act
Skin - Potential contribution to overall exposure possible via skin absorption
STEL - Short-term exposure limit; 15-min. time-weighted average
TLV - Threshold limit value
TWA - 8-hour time-weighted average

Chemical Group
Hoechst Celanese Corporation
*P.O. Box 819005/Dallas, Texas 75381-9005
*Information phone: 214 277 4000
Emergency phone: 800 424 9300 (CHEMTREC)

Figure 3.23 Continued

Section VI — Health Hazard Data

- Routes of entry
- Health hazards (acute or chronic)
- Carcinogenicity
- NTP (National Toxicological Program)
- IARC (International Agency for Research on Cancer) monographs
- OSHA regulated
- Signs and symptoms of exposure
- Medical conditions aggravated by exposure
- Emergency and first aid procedures

Section VII — Precautions for Safe Handling and Use

- Steps to be taken in case material has been released or spilled
- Waste disposal methods
- Handling and storing precautions
- Other precautions

Section VIII — Control Measures

- Respiratory protection
- Ventilation (local, mechanical, special, other)
- Protective gloves
- Eye protection
- Other protective clothing or equipment
- Work/hygienic practices

[NFPA 472: 2-2.1.9.2]

Section II is the area of the MSDS in which the first responder can find the name(s) of the hazardous material(s). The other sections describe, to the responder, the characteristics of the material and the actions that need to be taken. U.S. material safety data sheets are not accepted in Canada because they are slightly different from the Canadian MSDS.

First responders can avoid considerable confusion if the MSDS is used in pre-incident planning. This requires identifying the buildings that store or processes that use extremely toxic materials and then acquiring the MSDS to place with the pre-incident plan.

There are many other resources available to the responder for gathering information about a material and also for learning actions that should be taken to counteract a hazardous situation. The first responder should consult a minimum of two resources for information about a material. Consulting more than one resource allows the responder to verify the information found. If the resources used are not in agreement, the responder should take notice and use the information that is most conservative. If the various resources disagree, then the responder must do further research by using other available resources.

HAZARDOUS MATERIALS IDENTIFICATION AT FIXED FACILITIES

[29 CFR 1910.120(q)(6)(i)(c)]; [NFPA 472: 2-2.1];
[29 CFR 1910.120(q)(6)(i)(d)]; [NFPA 472: 2-2.2.3];
[NFPA 472: 3-2.1.3]

A hazardous material is most easily identified before it is released from its container. Much of the first responder's haz mat identification training can be performed at local commercial and industrial (fixed) facilities. Surveys at these sites allow the responder to document the location of hazardous materials and the physical layout of the plant (Figure 3.24). At these surveys, the responder becomes familiar with the colors, shapes, labels, and construction of many material containers. Responders should be particularly observant of materials commonly used in their response area and of those materials most likely to be encountered during an

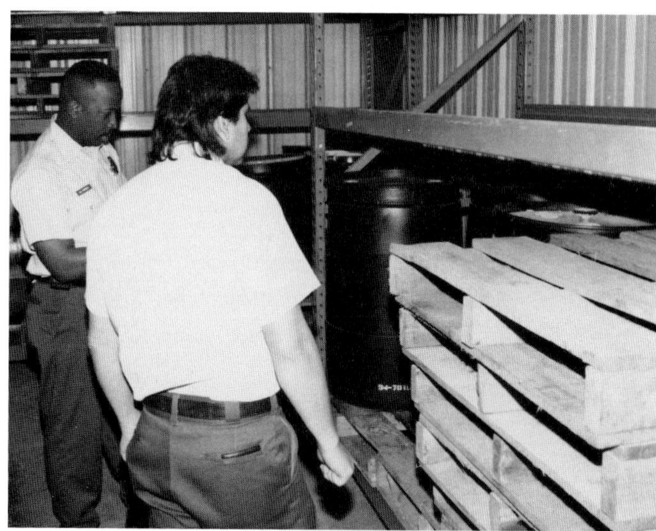

Figure 3.24 Inspect all occupancies for hazardous materials.

incident. Most important, the responder can research the characteristics of detected substances at length without the hazards, stress, and potential errors present in an emergency.

The information gathered when visiting fixed facilities is invaluable in performing pre-incident planning. Pre-incident plans are crucial when personnel face emergencies at facilities containing hazardous materials. First responders will often participate in the gathering of information for pre-incident plans, although they will typically be finalized by command staff.

Identifying The Fire Hazards Of Materials At Fixed Facilities

[NFPA 472: 2-2.2.1]

Hazardous materials that are manufactured, stored, processed, or used at a particular site are not subject to regulations affecting transported materials. Local agencies and county, city, and township governments may adopt their own identification system or use a widely recognized method such as that recommended by NFPA 704, *Standard System for the Identification of the Fire Hazards of Materials*. Some private companies and the military have created their own internal policies that regulate the marking of these materials.

[NFPA 472: 2-2.1.7(b)]

NFPA 704 is an acceptable identification system that can be implemented in an area where materials are regularly stored and used (Figure 3.25). The NFPA 704 system is an important aid in determining the actions and safety procedures to be used in the initial phases of a haz mat incident. The system is a widely recognized method for indicating the presence of hazardous materials at commercial, manufacturing, institutional, and other fixed-storage facilities. Use of this system is commonly required by local ordinances for all occupancies that contain hazardous materials. The NFPA 704 system is not designed to be used for the following:

- Transportation
- General public use
- Chronic (repeated, long-term) exposures
- Nonemergency occupational exposures

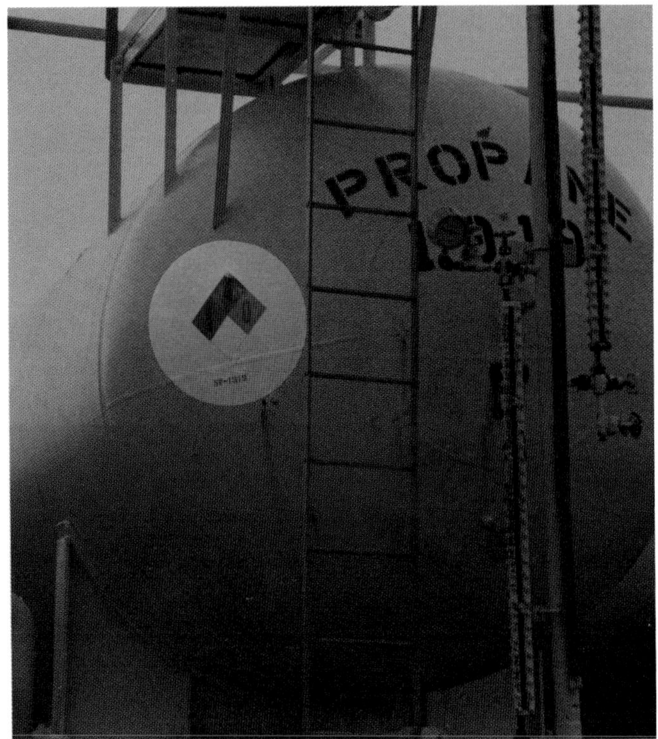

Figure 3.25 The 704 symbols are displayed conspicuously on the outside of the building or storage vessel.

The NFPA 704 system offers the following:

- It provides the appropriate signal or alert to first responders that hazardous materials are present. The first-arriving responder who sees the NFPA 704 marker on a structure can determine the hazards of a single material in a marked container or can determine the relative combined hazard severity of the collection of numerous materials in the occupancy (Figure 3.26).
- It identifies the general hazards and the degree of severity for health, flammability, and reactivity.

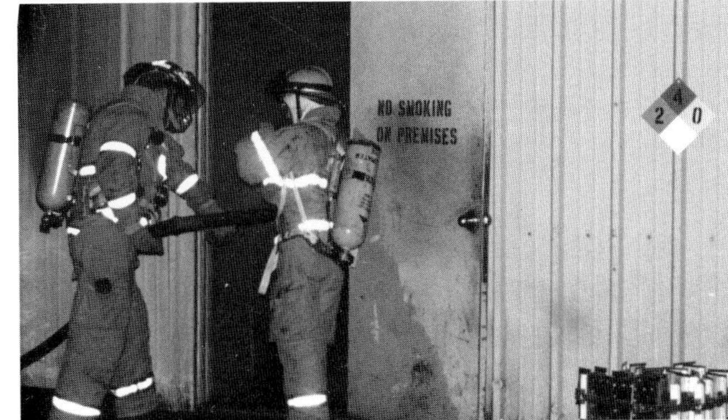

Figure 3.26 First responders should note the 704 symbol before entering a structure.

- It provides immediate information necessary to protect the lives of both the public and emergency response personnel.

The system does not identify the specific chemical or chemicals that may be present. Positive identification should be made through other means such as container markings, employee information, company records, and the pre-incident plan.

[NFPA 472: 2-2.1.7.1]; [NFPA 472: 2-2.1.8]

Specifically, the NFPA 704 system uses a rating system of zero (0) to four (4). A zero indicates there is no hazard present, and a four represents a severe hazard. The rating is assigned to three categories: health, flammability, and reactivity. The rating numbers are arranged on a diamond-shaped marker or sign (Figure 3.27). The health rating is located on a blue background at the nine o'clock position. The flammability hazard rating is positioned on a red background at the twelve o'clock position. The reactivity hazard rating appears on a yellow background and is positioned at three o'clock. As an alternative, the backgrounds for each of these rating positions may be any contrasting color, and the numbers (0 to 4) may be represented by the appropriate color (blue, red, and yellow) (Figure 3.28). Special hazards are located in the six o'clock position and have no specified background color; however, white is most commonly used.

Figure 3.27 Most 704 symbols have colored blocks and black or white numerals.

Figure 3.28 Some 704 symbols have white blocks and appropriately colored numerals.

The ratings for each hazard (health, flammability, and reactivity) are described in Table 3.3. This table also describes the special hazards that may be indicated on the NFPA 704 marker. The NFPA 704 system is used in conjunction with NFPA 49, *Hazardous Chemicals Data.* NFPA 49 describes the properties and hazards of various materials and provides information on personal protection and fire fighting when facing these specific chemicals. Valuable information is given on assigning appropriate ratings to the NFPA 704 markers at facilities that contain listed chemicals.

Identifying Containers At Fixed Facilities

[NFPA 472: 3-2.1.2]; [NFPA 472: 2-2.1.6]; [NFPA 472: 3-2.1.2.2]; [NFPA 472: 3-2.1.1]

First responders cannot depend exclusively on signs, symbols, labels, or other identification systems to ensure their safety. These systems are only as reliable as the people who place them on the containment system and maintain them. There is always the possibility of human error. Therefore, the first responder who faces a potential haz mat incident must evaluate each situation as it arises. The initial report may include information that hazardous materials are involved. The first responder may recognize the location of the incident or the type of occupancy as one that handles hazardous materials. The presence of certain storage vessels, containers, or vehicles may alert responding crews to hazardous materials. Certainly the employees of a plant should be questioned to deter-

TABLE 3.3
NFPA 704 Rating System

Identification of Health Hazard		Identification of Flammability		Identification of Reactivity	
Type of Possible Injury		Susceptibility of Materials to Burning		Susceptibility to Release of Energy	
Signal		Signal		Signal	
4	Materials that on very short exposure could cause death or major residual injury.	4	Materials that will rapidly or completely vaporize at atmospheric pressure and normal ambient temperature, or that are readily dispersed in air and that will burn readily.	4	Materials that in themselves are readily capable of detonation or of explosive decomposition or reaction at normal temperatures and pressures.
3	Materials that on short exposure could cause serious temporary or residual injury.	3	Liquids and solids that can be ignited under almost all ambient temperature conditions.	3	Materials that in themselves are capable of detonation or explosive decomposition or reaction but require a strong initiating source or which must be heated under confinement before initiation or which react explosively with water.
2	Materials that on intense or continued but not chronic exposure could cause temporary incapacitation or possible residual injury.	2	Materials that must be moderately heated or exposed to relatively high ambient temperatures before ignition can occur.	2	Materials that readily undergo violent chemical change at elevated temperatures and pressures or which react violently with water or which may form explosive mixtures with water.
1	Materials that on exposure would cause irritation but only minor residual injury.	1	Materials that must be preheated before ignition can occur.	1	Materials that in themselves are normally stable, but which can become unstable at elevated temperatures and pressures.
0	Materials that on exposure under fire conditions would offer no hazard beyond that of ordinary combustible material.	0	Materials that will not burn.	0	Materials that in themselves are normally stable, even under fire exposure conditions, and which are not reactive with water.

mine whether hazardous materials are involved. Pre-incident plans should also contain information on materials stored at a given facility.

In some cases, first responders may be able to detect the presence of hazardous materials by recognizing types of containers found at the facility. Typically, the first responder's scope of responsibility ends with the following:

- Recognition of the type of container
- Identification of the material in the container
- Transmission of this information to an appropriate authority

This information is then used to form an incident action plan.

Many types of containers are used to store hazardous and nonhazardous materials; however, the most common types can be easily grouped and recognized. In many cases, particular types of containers are almost always used to carry specific materials. Knowing this information, the first responder can make some basic assumptions about the nature of the material present in a given container. Some containers are so unusual that their presence strongly suggests that hazardous materials are involved.

CAUTION: Do not make a conclusive determination of the material within a container based solely on the type or shape of the container. Such an assumption could lead to incorrect action and place responders in unnecessary danger.

In general, aboveground storage tanks are divided into three major categories: atmospheric tanks, low-pressure tanks, and pressure vessels. Underground storage tanks are also common for some occupancies. The following sections highlight the features of each of these tanks.

ATMOSPHERIC TANKS

[NFPA 472: 3-2.1.1.4(a)]

Atmospheric storage tanks are designed to hold contents under little or no pressure. The maximum pressure under which any atmospheric tank is capable of holding its contents is 0.5 psig (4 kPa). There are six common types of atmospheric storage tanks:

- Ordinary cone roof tank
- Floating roof tank
- Lifter roof tank
- Internal floating roof tank
- Vapordome roof tank
- Horizontal tank

Ordinary Cone Roof Tank

An ordinary cone roof tank stores flammable, combustible, and corrosive liquids. It has a cone-shaped, pointed roof. It is designed with a weak roof-to-shell seam, intended to break when/if the container becomes overpressurized (Figure 3.29). A disadvantage of this type of tank is that when it is partially full, the remaining portion of the tank contains a potentially dangerous vapor space. The vapor space is explosive if the area is exposed to an ignition source.

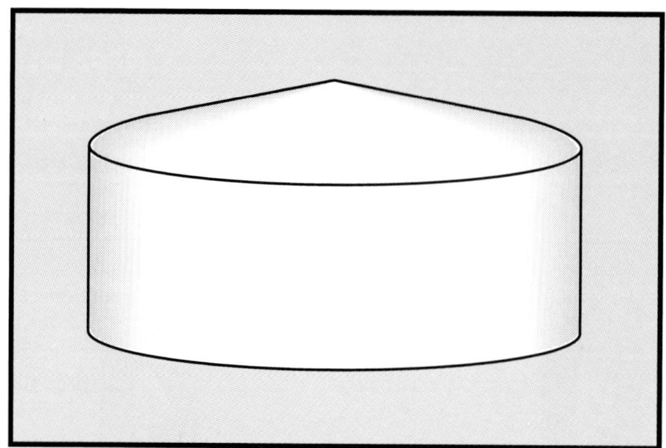

Figure 3.29 An ordinary cone roof tank.

Floating Roof Tank

A floating roof tank is a large-capacity, aboveground holding tank that is commonly used to store flammable and combustible liquids, particularly petroleum products. It stands vertically and is usually much wider than it is tall. It is designed so that the roof actually floats on the surface of the liquid (Figures 3.30 a and b). This eliminates the potentially dangerous vapor space. Evaporation and condensation buildup are also greatly reduced. The roof slides up and down the tank walls as the volume of the container changes. A fabric or rubber seal around the circumference of the roof provides a weather-tight seal (Figure 3.31). Its capacity can range from 50,000 gallons (200 000 L) to over 1,000,000 gallons (4 000 000 L).

Lifter Roof Tank

A lifter roof tank is similar in appearance to a floating roof tank, but there are some major differences between the two types. A lifter roof floats within a series of vertical guides that allow only a few feet of travel (Figure 3.32). The roof is either liquid- or fabric-sealed and moves up and down with changes in vapor pressure. The roof is designed so that when the vapor pressure exceeds a designated limit, the roof lifts up slightly and relieves the excess pressure. A lifter roof tank is usually used to store volatile liquids.

Internal Floating Roof Tank

The internal floating roof tank is a combination of the floating roof tank and the ordinary cone roof tank. It has a fixed cone roof and either a pan or deck-type float inside the tank that rides directly on

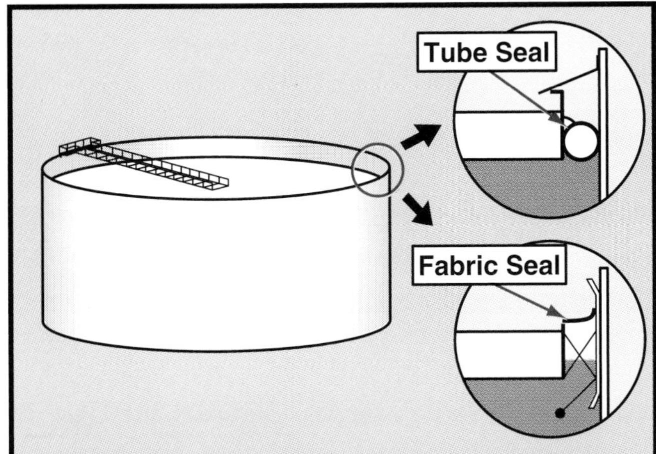

Figure 3.30a Floating roof tanks have different types of seals.

Figure 3.30b A typical floating roof tank.

Figure 3.31 A view of the floating roof tank seal.

the product surface (Figures 3.33 a and b). Vents are placed in the fixed cone roof to allow for the necessary adjustment in air pressures created by the movement of the float. These tanks are used primarily to store flammable liquids.

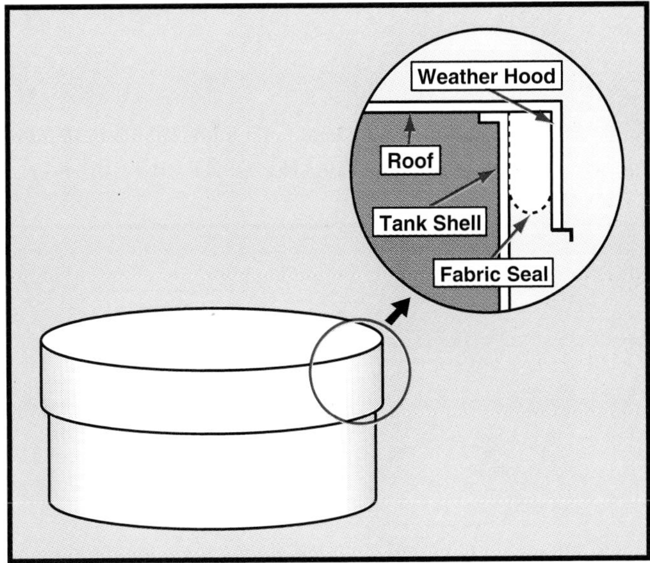

Figure 3.32 A lifter roof tank.

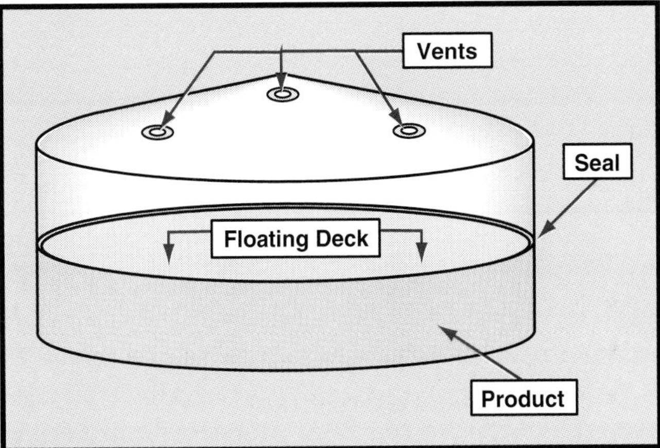

Figure 3.33a The internal floating roof tank has a fixed roof and a floating pan inside.

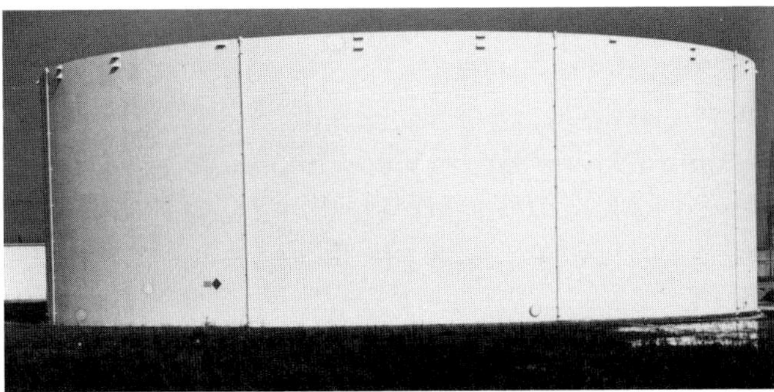

Figure 3.33b From the outside, the internal floating roof tank looks similar to an ordinary cone roof tank. However, note the air vents around the top rim of the tank.

Vapordome Roof Tank

A vapordome roof tank is a vertical storage tank that has a giant bulge or dome on its top (Figure 3.34). Attached to the underside of the dome is a flexible diaphragm that moves in conjunction with changes in vapor pressure. This design is used for combustible liquids of medium volatility. Some nonhazardous materials, such as molasses and fertilizer blends, may be stored in this style of tank. Vapordome roof tanks range in size up to a maximum of about 8,500,000 gallons (34 000 000 L).

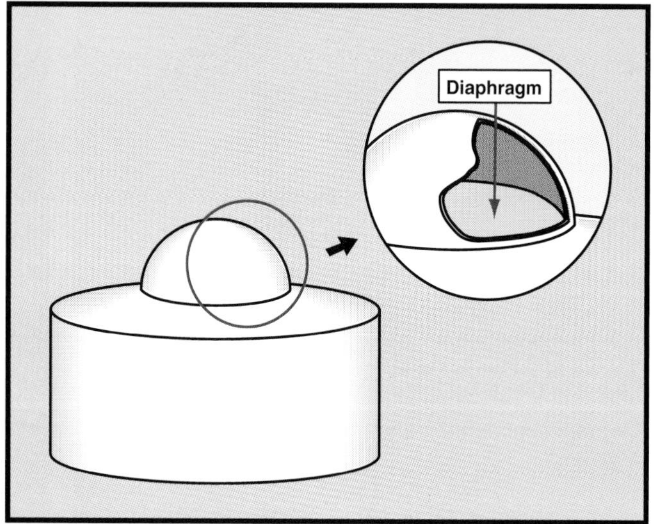

Figure 3.34 A vapordome tank.

Horizontal Tank

A horizontal tank is typically constructed of steel. Its capacities may range from a few thousand gallons (liters) to 20,000 gallons (80 000 L). It is commonly used for bulk storage in conjunction with fuel-dispensing operations. A horizontal tank supported by unprotected steel supports or stilts may fail quickly during fire conditions (Figure 3.35).

Figure 3.35 Horizontal tanks rest either on stilts or directly on the ground.

LOW-PRESSURE STORAGE TANKS

[NFPA 472: 3-2.1.1.4(b)]

A low-pressure storage tank is designed to have an operating pressure that ranges from 0.5 to 15 psig (4 kPa to 105 kPa). The two common types of low-pressure tanks are the spheroid and the noded spheroid.

Spheroid Tank

Know

A spheroid tank is designed to store liquid or gaseous commodities such as LPG, methane, propane, and other light gases (Figure 3.36). Certain flammable liquids, such as gasoline and crude oil, are also stored in these tanks. Some city water departments use this design for storage of domestic water. This tank can store 3,000,000 gallons (12 000 000 L) or more. A pressure-relief valve is located on top of the tank (Figure 3.37).

Noded Spheroid Tank

A noded spheroid tank is used for the same purpose as that of a spheroid tank. The two tanks differ in appearance and size. A noded spheroid tank can be substantially larger and flatter in shape, and it contains bulging, ribbed sections (Figure 3.38). Unlike the spheroid tank, the noded spheroid tank is held together by a series of internal ties and supports that reduce stress on the external shell. This tank also has a pressure-relief valve on its top.

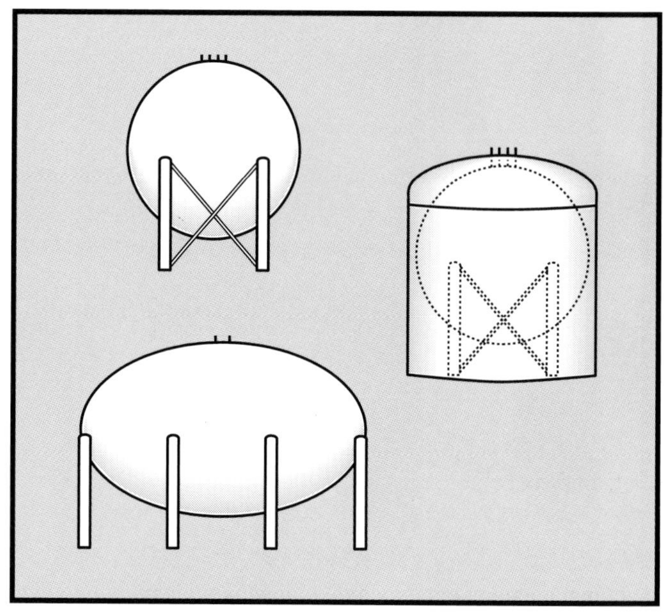

Figure 3.36 There are various types of low-pressure spheroid tanks.

Figure 3.37 A typical relief valve on a spheroid tank.

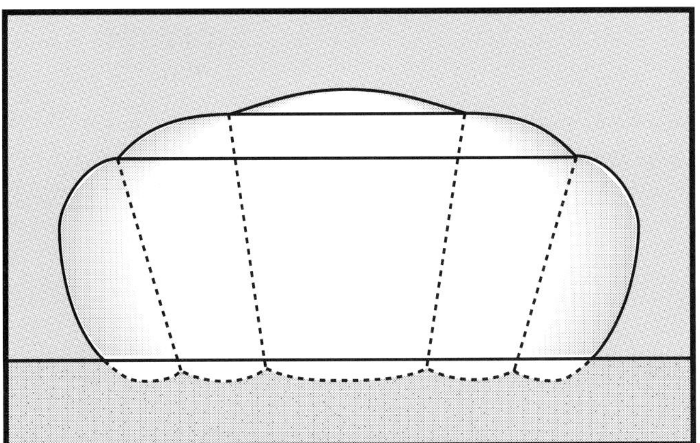

Figure 3.38 The basic design of a noded spheroid tank.

PRESSURE VESSELS

[NFPA 472: 3-2.1.1.4(b)]

A pressure vessel stores compressed or liquefied gases. It is designed to have an operating pressure above 15 psig (105 kPa). There are several types of pressure vessels: horizontal tanks, sphere tanks, and cryogenic-liquid storage tanks.

Horizontal Tank

A horizontal tank is readily identified by its rounded ends that signify a pressure vessel (Figure 3.39). Its capacities may range from 500 to over 40,000 gallons (2 000 L to over 160 000 L). Large industrial or storage locations may have several stored together (Figure 3.40). Horizontal tanks are commonly found at facilities that dispense fuel gases to the public. Substances commonly found in horizontal tanks include the following: propane, liquefied natural gas (LNG), compressed natural gas (CNG), butane, ethane, ammonia, sulfur dioxide, chlorine, and hydrogen chloride.

Figure 3.39 Horizontal pressure tanks are commonly found in a variety of locations.

Figure 3.40 Some occupancies may have more than one horizontal pressure tank on their property.

Sphere Tank

A sphere tank is a single-shell, noninsulated tank that has a capacity of up to 600,000 gallons (2 400 000 L). It can be identified by its round, ball-like appearance (Figure 3.41). The sphere tank is often supported off the ground by a series of concrete or steel legs. A sphere tank is usually painted white or some other highly reflective color. Painting the tank white reduces the heat level and the resulting amount of vaporization that occurs inside the tank. Liquefied petroleum gases are commonly stored in pressure sphere tanks.

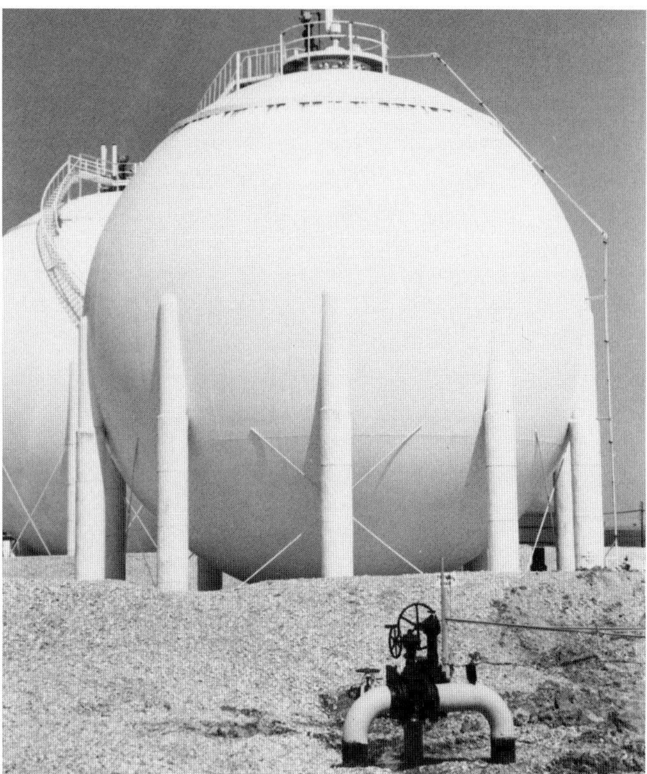

Figure 3.41 A typical high-pressure sphere tank.

Cryogenic-Liquid Storage Tank

A cryogenic-liquid storage tank is an insulated, vacuum-jacketed tank with safety relief valves and rupture disks. Its capacity can range from 300 to 400,000 gallons (1 200 L to 1 600 000 L). The pressure varies according to the material stored and its use (Figure 3.42).

Figure 3.42 Several large cryogenic storage tanks.

UNDERGROUND STORAGE TANKS

Underground storage tanks are found primarily at gasoline service stations and large private garages. This type of tank can be buried under a building or driveway or can be buried adjacent to the occupancy (Figure 3.43). Underground tanks are constructed of steel or fiberglass. This tank has fill and vent connections near the location of the tank (Figure 3.44).

Figure 3.43 One indication of an underground tank is the cover over the fill connection(s).

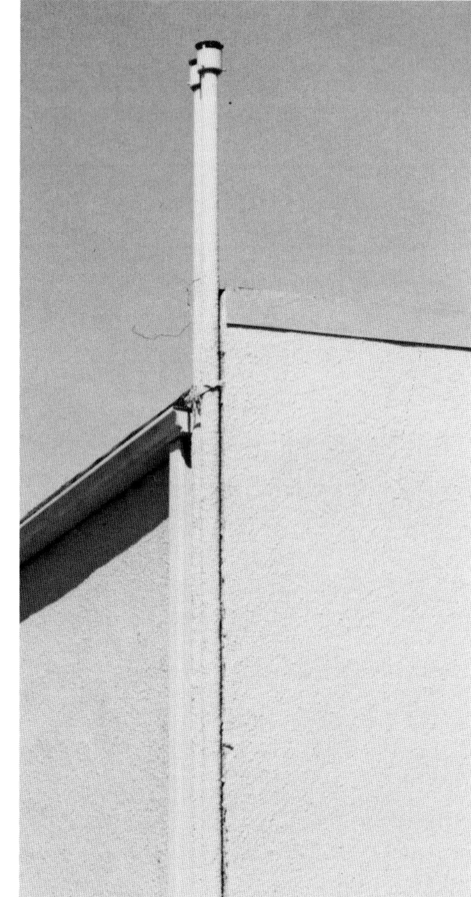

Figure 3.44 Underground tanks must be vented.

HAZARDOUS MATERIALS IDENTIFICATION IN TRANSPORTATION

[29 CFR 1910.120(q)(6)(i)(c)]; [NFPA 472: 2-2.1];
[29 CFR 1910.120(q)(6)(i)(d)]; [NFPA 472: 3-2.1.3]

Identifying the hazardous materials in a transportation accident can be a difficult task for the first responder. However, without first positively identifying the contents of a transportation vehicle, the first responder may inadvertently turn what should be a minor incident into a major catastrophe.

The federal Hazardous Materials Transportation Act (HMTA) gives the Secretary of Transportation regulatory authority to protect the public against the risks associated with the transportation of hazardous materials. The Department of Transportation (DOT) is the federal agency that develops, publishes, and enforces regulations for the safe transportation of hazardous materials. The DOT carries out these duties through a program of regulation, enforcement, emergency response education and training, and data collection and analysis.

[NFPA 472: 2-2.2.1]

Regulations that govern the transportation of roughly 500,000 haz mat shipments per day are extremely complex. The regulations establish safety standards for shippers and carriers and for the packaging, loading, and hauling of some 30,000 hazardous materials and wastes by air, water, highway, and rail. The complexity of the regulations is compounded by the different and overlapping responsibilities of federal, state, and local governments. At the federal level, several of the following agencies are involved in the regulation of hazardous materials and/or wastes:

- Department of Transportation (DOT) — In conjunction with the Transport Development Group has overall regulatory authority
- Transport Development Group (TDG) — In conjunction with the Department of Transportation has overall regulatory authority
- Nuclear Regulatory Commission (NRC) — Regulates the possession, use, and transport of radioactive materials
- Environmental Protection Agency (EPA) — Establishes requirements for the transportation of hazardous substances and wastes
- Occupational Safety and Health Administration (OSHA) — Regulates the health and safety of first responders and workers employed by shippers and carriers of hazardous materials
- Department of Energy (DOE) — Regulates companies that generate power (e.g., electric companies)

United Nations Classification System

Both the United States and Canada have adopted the United Nations (UN) system for classifying and identifying hazardous materials transported both internationally and domestically. Under this system, nine hazard classes are used to categorize hazardous materials. In addition to these nine classes, a separate category exists for other regulated materials (ORM-D). The nine hazard classes used for categorizing hazardous materials are as follows:

Class 1 — Explosives

Class 2 — Gases

Class 3 — Flammable Liquids

Class 4 — Flammable Solids

Class 5 — Oxidizers

Class 6 — Poisons and Infectious Substances

Class 7 — Radioactive Substances

Class 8 — Corrosives

Class 9 — Miscellaneous

[NFPA 472: 2-2.1.2]

The UN system forms the basis for the DOT regulations (Table 3.4). The DOT classifies hazardous materials according to their primary danger and assigns standardized symbols to identify the classes; this is similar to what the UN system has done. DOT regulations cover several other types of substances in addition to the nine classes identified in the UN system. The major classes and a brief description of each are given in the following sections (Figures 3.45 a and b).

TABLE 3.4
Examples Of Department Of Transportation Hazardous Materials Classes

Hazard Class	Product Example
1 Explosives	
1.1 Mass explosion hazard	Black powder
1.2 Projection hazard	Detonating cord
1.3 Fire hazard	Propellant explosives
1.4 No significant blast	Practice ammunition
1.5 Very insensitive	Prilled ammonium nitrate
1.6 Extremely insensitive	Fertilizer-fuel oil mixtures
2 Gases	
2.1 Flammable gas	Propane
2.2 Nonflammable gas	Anhydrous ammonia
2.3 Poisonous gas	Phosgene
2.4 Corrosive gas (Canada)	
3 Flammable Liquids	Gasoline, kerosene, diesel fuel
4 Flammable Solids, Spontaneously Combustible Materials, and Materials that are Dangerous When Wet	
4.1 Flammable solids	Magnesium
4.2 Spontaneously combustible	Phosphorus
4.3 Dangerous when wet	Calcium carbide
5 Oxidizers and Organic Peroxides	
5.1 Oxidizers	Ammonium nitrate
5.2 Organic peroxides	Ethyl ketone peroxide
6 Poisonous and Etiologic Materials	
6.1 Poisonous	Arsenic
6.2 Infectious (etiological agent)	Rabies, HIV, Hepatitis B
7 Radioactive Materials	Cobalt
8 Corrosives	Sulfuric acid, sodium hydroxide
9 Miscellaneous Hazardous Materials	
9.1 Miscellaneous (Canada Only)	PCBs, molten sulfur
9.2 Environmental Hazard (Canada Only)	PCB, asbestos
9.3 Dangerous Waste (Canada Only)	Fumaric acid
ORM-D (Other regulated materials)	Consumer commodities

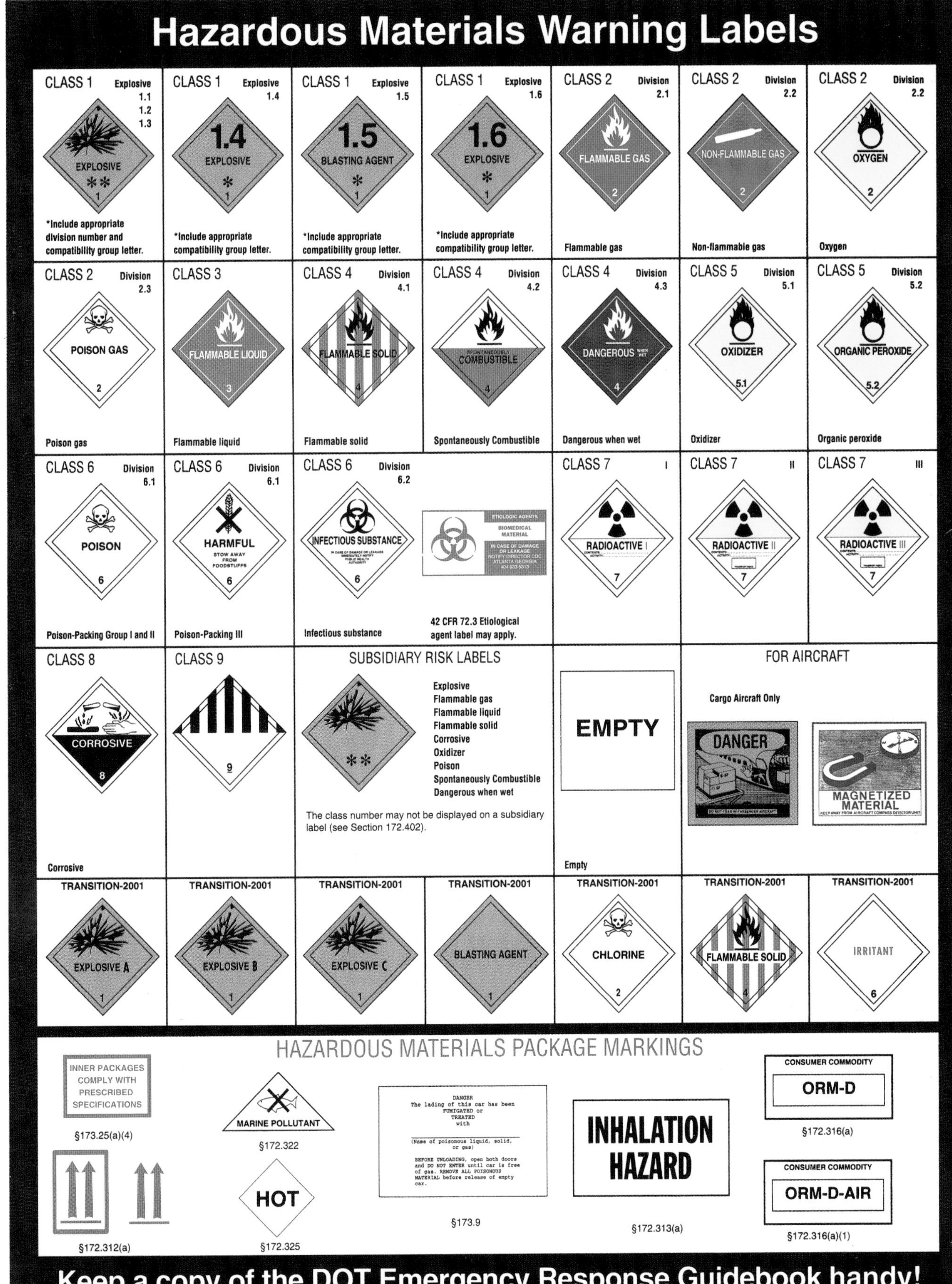

Figure 3.45a U.S. placards and markings related to the UN system.

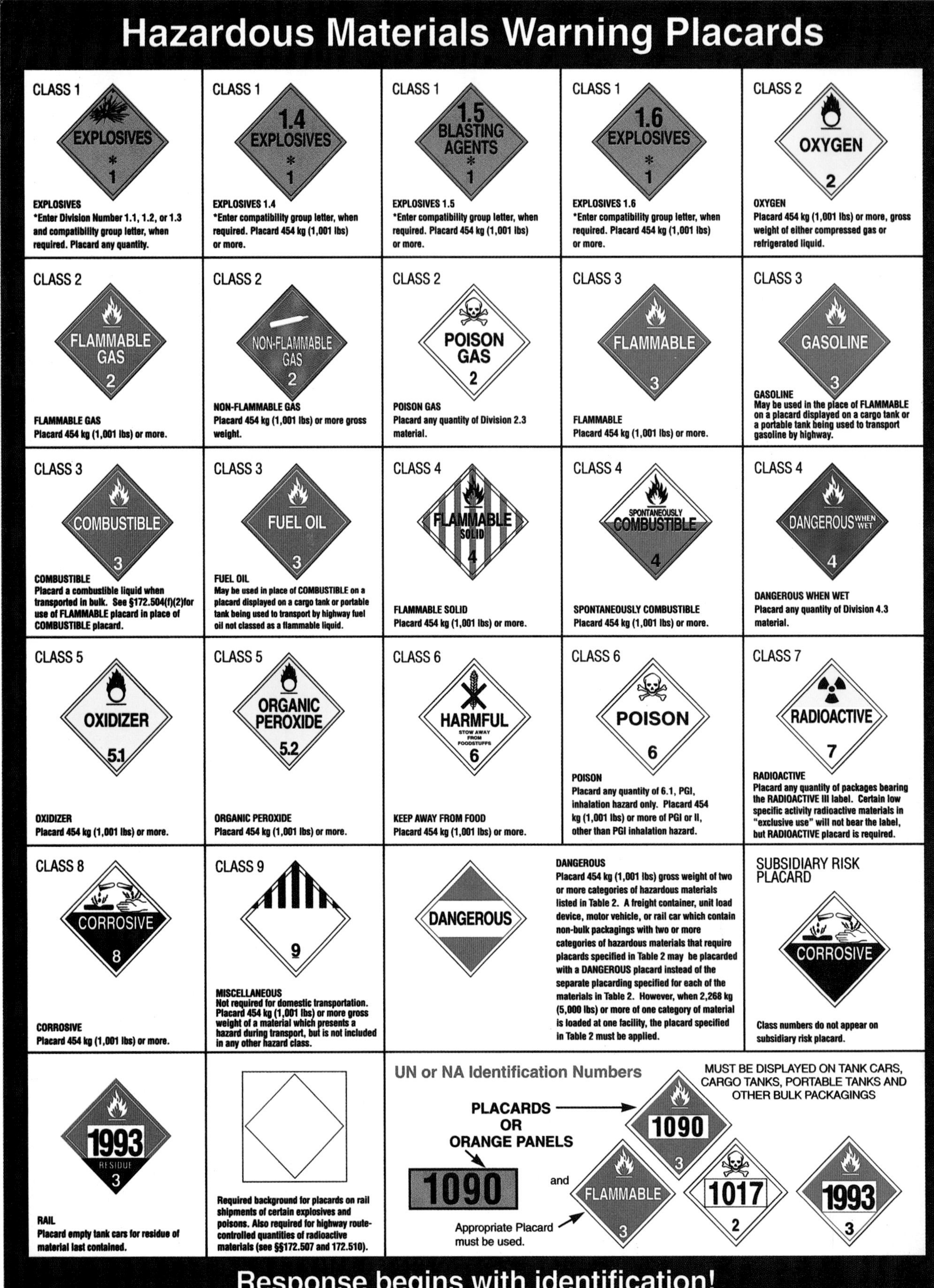

Figure 3.45a Continued.

General Guidelines on Use of Warning Labels and Placards

LABELS

See 49 CFR, Part 172, Subpart E for complete labeling regulations.

- Until October 1, 1993, all of the labels appearing on the Hazardous Materials Warning Labels chart may be used to satisfy the labeling requirements contained in Subpart E.
- On and after October 1, 1993, those labels in boxes marked "TRANSITION-2001" on the chart will not be authorized for use under Subpart E. (NOTE: these labels may be used IF they were affixed to a package offered for transportation and transported prior to October 1, 2001, and the package was filled with hazardous materials prior to October 1, 1991.)
- For classes 1,2,3,4,5,6 and 8, text indicating a hazard (e.g., "CORROSIVE") IS NOT required on a label. The label must otherwise conform to Subpart E [Section 172.405].
- Any person who offers a hazardous material for transportation MUST label the package, if required [Section 172.400(a)].
- The Hazardous Materials Table [Section 172.101] identifies the proper label(s) for the hazardous material listed.
- When required, labels must be printed on or affixed to the surface of the package near the proper shipping name [Section 172.406(a)].
- When two or more labels are required, they must be displayed next to each other [Section 172.406(c)].
- Labels may be affixed to packages when not required by regulations, provided each label represents a hazard of the material contained in the package [Section 172.401].

PLACARDS

See 49 CFR, Part 172, Subpart F for complete placarding regulations.

- All of the placards appearing on the Hazardous Materials Warning Placards chart may be used to satisfy the placarding requirements contained in Subpart F.
- Each person who offers for transportation or transports any hazardous material subject to the Hazardous Materials Regulations shall comply with all applicable requirements of Subpart F.
- Placards may be displayed for a hazardous material even when not required, if the placarding otherwise conforms to the requirements of Subpart F.
- For other than Class 7 or the OXYGEN placard, text indicating a hazard (e.g., "CORROSIVE") is not required on a placard [Section 172.519(b)].
- Any transport vehicle, freight container, or rail car containing any quantity of material listed in Table 1 (Section 172.504) must be placarded.
- When the gross weight of all hazardous materials covered in Table 2 is less than 454 kg (1,001 lbs), no placard is required on a transport vehicle or freight container [Section 172.504].

Effective October 1, 1994, and extending through October 1, 2001, these placards may be used for HIGHWAY TRANSPORTATION ONLY.

Illustration numbers in each square refer to Tables 1 and 2 below.

Poisonous Materials

§172.554

INHALATION HAZARD

§172.313

Materials which meet the inhalation toxicity criteria have additional "communication standards" prescribed by the HMR. First, the words "Poison-Inhalation Hazard" must be entered on the shipping paper, as required by Section 172.203(m)(3). Second, packagings must be marked "Inhalation Hazard" in accordance with Section 172.313(a). Lastly, transport vehicles, freight containers, portable tanks and unit load devices that contain a poisonous material subject to the "Poison-Inhalation Hazard" shipping description, must be placarded with a POISON or POISON GAS placard, as appropriate. This shall be in addition to any other placard required for that material in Section 172.504.

Table 1 (Placard any quantity)

Hazard class or division	Placard name
1.1	EXPLOSIVES 1.1
1.2	EXPLOSIVES 1.2
1.3	EXPLOSIVES 1.3
2.3	POISON GAS
4.3	DANGEROUS WHEN WET
6.1 (PGI, PIH only)	POISON
7 (Radioactive Yellow III)	RADIOACTIVE

Table 2 (Placard 1,001 pounds or more)

Hazard class or division	Placard name
1.4	EXPLOSIVES 1.4
1.5	EXPLOSIVES 1.5
1.6	EXPLOSIVES 1.6
2.1	FLAMMABLE GAS
2.2	NON-FLAMMABLE GAS
3	FLAMMABLE
Combustible Liquid	COMBUSTIBLE
4.1	FLAMMABLE SOLID
4.2	SPONTANEOUSLY COMBUSTIBLE
5.1	OXIDIZER
5.2	ORGANIC PEROXIDE
6.1 (PGI or II, other than PGI PIH)	POISON
6.1 (PGIII)	KEEP AWAY FROM FOOD
6.2	NONE
8	CORROSIVE
9	CLASS 9
ORM-D	NONE

For complete details, refer to one or more of the following:

- Code of Federal Regulations, Title 49, Transportation. Parts 100-199. [All modes]
- International Civil Aviation Organization (ICAO) Technical Instructions for Safe Transport of Dangerous Goods by Air [Air]
- International Maritime Organization (IMO) Dangerous Goods Code [Water]
- "Transportation of Dangerous Goods Regulations" of Transport Canada. [All Modes]

U.S. Department of Transportation
Research and Special Programs Administration

Copies of this Chart can be obtained by writing OHMIT/DHM-51, Washington, D.C. 20590

CHART 10
REV. FEBRUARY 1994

Figure 3.45a Continued.

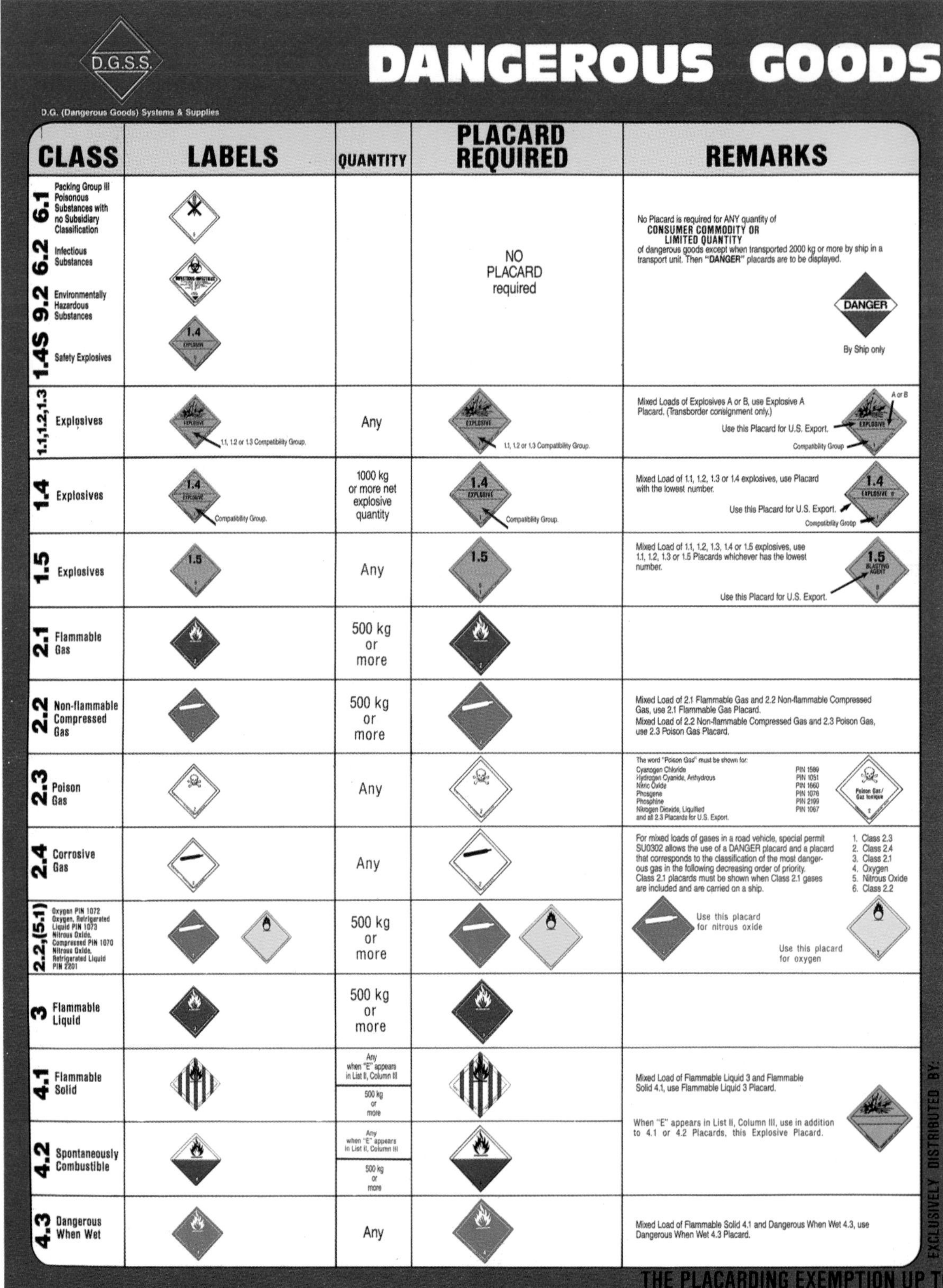

D.G.S.S.

D.G. (Dangerous Goods) Systems & Supplies

DANGEROUS GOODS

CLASS	LABELS	QUANTITY	PLACARD REQUIRED	REMARKS
6.1 Packing Group III Poisonous Substances with no Subsidiary Classification **6.2** Infectious Substances **9.2** Environmentally Hazardous Substances **1.4S** Safety Explosives			NO PLACARD required	No Placard is required for ANY quantity of **CONSUMER COMMODITY OR LIMITED QUANTITY** of dangerous goods except when transported 2000 kg or more by ship in a transport unit. Then "**DANGER**" placards are to be displayed. DANGER By Ship only
1.1,1.2,1.3 Explosives	1.1, 1.2 or 1.3 Compatibility Group.	Any	1.1, 1.2 or 1.3 Compatibility Group.	Mixed Loads of Explosives A or B, use Explosive A Placard. (Transborder consignment only.) A or B Use this Placard for U.S. Export. Compatibility Group
1.4 Explosives	Compatibility Group.	1000 kg or more net explosive quantity	Compatibility Group.	Mixed Load of 1.1, 1.2, 1.3 or 1.4 explosives, use Placard with the lowest number. Use this Placard for U.S. Export. Compatibility Group
1.5 Explosives		Any		Mixed Load of 1.1, 1.2, 1.3, 1.4 or 1.5 explosives, use 1.1, 1.2, 1.3 or 1.5 Placards whichever has the lowest number. Use this Placard for U.S. Export.
2.1 Flammable Gas		500 kg or more		
2.2 Non-flammable Compressed Gas		500 kg or more		Mixed Load of 2.1 Flammable Gas and 2.2 Non-flammable Compressed Gas, use 2.1 Flammable Gas Placard. Mixed Load of 2.2 Non-flammable Compressed Gas and 2.3 Poison Gas, use 2.3 Poison Gas Placard.
2.3 Poison Gas		Any		The word "Poison Gas" must be shown for: Cyanogen Chloride PIN 1589 Hydrogen Cyanide, Anhydrous PIN 1051 Nitric Oxide PIN 1660 Phosgene PIN 1076 Phosphine PIN 2199 Nitrogen Dioxide, Liquified PIN 1067 and all 2.3 Placards for U.S. Export. Poison Gas/ Gaz toxique
2.4 Corrosive Gas		Any		For mixed loads of gases in a road vehicle, special permit SU0302 allows the use of a DANGER placard and a placard that corresponds to the classification of the most dangerous gas in the following decreasing order of priority. Class 2.1 placards must be shown when Class 2.1 gases are included and are carried on a ship. 1. Class 2.3 2. Class 2.4 3. Class 2.1 4. Oxygen 5. Nitrous Oxide 6. Class 2.2
2.2,(5.1) Oxygen PIN 1072 Oxygen, Refrigerated Liquid PIN 1073 Nitrous Oxide, Compressed PIN 1070 Nitrous Oxide, Refrigerated Liquid PIN 2201		500 kg or more		Use this placard for nitrous oxide Use this placard for oxygen
3 Flammable Liquid		500 kg or more		
4.1 Flammable Solid		Any when "E" appears in List II, Column III 500 kg or more		Mixed Load of Flammable Liquid 3 and Flammable Solid 4.1, use Flammable Liquid 3 Placard. When "E" appears in List II, Column III, use in addition to 4.1 or 4.2 Placards, this Explosive Placard.
4.2 Spontaneously Combustible		Any when "E" appears in List II, Column III 500 kg or more		
4.3 Dangerous When Wet		Any		Mixed Load of Flammable Solid 4.1 and Dangerous When Wet 4.3, use Dangerous When Wet 4.3 Placard.

EXCLUSIVELY DISTRIBUTED BY:

THE PLACARDING EXEMPTION UP T

Figure 3.45b The *Canadian Dangerous Goods Placarding Guide. Courtesy of Markmaster-Monette.*

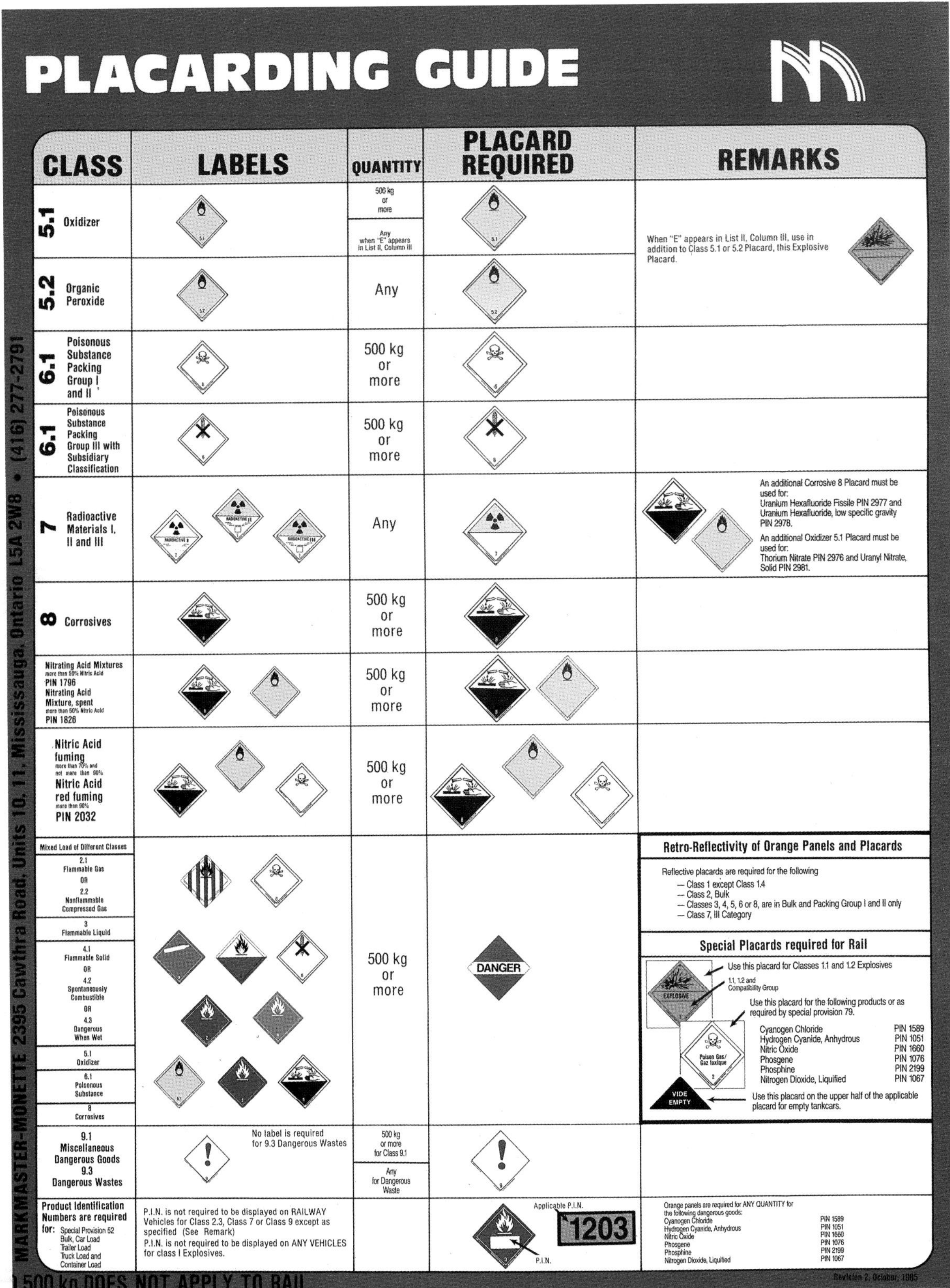

PLACARDING GUIDE

MARKMASTER-MONETTE 2395 Cawthra Road, Units 10, 11, Mississauga, Ontario L5A 2W8 • (416) 277-2791

CLASS	LABELS	QUANTITY	PLACARD REQUIRED	REMARKS
5.1 Oxidizer		500 kg or more / Any when "E" appears in List II, Column III		When "E" appears in List II, Column III, use in addition to Class 5.1 or 5.2 Placard, this Explosive Placard.
5.2 Organic Peroxide		Any		
6.1 Poisonous Substance Packing Group I and II '		500 kg or more		
6.1 Poisonous Substance Packing Group III with Subsidiary Classification		500 kg or more		
7 Radioactive Materials I, II and III		Any		An additional Corrosive 8 Placard must be used for: Uranium Hexafluoride Fissile PIN 2977 and Uranium Hexafluoride, low specific gravity PIN 2978. An additional Oxidizer 5.1 Placard must be used for: Thorium Nitrate PIN 2976 and Uranyl Nitrate, Solid PIN 2981.
8 Corrosives		500 kg or more		
Nitrating Acid Mixtures more than 50% Nitric Acid PIN 1796 Nitrating Acid Mixture, spent more than 50% Nitric Acid PIN 1826		500 kg or more		
Nitric Acid fuming more than 70% and not more than 90% Nitric Acid red fuming more than 90% PIN 2032		500 kg or more		
Mixed Load of Different Classes 2.1 Flammable Gas OR 2.2 Nonflammable Compressed Gas 3 Flammable Liquid 4.1 Flammable Solid OR 4.2 Spontaneously Combustible OR 4.3 Dangerous When Wet 5.1 Oxidizer 6.1 Poisonous Substance 8 Corrosives		500 kg or more	DANGER	**Retro-Reflectivity of Orange Panels and Placards** Reflective placards are required for the following — Class 1 except Class 1.4 — Class 2, Bulk — Classes 3, 4, 5, 6 or 8, are in Bulk and Packing Group I and II only — Class 7, III Category **Special Placards required for Rail** EXPLOSIVE — Use this placard for Classes 1.1 and 1.2 Explosives; 1.1, 1.2 and Compatibility Group Poison Gas/ Gaz toxique 2 — Use this placard for the following products or as required by special provision 79. Cyanogen Chloride PIN 1589 Hydrogen Cyanide, Anhydrous PIN 1051 Nitric Oxide PIN 1660 Phosgene PIN 1076 Phosphine PIN 2199 Nitrogen Dioxide, Liquified PIN 1067 VIDE EMPTY — Use this placard on the upper half of the applicable placard for empty tankcars.
9.1 Miscellaneous Dangerous Goods 9.3 Dangerous Wastes	No label is required for 9.3 Dangerous Wastes	500 kg or more for Class 9.1 / Any for Dangerous Waste		
Product Identification Numbers are required for: Special Provision 52 Bulk, Car Load Trailer Load Truck Load and Container Load	P.I.N. is not required to be displayed on RAILWAY Vehicles for Class 2.3, Class 7 or Class 9 except as specified (See Remark) P.I.N. is not required to be displayed on ANY VEHICLES for class I Explosives.		Applicable P.I.N. 1203 P.I.N.	Orange panels are required for ANY QUANTITY for the following dangerous goods: Cyanogen Chloride PIN 1589 Hydrogen Cyanide, Anhydrous PIN 1051 Nitric Oxide PIN 1660 Phosgene PIN 1076 Phosphine PIN 2199 Nitrogen Dioxide, Liquified PIN 1067

) 500 kg DOES NOT APPLY TO RAIL

Revision 2, October, 1985

Figure 3.45b Continued

[NFPA 472: 2-2.1.3]; [NFPA 472: 3-2.2.1]

CLASS 1 — DIVISION 1.1, 1.2, 1.3, 1.4, 1.5, 1.6

An *explosive* is any substance or article, including a device, that is designed to function by explosion (i.e., an extremely rapid release of gas and heat) or that, by chemical reaction within itself, is able to function in a similar manner even if not designed to function by explosion, unless the substance or article is otherwise classed.

Explosives in Class 1 are divided into six divisions as follows:

- 1.1 — *Explosives*. Consists of explosives that have a mass explosion hazard. A mass explosion is one that affects almost the entire load instantaneously.
- 1.2 — *Explosives*. Consists of explosives that have a projection hazard but not a mass explosion hazard.
- 1.3 — *Explosives*. Consists of explosives that have a fire hazard and either a minor blast hazard or a minor projection hazard or both but not a mass explosion hazard.
- 1.4 — *Explosives*. Consists of explosives that present a minor explosion hazard. The explosive effects are largely confined to the package, and no projection of fragments of appreciable size or range is to be expected. An external fire must not cause instantaneous explosion of almost the entire contents of the package.
- 1.5 — *Explosives*. Consists of very insensitive explosives. This division is composed of substances that have a mass explosion hazard but are so insensitive that there is very little probability of initiation or of transition from burning to detonation under normal conditions of transport.
- 1.6 — *Explosives*. Consists of extremely insensitive articles that do not have a mass explosive hazard. This division is composed of articles that contain only extremely insensitive detonating substances and that demonstrate a negligible probability of accidental initiation or propagation.

CLASS 2 — DIVISION 2.1, 2.2, 2.3, (2.4 CANADA ONLY)

- 2.1 — *Flammable gas* is any material that is a gas at 68°F (20°C) or less at normal atmospheric pressure or a material that has a boiling point of 68°F (20°C) or less at normal atmospheric pressure and that:
  (1) Is ignitable at normal atmospheric pressure when in a mixture of 13 percent or less by volume with air or,
  (2) Has a flammable range at normal atmospheric pressure with air at least 12 percent, regardless of the lower limit.
- 2.2 — *Nonflammable compressed gas* (nonflammable, nonpoisonous compressed gas, liquefied gas, pressurized cryogenic gas, and compressed gas in solution) is any material (or mixture) that exerts in the package an absolute pressure of 41 psia (280 kPa) at 68°F (20°C) and does not meet the definition of Division 2.1 or 2.3.
- 2.3 — *Poisonous gas* (gas poisonous by inhalation) is a material that is a gas at 68°F (20°C) or less at normal atmospheric pressure, that has a boiling point of 68°F (20°C) or less at normal atmospheric pressure and that:
  (1) Is known to be so toxic to humans as to pose a hazard to health during transportation or,
  (2) In the absence of adequate data on human toxicity, is presumed to be toxic to humans because when tested on laboratory animals it has an LC_{50} value of not more than 5000 ml/m^3.
- 2.4 (Canada Only) — Corrosive Gases

CLASS 3

- *Flammable liquid* is:
  (1) A liquid having a flash point of not more than 141°F (61°C) or any material in a liquid phase with a flash point at or above 100°F (38°C) that is intentionally heated (mixtures and solutions) and offered for transportation or transported at or above its flash point in a bulk packaging.
  (2) A distilled spirit of 140 proof or lower is considered to have a flash point of no lower than 73°F (23°C).

- *Combustible liquid* is:
  (1) Any liquid that does not meet the definition of any other hazard class, except Class 9, and has a flash point above 141°F (61°C) and below 200°F (93°C).
  (2) A flammable liquid with a flash point at or above 100°F (38°C) that does not meet the definition of any other hazard class, except Class 9, may be reclassed as a combustible liquid.

NOTE: Specific flash points for the definition of a flammable liquid may vary depending on the regulation or standard referenced. For example, transportation regulations and codes for fixed facilities, such as NFPA 30, *Flammable and Combustible Liquids Code*, have slightly different definitions. These differences are important from both the inspection and enforcement perspectives. However, as a general guideline, the first responder should consider anything with a flash point below 100°F (38°C) as a flammable liquid.

CLASS 4 — DIVISION 4.1, 4.2, 4.3

- 4.1 — *Flammable solid* is any of the following three types of materials:
  (1) Wetted explosives
  (2) Self-reactive materials
  (3) Readily combustible solids
- 4.2 — *Spontaneously combustible materials* are the following:
  (1) A pyrophoric material, liquid or solid, that even in small quantities and without an external ignition source can ignite within five minutes after coming into contact with air.
  (2) A self-heating material that when in contact with air and without an energy supply is liable to self-heat.
- 4.3 — *Dangerous when wet material* is a material that, by contact with water, is liable to become spontaneously flammable or to give off flammable or toxic gas at a rate greater than 1 liter per kilogram of the material per hour.

CLASS 5 — DIVISION 5.1, 5.2

- 5.1 — *Oxidizer* is a material that may, generally by yielding oxygen, cause or enhance the combustion of other materials.
- 5.2 — *Organic peroxide* is any organic compound containing oxygen in the bivalent -O-O- structure and that may be considered a derivative of hydrogen peroxide, where one or more of the hydrogen atoms have been replaced by organic radicals.

CLASS 6 — DIVISION 6.1, 6.2

- 6.1 — *Poisonous material* is a material, other than a gas, that is known to be so toxic to humans as to afford a hazard to health during transportation.
- 6.2 — *Infectious substance* (etiologic agent) is a viable microorganism, or its toxin, that causes or may cause disease in humans or animals and includes those agents listed in the regulations of the Department of Health and Human Services or any other agent that causes or may cause severe, disabling, or fatal disease.

CLASS 7

- *Radioactive material* is any material having a specific activity greater than 0.002 microcuries per gram (uCi/g).

CLASS 8

- *Corrosive material* is a liquid or solid that causes visible destruction or irreversible alterations to human skin tissue at the site of contact or is a liquid that has a severe corrosion rate on steel or aluminum.

CLASS 9

- *Miscellaneous hazardous materials*
  (a) A material that has an anesthetic, noxious or other similar property that could cause extreme annoyance or discomfort to a flight crew member so as to prevent the correct performance of assigned duties.
  (b) Meets the definitions in 49 CFR 171.8 for a hazardous substance or a hazardous waste.

(c) Meets the definition in 49 CFR 171.8 for an elevated temperature material.

In Canada, Class 9 has three divisions:

9.1 — Miscellaneous Dangerous Goods

9.2 — Environmental Hazard

9.3 — Dangerous Waste

OTHER REGULATED MATERIALS — ORM-D

- ORM-D is a material, such as a consumer commodity, that presents a limited hazard during transportation due to its form, quantity, and packaging.

Number Identification System

[NFPA 472: 2-2.1.7(a)]; [NFPA 472: 2-2.1.8]

The United Nations (UN) has developed a system of numbers and reference materials that the United States and Canada use in conjunction with illustrated placards. The UN system provides uniformity in recognizing hazardous materials in international transport. By using the UN numbers and reference materials, the first responder can determine information on the general hazard class and the identity of certain predetermined commodities.

A haz mat identification number is a four-digit number assigned to each material listed in the Hazardous Materials Table appearing in both the U.S. DOT *Emergency Response Guidebook (ERG)* and Transport Canada *Dangerous Goods Guide to Initial Emergency Response (IERG)*. This four-digit number may appear on labels, placards, orange panels, or white square-on-point configurations in association with materials being transported in cargo tanks, portable tanks, or tank cars (Figure 3.46). The identification number assists first responders in identifying materials and in correctly referencing the material in the *ERG*/*IERG*. Apart from a full description of a material on a shipping document, the identification number may be the first responder's most valuable source of information for properly identifying a material and establishing initial protective actions.

Communication Of Hazards

[NFPA 472: 2-2.2.2]

The vast majority of hazardous materials are transported safely. Yet, because the potential for

Figure 3.46 Many placards have the UN number on them.

catastrophe is so great, haz mat transportation incidents present immense challenges to first responders. An inappropriate response to an incident can endanger first responders, the surrounding community, and the environment. In the interest of first responder safety, DOT has established regulations that require shippers and carriers of hazardous materials to "communicate" the hazards of their cargo by providing four essential pieces of information:

- Shipping papers
- Marking
- Labels
- Placards

These requirements furnish essential information about the cargo to emergency response personnel who respond to an incident that involves the release of hazardous materials. Both the U.S. Department of Transportation and Transport Canada (TC) have placarding and labeling systems that generally identify hazardous cargoes by classes. The first responder must, however, realize that in some cases the hazardous substance or its conveyance will be unmarked or mismarked. Regulations on placarding and labeling hazardous materials are complex, and some shippers and their drivers are unaware of the rules or are indifferent to them. Some carriers take shortcuts or ignore regulations. The following sections cover formal ways that the first responder can identify hazardous cargoes in transit.

SHIPPING PAPERS

[NFPA 472: 2-2.1.9.3]

Most shipments of hazardous materials must be accompanied by shipping papers that describe the hazardous material. For most shipments, DOT does not specify the use of a particular type of document. The information can be provided on a bill of lading, waybill, or similar document (Figure 3.47). The exceptions are hazardous waste shipments, which must be accompanied by a specific document called a *Uniform Hazardous Waste Manifest*. Instructions for describing hazardous materials are provided in the Department of Transportation/Transport Canada regulations. These descriptions include the following:

- Proper shipping name of the material
- Hazard class represented by the material
- Packing group assigned to the material
- Quantity of material

In addition, special description requirements apply to certain types of materials (for example, those that cause poison by inhalation, radioactive materials, and hazardous substances) and modes of transportation. The following example shows how a hazardous material is described on a shipping paper:

"Gasoline, 3, UN1203, PGII, 2000 lbs"

Other information is also required on shipping papers. See Table 3.5 for a list of the minimum requirements on shipping papers.

[NFPA 472: 2-2.1.9.4]; [NFPA 472: 2-2.1.9.5]; [NFPA 472: 2-2.1.9.7]; [NFPA 472: 2-2.1.9.6]

When first responders know that a close approach to an incident is safe, they should then examine the cargo shipping papers. The location and type of paperwork changes according to the mode of transport (Table 3.6). In each of the follow-

TABLE 3.5
Required Information On Shipping Papers

Information	Regulation
Shipper's name and address	172.201(b)
Receiver's name and address	172.201(c)
Proper shipping names of materials	172.202(a)(1)
Hazard class of materials	172.202(a)(2)
Identification number (UN/NA number)	172.202(a)(3)
Packing group (in Roman numerals)	172.202(a)(4)
Gross weight or volume of materials shipped	172.202(a)(5)
First-listed order of hazardous materials on the shipping papers	172.201(a)(1)(i)
"X" placed before the shipping name in the column captioned "HM" for hazardous material ("X" may be replaced with an "RQ" when the hazardous material is considered a reportable quantity.)	172.201(a)(1)(iii)
Emergency response telephone number	172.201(d)

TABLE 3.6
Shipping Paper Identification

Transportation Mode	Shipping Paper Name	Location of Papers	Party Responsible
Air	Air bill	Cockpit	Pilot
Highway	Bill of lading	Cab of vehicle	Driver
Rail	Waybill/consist	Engine or caboose	Conductor
Water	Dangerous cargo manifest	Bridge or pilot house	Captain or master

THIS MEMORANDUM is an acknowledgement that a Bill of Lading has been issued and is not the Original Bill of Lading nor a copy or duplicate, covering the property named herein, and is intended solely for filing or record.
RECEIVED, subject to the classifications and lawfully filed tariffs in effect on the date of the receipt by the carrier of the property described in the Original Bill of Lading.
SHELL CHEMICAL COMPANY

BILLING OFFICE MEMORANDUM 1

SC-6034.4 (REV. 2-91)

AT	PAGE	OF	SHIPPER'S I.D. NO.	B/L SEQ. NO.	CARRIER NAME OR RAIL CAR INITIALS AND NO.	SCAC	SHIPPING DATE
DEER PARK TX			141 A04602	978	MATLACK 6189	MTLK	11-05-94

SHIP VIA

FOR CHEMICAL EMERGENCY
Spill, Leak, Fire, Exposure or Accident
CALL CHEMTREC Day or night
800-424-9300

ROUTE CODE | ROUTE

SEAL NO.'S 188288-90

Subject to Section 7 of Conditions of applicable bill of lading; if this shipment is to be delivered to the consignee without recourse on the consignor, the consignor shall sign the following statement:
The carrier shall not make delivery of this shipment without payment of freight and all other lawful charges.
SHELL CHEMICAL COMPANY.
(Signature of Consignor)

CUSTOMER NO. 39300001 CITY 100 STATE 09 DATE ORDERED CUSTOMER ORDER NO. 90576 R-4

CONSIGNED TO:
JOHN OGORMAN
1123 DATELOG WAY
HOUSTON, TX 77090

BILL TO:
JPO INDUSTRIES
P. O. BOX 90674
HOUSTON, TX 77090

FREIGHT TERMS
CODE 2

SR DIST.	CR DIST.	ORDERED BY AND DATE DIST.	MO	DAY	YR.	SUGGESTED SHIPPING DATE			REQUESTED DELIVERY DATE		
82	82	47	11	01	93	11	05	93	11	05	AM

ORDERER'S INITIALS TELEPHONE & TWX: AMH 713-444-2430
SPLC CODES: ORIGIN | DEST.

DESCRIPTION OF ARTICLES, SPECIAL MARKS AND EXCEPTIONS

HM	NO. & KIND PACKAGES	TRUCK/ TRAILER NUMBER	TRUCK/ TRAILER OR T/C COMPT	CAPACITY	GALLONS OUTAGE AT LOADING TEMP.	LOADING TEMP. °F	LB./GAL AT LOAD TEMP.	GALLON CORR. FACTOR	NET WEIGHT OR GALLONS AT 60°F	GROSS WEIGHT (Subject to correction)
X	1 T/T	EPICHLOROHYDRIN, 6.1, UN 2023, II, RQ, GUIDE 30								

LINE 01 4,500 GAL EPICHLOROHYDRIN
BULK
CA 32410

GROSS	TARE	NET
70,240	25,960	44,280

SPECIAL INSTRUCTIONS: ANY UNLOADING DETENTION CHARGES BILL TO CONSIGNEE EQUIP. T/T WITH 2" CAMLOCK FITTING FOR UNLOADING AND 2" MALE CAMLOCK FITTING FOR VENTING VAPORS. DELIVER 10 AM - 3 PM 11/5.

IF SHIPMENT IS PREPAID, MAIL FREIGHT BILL IN DUPLICATE WITH NO. 4 COPY OF B/L TO: Shell Chemical Company, P.O. Box 1876, Houston, Texas 77251. Attention: Chemical Products Accounting, Freight Accounting

SHIPMENTS VIA MOTOR CARRIER — D.O.T. HAZARDOUS MATERIALS PLACARDS FURNISHED BY: ☐ SHIPPER ☐ CARRIER

1. The agreed or declared value of the property is hereby specifically stated by the shipper to be not exceeding 50 cents per pound where the rate is dependent on value. 2. Description and gross weight thereof as shown herein are correct, per Agreement filed with Weighing and Inspection Bureau, if applicable. 3. This is to certify that the above-named materials are properly classified, described, packaged, marked and labeled and are in proper condition for transportation according to the applicable regulations of the Department of Transportation.
SHELL CHEMICAL COMPANY, Shipper — is a division of, and means herein, SHELL OIL COMPANY, a Delaware Corporation

Per

Carrier certifies that the container supplied by Carrier for this shipment is a proper container for transportation of the Materials as above described.
Carrier
Per Agent

DELIVERY RECEIPT — Received in good order Customer/Customer's Carrier certifies that the container supplied by it for this shipment is a proper container for transportation
For
By

Figure 3.47 A typical shipping paper. *Courtesy of Shell Oil Company.*

ing modes, the general area is given. However, the exact location of the documents will vary. Responders may need to check the entire general area in order to locate these documents. In trucks and airplanes, these papers are placed near the driver or pilot. On ships and barges, the papers are placed on the bridge or in the pilothouse of a controlling tugboat. On trains, the waybills and consist may be placed in the engine, caboose, or both. During pre-incident planning, the location of the papers for a specific rail line can be determined.

MARKING

The DOT requires that shippers provide marking on a container (Figure 3.48). The marking gives certain information on the contents and the qualifications of the packaging used to contain the material. This marking includes the following information:

- Proper shipping name
- Identification number (UN/NA)
- Name and address of the consignee

Marking regulations are designed to provide first responders with a method for identifying a haz mat package or container in the event that it becomes separated from its associated shipping document.

Figure 3.48 Even small containers should have DOT markings on them.

LABELS

Basically, labels are small replicas of vehicle placards (Figure 3.49). Labels on packages communicate the hazards posed by the material in the event the package is spilled from the transport vehicle. Labels may or may not have written text that identifies the hazardous material within the packaging. First responders must be familiar with the pictogram and hazard class or division number for the materials. (NOTE: Class 7 Radioactive I, II, and III labels *must always* contain text.) For materials that meet the definition of more than one hazard class, the packaging contains a primary label and a subsidiary label. The regulations governing the use of labels are contained in 49 CFR, Subpart E.

PLACARDS

[NFPA 472: 2-2.1.8]

If a shipping document is not available, a placard is an important tool for first responders. A placard identifies the hazards posed by a hazardous material at a transportation incident site. This 10¾-inch square, on-point, diamond-shaped sign communicates the hazards represented by the materials contained within the following conveyances:

- Bulk package
- Freight container
- Unit load device (aircraft)
- Motor vehicle (Figure 3.50)
- Rail car

Figure 3.49 DOT labels are placed on small packages.

Figure 3.50 Placards should be plainly visible on the vehicle.

Tables 3.7 and 3.8 identify the required placard that must be used in conjunction with shipments of each hazard class or division.

Placards are required when *any* quantity of materials identified in Table 3.7 is ready for transportation. For materials identified in Table 3.8, a placard is not required when the gross weight of any of the materials covered in the table is less than 1,001 pounds (454 kg) for any one shipment. The following is a list of important facts that relate to placarding:

- A placard is not required for shipment of infectious substances (Class 6.2), ORM-D, limited quantities, small-quantity packages, or combustible liquids in nonbulk packaging.
- A freight container, unit load device, transport vehicle, or rail car that contains nonbulk packaging with two or more categories of hazardous materials may be placarded DANGEROUS (in the U.S.) or DANGER (in Canada).
- The word GASOLINE may be used instead of the word FLAMMABLE on a placard displayed on a cargo tank or on a portable tank used to transport gasoline by highway.
- The words FUEL OIL may be used in place of the word COMBUSTIBLE on a placard displayed on a cargo tank or on a portable tank used for highway transport.
- A tank car that contains a residue from either a Division 2.3, Hazard Zone A material or a Division 6.1, Packing Group I, Hazard Zone A material requires that the words POISON GAS-RESIDUE or POISON-RESIDUE be placed on a placard with a square background. This placard must be affixed to a rail car.
- Each transport vehicle and freight container containing lading that has been fumigated or treated with poisonous liquid, solid, or gas and that is offered for transportation by rail must have the FUMIGATION placard affixed on or near each door.
- Some states do not require placarding of haz mat shipments when the origin and destination are both within the state.

Improperly marked, unmarked, and illegal shipments are common occurrences.

TABLE 3.7
Materials Requiring Placarding Regardless Of Quantity

Hazard Class or Division	Placard Name
1.1	Explosives 1.1
1.2	Explosives 1.2
1.3	Explosives 1.3
2.3	Poison Gas
4.3	Dangerous When Wet
6.1 (PGI, PIH only)	Poison
7 (Radioactive Yellow III)	Radioactive

TABLE 3.8
Materials Requiring Placarding In Excess Of 1,000 Pounds

Hazard Classes or Division	Placard Name
1.4	Explosives 1.4
1.5	Explosives 1.5
1.6	Explosives 1.6
2.1	Flammable Gas
2.2	Non-Flammable Gas
3	Flammable
Combustible Liquid	Combustible
4.1	Flammable Solid
4.2	Spontaneously Combustible
5.1	Oxidizer
5.2	Organic Peroxide
6.1 (PG I or II, other than PGI PIH)	Poison
6.1 (PGIII)	Keep Away From Food
6.2	None
8	Corrosive
9	Class 9
ORM-D	None

HAZARDOUS MATERIALS IDENTIFICATION BY TRANSPORTATION CONTAINER

[NFPA 472: 3-2.1.2]; [NFPA 472: 2-2.1.6]; [NFPA 472: 3-2.1.1]

First responders encounter a wide variety of containers used to transport both hazardous and nonhazardous materials (Table 3.9). As with containers in fixed facilities, the recognition of the container may give some indication of what material is inside it. This piece of information will be of great assistance in sizing up the emergency scene. It is important that first responders recognize the most common types of transportation containers. The following types of vehicles routinely transport hazardous materials:

- Cargo tank trucks
- Compressed-gas trailer
- Dry bulk carrier
- Elevated-temperature materials carrier
- Railroad tank cars
- Dedicated railcars
- Non-tank railcars
- Intermodal tank containers/cylinders

Cargo Tank Trucks

[NFPA 472: 3-2.1.2.1(c)]

Cargo tank trucks are recognizable because they have construction features, fittings, attachments, or shapes characteristic of their use. Of course, many of these vehicles also transport nonhazardous materials. (**NOTE:** Even if first responders recognize one of the cargo tanks described in this section, the process of positive identification must proceed from placards, shipping papers, or other formal sources of information.)

MC-306/DOT-406 NONPRESSURE (ATMOSPHERIC PRESSURE) CARRIER

[NFPA 472: 3-2.1.1.3(a)]; [NFPA 472: 3-4.4.3]

This tank carries flammable liquids (such as gasoline and alcohol), combustible liquids (such as fuel oil), Division 6.1 poisons, and liquid food products. All of these materials are maintained at a vapor pressure under 3 psi (20 kPa). Distinguishing features of this tank include the following characteristics (Figure 3.51):

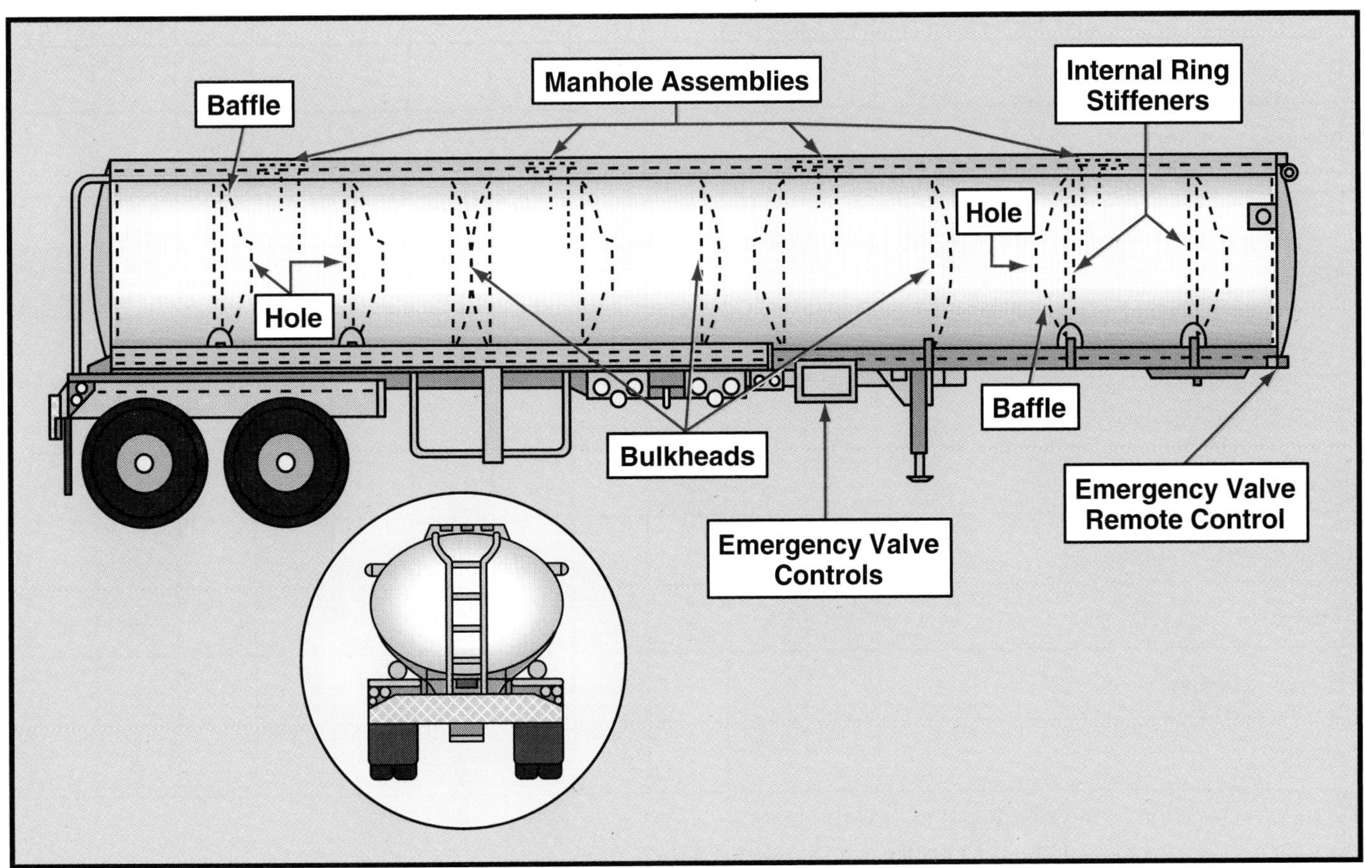

Figure 3.51 MC 306 tanks are designed to carry materials that are not under pressure.

TABLE 3.9
Table Of Shipping Containers

The following table shows the various containers that are used to transport hazardous materials and the Department of Transportation hazard classes. An "X" at an intersection of vertical and horizontal columns means that the hazard class is shipped in that container.

	Hazard Class										
	Explosive	Compressed gas	Flammable & Combustible liquid	Flammable solid	Oxidizer	Organic peroxide	Poisonous material	Etiologic agent	Radioactive material	Corrosive material	Other regulated mat.
Pail			x	x	x	x	x				
Glass carboy in plywood drum or box										x	
Fiberboard box**	x		x	x	x	x		x	x	x	x
Wooden box**	x		x	x	x	x	x	x	x	x	x
Mailing tubes***	x			x	x	x	x	x			
Wooden barrel											x
Bag	x			x	x	x	x				x
Cylinder		x	x				x		x	x	
Fiberboard drum	x		x	x	x	x	x		x	x	
Metal drum	x		x	x	x	x	x		x	x	x
Metal keg	x									x	
Polystyrene case**										x	
Lead-shielded container**									x		
Portable tank		x	x		x	x	x			x	
Tank truck	x	x	x		x	x*	x			x	x
Tank car	x	x	x	x	x		x			x	x
Tanker (Marine)		x	x		x		x			x	
Barge		x	x	x	x		x			x	

* Under exemption from Department of Transportation.

** Indicates outside package for inside containers.

*** Indicates shape of package only. They are *not* used to ship hazardous materials through the mail.

- Elliptical aluminum tank construction (older vehicles may be constructed of steel)
- Longitudinal rollover protection
- Valving and unloading control box under tank
- Vapor recovery piping on right side and rear (not all MC-306 have vapor recovery), manhole assemblies, and vapor recovery valves on top for each compartment
- Possible permanent markings for compartment capacities, materials, or ownership that are locally identifiable (Hess, Shell, Chevron, etc.) (Figure 3.52)

MC-307/DOT-407 LOW-PRESSURE CARRIER

[NFPA 472: 3-2.1.1.3(b)]

This tank is designed to carry various chemicals with pressures not to exceed 40 psi (280 kPa) at 70°F (21°C). It may carry flammables, corrosives, or poisons. This tank can be recognized by the following characteristics (Figure 3.53):

Figure 3.52 A typical MC 306 trailer.

- Single (or double) top manhole assembly protected by a flash box that also provides rollover protection
- Circumferential rollover protection at each end
- Single outlet discharge piping at midship or rear
- Double shell with covered ring stiffeners, although some have external ring stiffeners
- Fusible plugs, frangible disks, or "Christmas tree" vents (a combination vacuum breaker

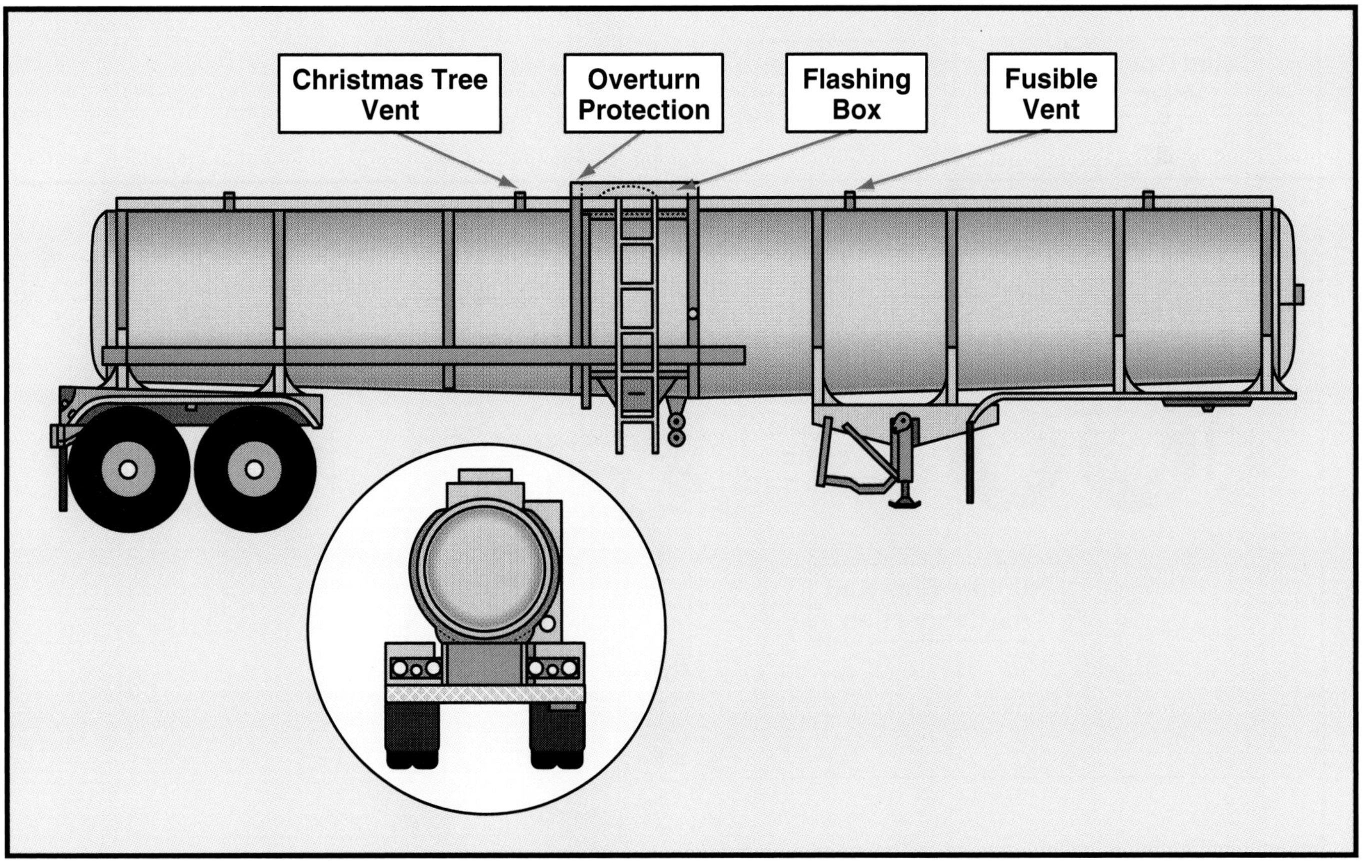

Figure 3.53 MC 307 trailers carry materials that are under low pressures.

and relief device) outside the flash box on top of the tank

- Drain hose from the flash box down the side of the tank
- Rounded ends
- Permanent ownership markings that are locally identifiable (Figure 3.54)

Figure 3.54 A typical MC 307 trailer.

MC-312/DOT-412 CORROSIVE LIQUID CARRIER

[NFPA 472: 3-2.1.1.3(c)]

This tank carries corrosive liquids, usually acids. Pressures may range up to 75 psi (517 kPa). The tank can be identified by the following characteristics (Figure 3.55):

- Small-diameter "cigar" shape
- Exterior stiffening rings
- Rear or middle top-loading/unloading station with exterior piping extending to the bottom of the tank
- Splash guard serving as rollover protection around valving
- Additional circumferential rollover protection at front of tank
- Flange-type rupture disk vent either inside or outside the splash guard
- Discoloration around loading/unloading area, or area may be painted or coated with corrosive-resistant material
- Permanent ownership markings that are locally identifiable (Figure 3.56)

MC-331 HIGH-PRESSURE CARRIER

[NFPA 472: 3-2.1.1.3(d)]; [NFPA 472: 3-4.4.3]

This tank carries gases that have been liquefied by increasing the pressure and compressing them

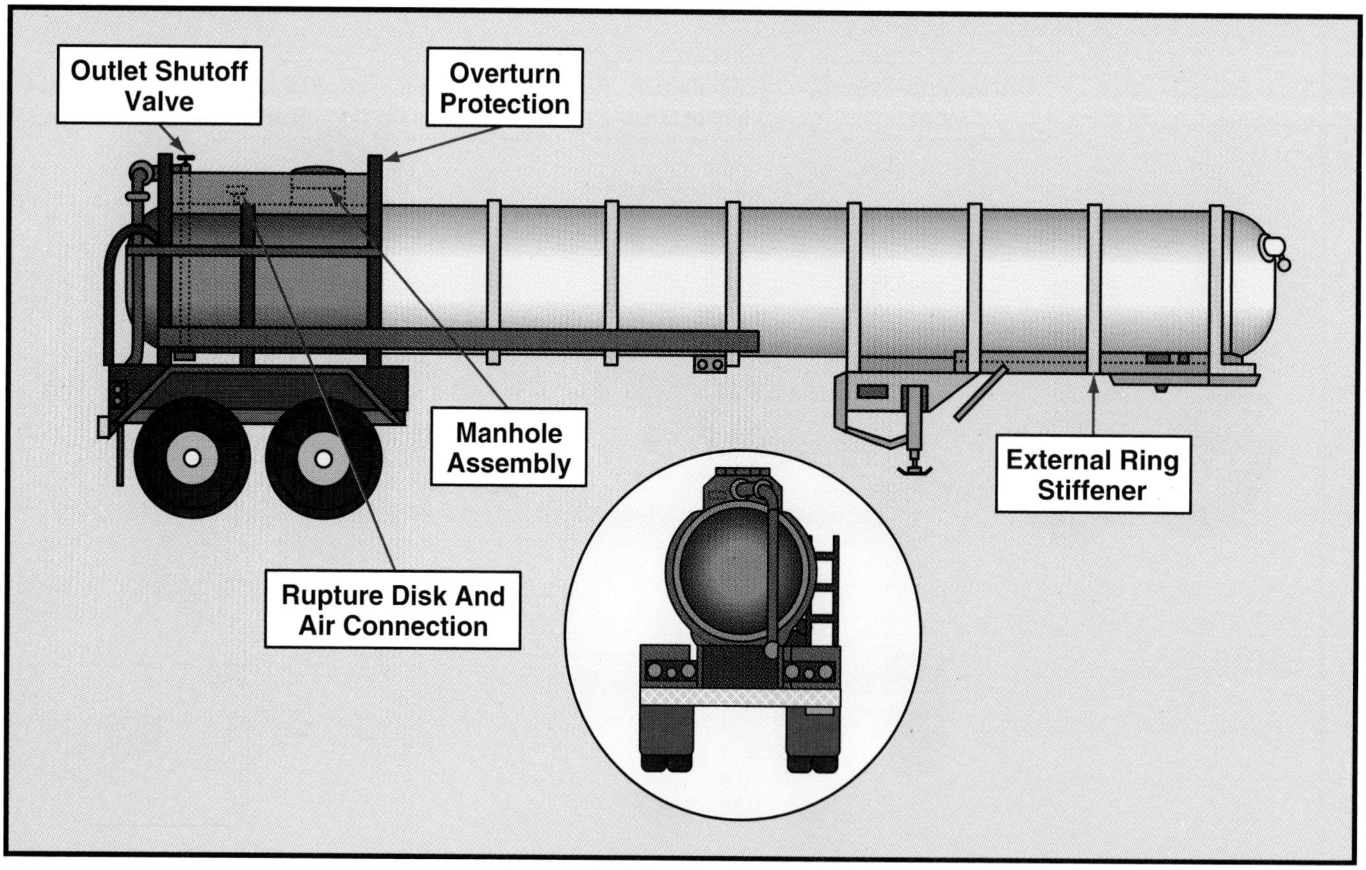

Outlet Shutoff Valve
Overturn Protection
Manhole Assembly
External Ring Stiffener
Rupture Disk And Air Connection

Figure 3.55 MC 312 tanks are designed to carry corrosive materials.

Figure 3.56 A typical MC 312 trailer.

into the liquid state. Examples of materials that are carried include propane, butane, and anhydrous ammonia. The tank can be identified by the following characteristics (Figure 3.57):

- Large hemispherical heads on both ends
- Bolted manhole at the rear
- Guard cage around the bottom loading/unloading piping
- Uninsulated tanks, single-shell vessels usually painted white
- Permanent markings such as FLAMMABLE GAS, COMPRESSED GAS, or an identifiable manufacturer or distributor name (Figure 3.58)

Figure 3.58 A typical MC 331 trailer.

MC-338 CRYOGENIC LIQUID CARRIER

[NFPA 472: 3-2.1.1.3(e)]

This tank carries gases that have been liquefied by temperature reduction. Typical materials that are carried include liquid oxygen (LOX), nitrogen, hydrogen, and carbon dioxide. The tank can be identified by the following characteristics (Figure 3.59):

Shutoff Valve Remote Control
Fill Pipe
Vents
Shutoff Valve Remote Control
Liquid Outlet
Rotary Gauge
Vapor Pipe
Outlets
Fitting Guard
Internal Shutoff Valve and Excess Flow Valves

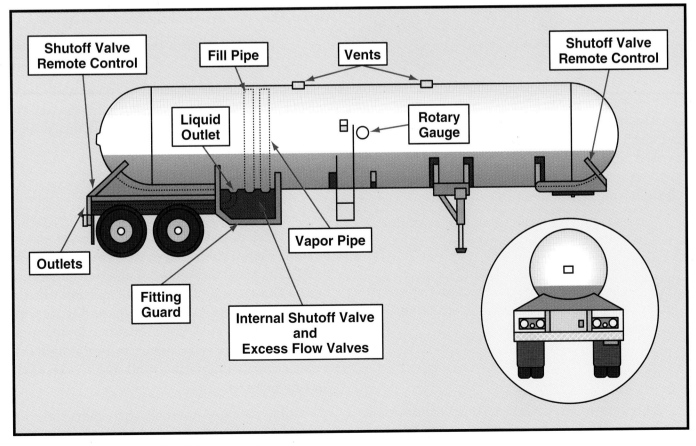

Figure 3.57 MC 331 tanks carry liquefied compressed gases.

- Large and bulky double shelling and heavy insulation
- Ends that are basically flat
- Loading/unloading station attached either at the rear or in front of the rear dual wheels
- Permanent markings, such as REFRIGERATED LIQUID, or an identifiable manufacturer name such as Air Products or Union Carbide (Figure 3.60)

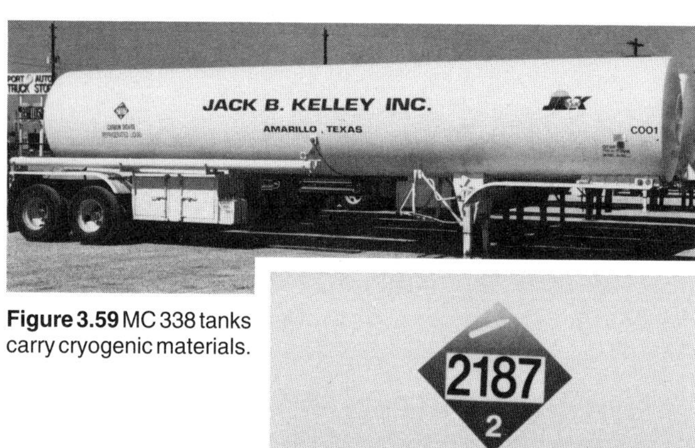

Figure 3.59 MC 338 tanks carry cryogenic materials.

Figure 3.60 The material name may be stenciled on the MC 338 tank.

Compressed-Gas Trailer (Tube Trailer)

This trailer carries compressed gases; it does not carry liquefied gases. The materials that are carried include air, argon, helium, hydrogen, nitrogen, oxygen, and refrigerant gases. This tank has the following characteristics:

- Several horizontal tubes on a trailer or intermodal unit (Figure 3.61)
- Manifold enclosed at the rear
- Permanent marking for the material or ownership (AIRCO, Liquid Air, etc.) that is locally identifiable

Dry Bulk Carrier

[NFPA 472: 3-2.1.1.3(f)]

This vehicle carries various types of hazardous materials in dry bulk and slurry forms that can burn and release toxic products of combustion. The container is identifiable by the following characteristics (Figure 3.62):

Figure 3.61 Tube trailers carry gases that are compressed at high pressures.

- Large, sloping, V-shaped bottom-unloading compartments
- Rear-mounted, auxiliary-engine-powered compressor or tractor-mounted power-take-off air compressor
- Exterior loading and bottom unloading pipes
- Top manhole assemblies

Elevated-Temperature Materials Carrier

When elevated-temperature materials are ready for transportation or are transported in bulk packaging, they are in one of the following forms:

- A liquid state at a temperature at or above 212°F (100°C)
- A liquid state with a flash point at or above 100°F (38°C) that is intentionally heated and ready for transportation or is transported at or above its flash point
- A solid state at a temperature of 464°F (240°C) or above

Elevated-temperature materials are carried in a tank truck or in one or more large metal pots on a truck (Figure 3.63). These pots are heavily insulated and keep the material molten for several hours. The major problems a spill presents are that exposed combustibles can ignite, and if they are attacked with water, huge amounts of steam are generated. It may be necessary to dam or dike with noncombustible materials well in advance of the flowing material.

When elevated-temperature materials are transported in bulk packaging, the marking HOT will be affixed to the packaging, except for pack-

Figure 3.62 A typical dry bulk carrier.

Figure 3.63 A typical elevated-temperature carrier.

ages containing molten aluminum or molten sulfur. Packaging containing these materials are marked either ALUMINUM MOLTEN or MOLTEN SULFUR.

Railroad Tank Cars

[NFPA 472: 3-2.1.2.1(a)]

Some railroad tank cars have capacities in excess of 30,000 gallons (120 000 L). Because of the large quantities these cars hold, a pressurized/liquefied material that was released suddenly would overwhelm the capabilities of most responding agencies.

By recognizing distinctive railroad cars, the first responder can begin the identification process from the greatest possible distance. Informal identification of hazardous materials in railroad tank cars can be based on some or all of the following criteria:

- Recognition of pressure/nonpressure cars
- Recognition of "dedicated" cars
- Stenciled material names
- Material packaging
- Construction features and color codings

Tank cars carry the bulk of the hazardous materials transported by rail. These tank cars are divided into four categories: pressure cars, nonpressure cars, cryogenic-liquid cars, and miscellaneous cars.

PRESSURE TANK RAILCAR

[NFPA 472: 3-2.1.1.1(b)]

A pressure tank railcar carries flammable and nonflammable liquefied gases, poisons, and other hazardous liquids (Figure 3.64). It may also carry other commodities such as ethylene oxide, sodium metal, anhydrous hydrofluoric acid, and motor-fuel antiknock compounds. This type of car is easy to recognize because of the protective housing around the manhole, valves, gaging rod, and sampling well. This car may have a shell capacity in excess of 30,000 gallons (152 000). The gas that a car is carrying is actually 80 percent liquid. When a pressurized tank railcar transports flammable gases, it is covered by a thermal jacket or by a sprayed-on protective coating. Some pressurized tank railcars are insulated. The primary visual characteristics of a pressure tank railcar are the valve enclosure at the top of the car and the lack of bottom-unloading piping under the car (Figure 3.65). The ends of an insulated/jacketed car are less round than the ends of a thermally protected or single-shell car. Pressurized tank cars are classified by DOT as DOT-105, 109, 112, and 114.

Figure 3.64 A typical pressure tank railcar.

Figure 3.65 The relief valve is on the top of the vehicle.

NONPRESSURE TANK RAILCAR

[NFPA 472: 3-2.1.1.1(a)]

A nonpressure tank rail car that transports hazardous materials is also known as a general service or a low-pressure tank car. DOT classifies this type of car by the designation of DOT 103, 104, 111, and 115 (Figures 3.66 a and b). A general service tank car can carry flammable and combustible liquids, flammable solids, oxidizers, Division 6.1 poisons, organic peroxides, molten solids, and corrosives. It can also transport nonhazardous materials such as fruit juices, molasses, tomato paste, and tallow.

General service tank cars are not distinguishable from those transporting nonhazardous materials, unless the tank railcars are required to carry commodity names or to display placards. A general service tank car may be lined, insulated, or single shelled. It may have capacities in excess of 30,000 gallons (120 000 L). This car is usually distinguished from the pressure cars by the visible fittings or by an expansion dome (Figure 3.67).

A tank railcar that carries corrosives is usually smaller than other types of general service cars because of the density and weight of the materials. A tank railcar that carries sulfuric acid may be limited to 14,000 gallons (56 000 L). Most tank cars that carry corrosives are loaded and unloaded from the top and have no plumbing underneath. The top valving arrangement is usually a means of visual identification. Cars that are bottom loading may be retrofitted with skid protectors to prevent damage to valves. Another distinctive feature of many corrosives cars is the protective black band around the car under the dome area.

A multicompartmented nonpressure tank railcar may have up to six compartments, with each compartment being considered a separate tank. Each compartment may be of a different capacity, and it will have its own loading and unloading fittings.

CRYOGENIC-LIQUID TANK RAILCAR

[NFPA 472: 3-2.1.1.1(c)]

This type of tank railcar is a large, nonpressure tank that is heavily insulated with a vacuum pulled on the space between the inner and outer shells

Figure 3.66a A DOT 103 railcar.

Figure 3.66b A DOT 111 railcar.

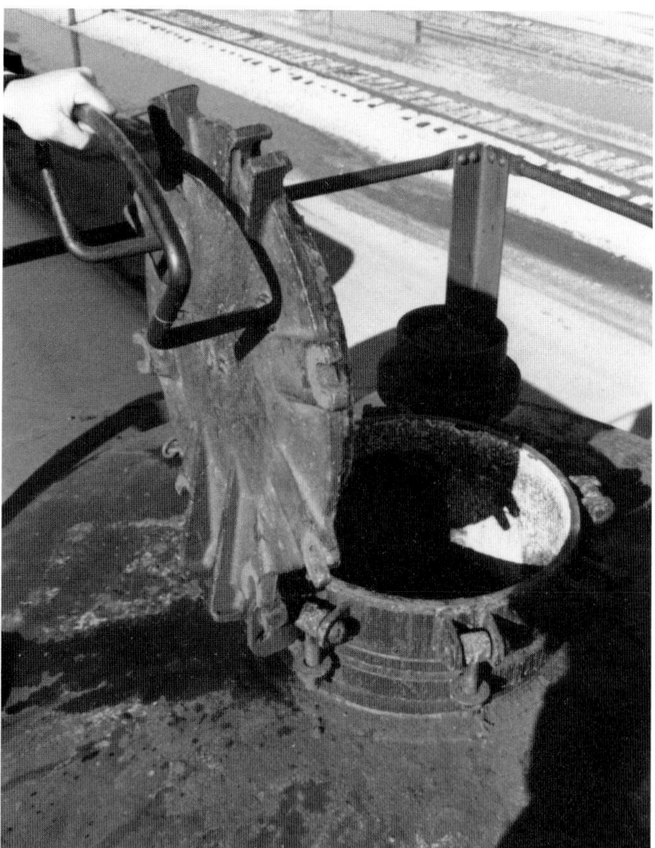

Figure 3.67 Nonpressure cars have dome covers on their tops.

(Figure 3.68). This tank may have capacities of 15,000 to 30,000 gallons (60 000 L to 120 000 L). It is not transported completely full, depending on material outage requirements. This tank railcar is equipped with a relief valve for overpressure protection and a safety vent that is set to dump the contents at the tank test pressure. There is also a safety vent for the space between the tank and the outer jacket. This safety vent is set at 16 psi (110 kPa), with the same flow capability as the inner tank relief valve. Cryogenic tanks may be enclosed in a boxcar (Figure 3.69). A cryogenic tank carries such commodities as argon, nitrogen, hydrogen, and oxygen.

Figure 3.68 A typical cryogenic railcar.

Figure 3.69 Some cryogenic tanks are enclosed within a boxcar. *Courtesy of Charles J. Wright.*

MISCELLANEOUS RAIL TANK CARS

Several railcars are used for special transportation needs such as the transport of radioactive materials, multi-unit tanks of one-ton (900 kg) cylinders, and high-pressure steel cylinders or tubes (Figure 3.70). Containers for rail flatcars and intermodal shipping are also used to carry hazardous materials of the liquid, gas, or solid variety.

Dedicated Railcars

A dedicated railcar is set aside by the material manufacturer to transport a single material. The name of the material is painted on the car, and the

Figure 3.70 Tube railcars are similar to tube trailers. *Courtesy of Charles J. Wright.*

manufacturer's name may also be on the car. The manufacturer's name or logo is typically the largest printing on the container and is easier to see from a distance. Some dedicated cars are also a distinctive color, which is determined by the owner and is not subject to regulation (Figure 3.71).

Typically, color alone cannot be used for material identification with rail cars. However, a car that is white with a horizontal red stripe around it and two vertical red stripes 3 feet (1 m) from each end always hauls hydrogen cyanide (hydrocyanic acid). Hydrogen cyanide can also be shipped in a noncolor-coded car.

A number of hazardous materials transported by rail are required to have their names stenciled on the sides of the railcar in 4-inch (100 mm) letters (Figure 3.72). These chemicals are listed in Table 3.10.

Figure 3.71 A dedicated railcar has the product name stenciled on it.

Figure 3.72 A close-up of the product name.

TABLE 3.10
Stenciled Commodity Names

ACROLEIN	LIQUEFIED PETROLEUM GAS (may also be stenciled PROPANE, BUTANE, PROPYLENE, ETHYLENE)
ANHYDROUS AMMONIA	METHYL ACETYLENE PROPADIENE STABILIZED
BROMINE	METHYL CHLORIDE
BUTADIENE	METHYL MERCAPTAN
CHLORINE	METHYL CHLORIDE - METHYLENE CHLORIDE MIXTURE
CHLOROPRENE (when transported in DOT 115A specification tank car)	MONOMETHYLAMINE, ANHYDROUS
DIFLUOROETHANE*	MOTOR FUEL ANTIKNOCK COMPOUND OR ANTI-KNOCK COMPOUND
DIFLUOROMONOCHLOROMETHANE*	NITRIC ACID
DIMETHYLAMINE, ANHYDROUS	NITROGEN TETROXIDE
DIMETHYL ETHER (transported only in ton cylinders)	NITROGEN TETROXIDE - NITRIC OXIDE MIXTURE
ETHYLENE IMINE	PHOSPHORUS
ETHYLENE OXIDE	SULFUR TRIOXIDE
FORMIC ACID	TRIFLUOROCHLOROETHYLENE*
FUSED POTASSIUM NITRATE AND SODIUM NITRATE	TRIMETHYLAMINE, ANHYDROUS
HYDROCYANIC ACID	VINYL CHLORIDE
HYDROFLUORIC ACID	VINYL FLUORIDE INHIBITED
HYDROGEN	VINYL METHYL ETHER INHIBITED
HYDROGEN CHLORIDE (by exemption from DOT)	
HYDROGEN FLUORIDE	
HYDROGEN PEROXIDE	
HYDROGEN SULFIDE	
LIQUEFIED HYDROGEN	
LIQUEFIED HYDROCARBON GAS (may also be stenciled PROPANE, BUTANE, PROPYLENE, ETHYLENE)	

*May be stenciled DISPERSANT GAS or REFRIGERANT GAS in lieu of name. Only *flammable* refrigerant or dispersant gases are stenciled.

Non-Tank Railcars

Some hazardous materials are transported in trailers on flatcars (TOFC) and containers on flatcars (COFC) (Figures 3.73 a and b). These TOFC and COFC shipments are placarded according to the regulations for their highway transportation. The waybill has to indicate the contents for these cars as it would for a normal railroad car.

Hazardous materials transported by boxcar, hopper car, and other freight cars cannot be identified by the outer appearance of the car. Refrigerated boxcars carry up to 500 gallons (2 000 L) of diesel fuel in a tank that is used for cooling system power generation equipment. Some refrigerated cars now use a refrigerated liquid cooling system. There will be signs on the car and a tank under the body (Figure 3.74). Other boxcars have alcohol-burning heaters to protect materials from freezing temperatures. First responders should immediately contact the conductor for help in identifying the contents of the cars involved.

Intermodal Tank Containers

[NFPA 472: 3-2.1.2.1(b)]

An intermodal tank is a freight container that is designed and constructed to be used interchangeably in multiple modes of transport such as both rail and highway transport (Figure 3.75). The tank of an intermodal container is generally a cylinder enclosed at both ends. The first responder may see

Figure 3.73a A trailer on flatcar.

Figure 3.73b A container on flatcar.

Figure 3.74 A typical refrigerated boxcar.

Figure 3.75 Intermodal (IM) containers are carried on either railcars or trailers.

tube modules, compartmented tanks, or other shapes. The tank container is placed in frames to protect it and to provide for stacking, lifting, and securing. The two types of basic frames are the box type and the beam type. The box type encloses the tank in a cage, and the beam type uses frame structures only at the ends of the tank (Figures 3.76 a and b). The capacity of these containers ordinarily will not exceed 6,340 gallons (24 000 liters). The two general classifications of intermodal containers are nonpressure and pressure (Table 3.11).

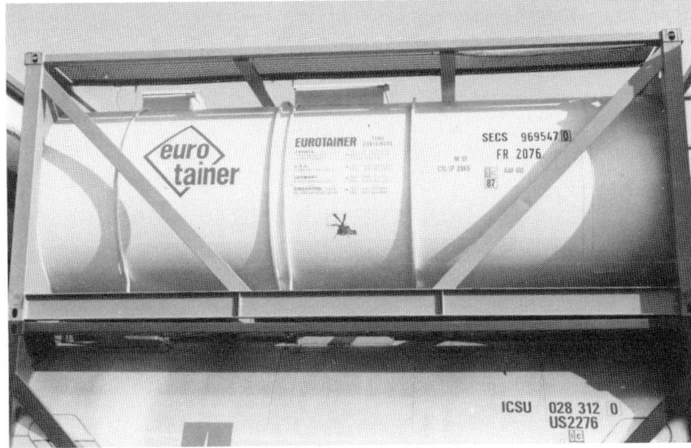

Figure 3.76a A box-type intermodal container.

Figure 3.76b A beam-type intermodal container.

TABLE 3.11
Intermodal Tank Containers

Specification	Capacity	Design Pressure
IM 101	up to 6,300 gallons	25.4 to 100 psig
IM 102	up to 6,300 gallons	14.5 to 25.4 psig
Spec 51	up to 5,500 gallons	100 to 500 psig

An intermodal tank is required to be placarded in accordance with DOT hazardous materials regulations. An intermodal tank is also marked with initials (reporting marks) and a tank number (Figure 3.77). These markings are generally found on the right-hand side of the tank as the first responder faces it from either the sides or the ends. The markings will either be on the tank or on the frame. First responders can use this information, in conjunction with the shipping papers or computer data, to identify the contents of the tank. The tank will also display a country code and a size/type code (Figure 3.78). The four-digit size/type code follows the country code. The first two numbers identify the container length and height; the second two numbers indicate the pressure range. See Tables 3.12 and 3.13.

Figure 3.77 The intermodal container has much information stenciled on it.

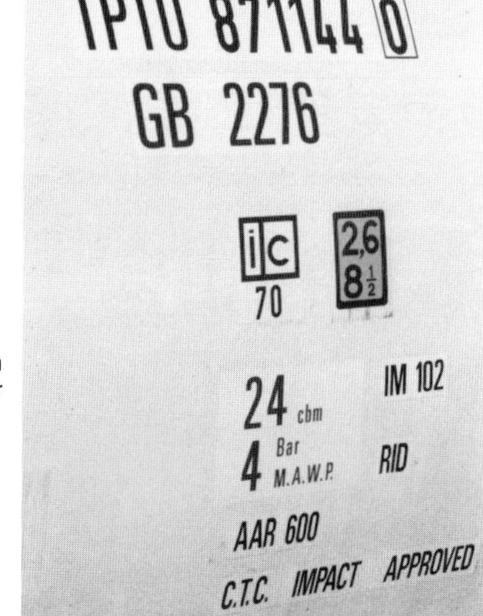

Figure 3.78 The "GB" on this container stands for Great Britain.

TABLE 3.12
Common Intermodal Container Country Codes

Code	Country
BM (BER)	Bermuda
CH (CHS)	Switzerland
DE	West Germany
DKX	Denmark
FR (FXX)	France
GB	Great Britain
HKXX	Hong Kong
ILX	Israel
IXX	Italy
JP (JXX)	Japan
KR	Korea
LIB	Liberia
NLX	Netherlands
NZX	New Zealand
PA (PNM)	Panama
PIX	Phillipines
PRC	People's Republic of China States
RCX	People's Republic of China (Taiwan)
SGP	Singapore
SXX	Sweden
US (USA)	United States

NONPRESSURE INTERMODAL TANK CONTAINER

[NFPA 472: 3-2.1.1.2(a)]

This is the most common type of intermodal tank used in transportation; it is also called an intermodal portable tank or an IM portable tank. The two common groups of nonpressure intermodal tank containers are IM 101 and IM 102 portable tanks.

An IM 101 intermodal tank is built to withstand a working pressure of 25.4 to 100 psig (177 kPa to 700 kPa). It transports both hazardous and nonhazardous materials. Internationally, an IM 101 portable tank is called an IMO Type 1 tank container (Figure 3.79).

An IM 102 intermodal tank is designed to handle maximum allowable working pressures of 14.5 to

TABLE 3.13
Common Intermodal Tank Size And Type Codes

Common Size Codes

20 - 20' (8 " high)

22 - 20' (8'6" high)

24 - 20' (>8'6" high)

Common Type Codes - Tanks

Maximum allowable working pressure
Nonhazardous commodities

70 — <0.44 Bar test pressure

71 — 0.44 — 1.47 Bar test pressure

72 — 1.47 — 2.94 Bar test pressure

73 — spare

Hazardous

74 — <1.47 Bar test pressure

75 — 1.47 — 2.58 Bar test pressure

76 — 2.58 — 2.94 Bar test pressure

77 — 2.94 — 3.93 Bar test pressure

78 — >3.93 Bar test pressure

79 — spare

Conversion Table

0.44 Bar = 6.4 psig	2.94 Bar = 42.6 psig
1.47 Bar = 21.3 psig	3.93 Bar = 57.0 psig
2.58 Bar = 37.4 psig	4.00 Bar = 58.0 psig

Figure 3.79 An IM 101 container.

25.4 psig (100 kPa to 177 kPa). It transports materials such as alcohols, pesticides, resins, industrial solvents, and flammables with a flash point between 32°F and 140°F (0°C to 60°C). Most commonly, it transports nonregulated materials such as food commodities. Internationally, an IM 102 tank is called an IMO Type 2 tank container (Figure 3.80).

Figure 3.80 An IM 102 container.

PRESSURE INTERMODAL TANK CONTAINER

[NFPA 472: 3-2.1.1.2(b)]

A pressure intermodal tank container is less common in transport. It is designed for working pressures of 100 to 500 psig (700 kPa to 3 500 kPa) and usually transports liquefied gases under pressure. DOT classifies this tank as Spec 51, while internationally it is known as an IMO Type 5 tank container (Figure 3.81).

Figure 3.81 A Type 5 pressure intermodal container.

SPECIALIZED INTERMODAL TANK CONTAINER

There are several types of specialized intermodal tank containers. Cryogenic liquid tank containers carry refrigerated liquid gases, argon, oxygen, and helium. Another type is the tube module, which transports gases in high-pressure cylinders (3,000 to 5,000 psi [21 000 kPa to 35 000 kPa]).

Bulk Packaging And Nonbulk Packaging

Bulk packaging refers to a packaging, other than a vessel or a barge, in which hazardous materials are loaded with no intermediate form of containment. This includes a transport vehicle or freight container such as a cargo tank, railcar, or portable tank. To be considered bulk packaging, one of the following criteria must be met:

- Maximum capacity greater than 119 gallons (450 L) as a receptacle for a liquid
- Maximum net mass greater than 882 pounds (400 kg) or a maximum capacity greater than 119 gallons (450 L) as a receptacle for a solid
- Liquid capacity greater than 1,000 pounds (454 kg)as a receptacle for a gas

Nonbulk packaging is packaging that is smaller than the minimum criteria established for bulk packaging. Common types of nonbulk packaging include the following containers:

- Drums (Figure 3.82)
- Boxes (Figure 3.83)
- Bags (Figure 3.84)
- Bottles (Figure 3.85)
- Carboys (Figure 3.86)
- Wooden or cardboard barrels (Figure 3.87)
- Portable tanks and bins

Cylinders

To transport, store, and use large volumes of gaseous materials economically, industry compresses them under high pressure. Compressed-gas cylinders range in size from pint (liter) size to railcar size and have equally varying pressures (Figure 3.88). All approved cylinders, with the exception of some that store poisons, are equipped with safety relief devices. These devices may be spring-loaded valves that reclose after operation, heat-fusible plugs, or pressure-activated bursting disks that completely empty the container. All fittings and threads are standardized according to the material stored in the cylinder. As yet, there is no nationally regulated color code that permits visual identification of the material. Some manufacturers use a single color for all their cylinders while other manufacturers have their own color-

Figure 3.82 A drum.

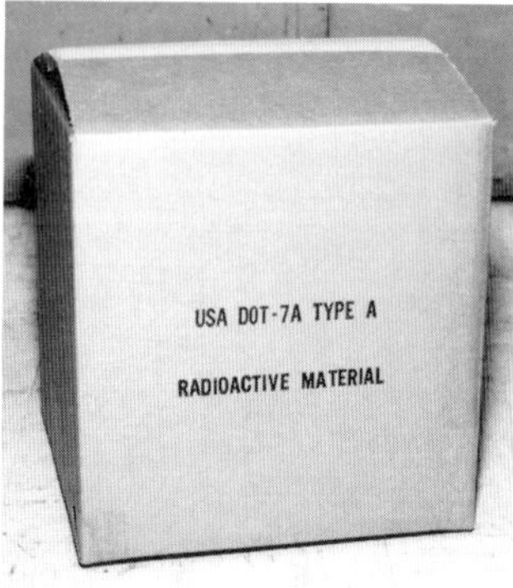

Figure 3.83 A box. *Courtesy of Lawrence Livermore National Laboratory.*

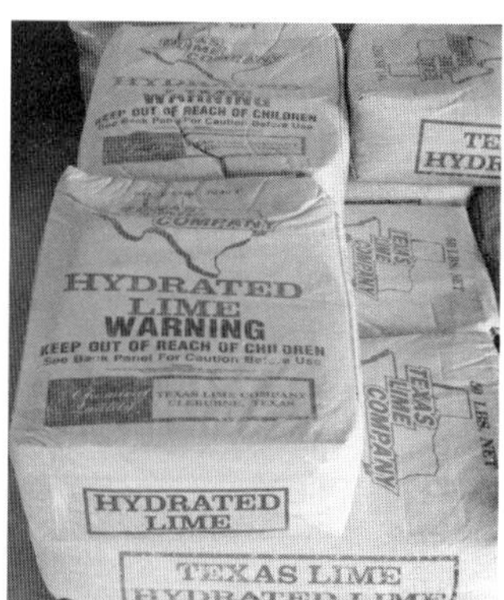

Figure 3.84 A bag.

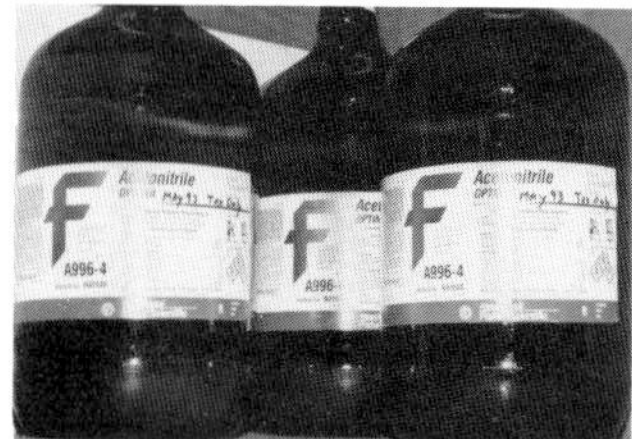

Figure 3.85 A bottle.

Figure 3.86 A carboy.

Figure 3.87 A cardboard barrel.

Figure 3.88 A compressed gas cylinder.

coding system. If local manufacturers and distributors use an identification system, it should be identified in pre-incident plans.

IDENTIFICATION OF SPECIALIZED HAZARDOUS MATERIALS MARKING SYSTEMS

First responders must be knowledgeable of some of the specialized marking systems for hazardous materials that they may encounter. The following sections highlight some of the more common specialized systems.

Military Markings

[NFPA 472: 2-2.1.7(c)]

The United States military and the Canadian military have established their own marking systems for hazardous materials and chemicals. The first responder who approaches a military vehicle involved in an accident or fire, on or off base, should exercise extreme caution. The first responder must remember that nearly all military ordnance are designed to inflict great bodily harm and/or heavy property damage.

Figure 3.89 The U.S. military special hazard symbols.

WARNING

Military drivers may be under orders not to identify their cargoes. If the military driver rapidly abandons the vehicle, the first responder should withdraw immediately.

The military system identifies detonation, fire, and special hazards. This system consists of seven symbols that can be found on structures containing hazardous materials. The three special hazard symbols include the following (Figure 3.89):

- Chemical hazard
- Apply no water
- Wear protective mask or breathing apparatus

The other four symbols give the relative detonation and fire hazards (Figure 3.90):

- Class 1 — Mass detonation
- Class 2 — Explosion with fragments

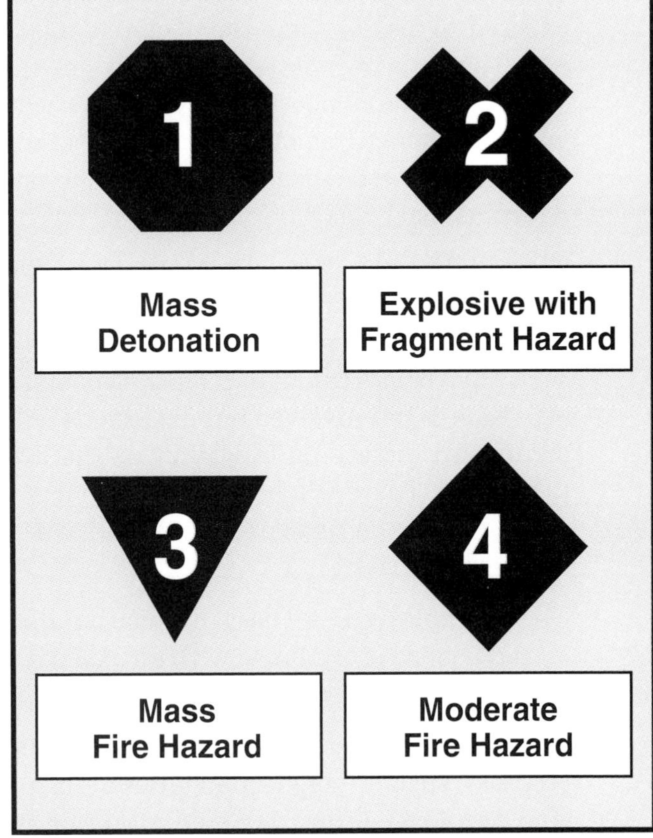

Figure 3.90 The U.S. military detonation hazard symbols.

- Class 3 — Mass fire hazard
- Class 4 — Moderate fire hazard

The Canadian Forces use a similar system (Figure 3.91):

- Class 1 Explosives — Those that must be expected to explode or detonate en masse very soon after fire reaches them.
- Class 2 Explosives — Those that are readily ignited and burn with great violence without necessarily exploding.
- Class 3 Explosives — Those that may explode en masse but, compared with Class 1 explosives, may be exposed to fire for some time before exploding. There will be a blast and fragment hazard.
- Class 4 Explosives — Those that burn fiercely and give off dense smoke with toxic effects in some instances. There is no risk of mass explosion.
- Class 5 Explosives — Those containing toxic substances.
- Class 6 Explosives — Those that may be exposed to fire for some time before exploding. The risk of mass explosion is not involved, but small sporadic explosions occur with increasing frequency as the fire takes hold. There will be a fragment hazard but not a serious blast risk.
- Class 7 Explosives — Those that involve combined flammable, toxic, and corrosive hazards. These may be exposed to fire for some time before exploding. The risk of mass explosion is not involved but explosions will occur with increasing frequency as the fire takes hold. There will be a fragment hazard arising from the pressure bursts but not a serious blast risk.
- Class 8 Explosives — Those in which a radiological hazard is combined with an explosive hazard.
- Class MP — Substance containing metallic powders, such as magnesium, aluminum, or zinc powders, either in ammunition or in bulk facility storage.

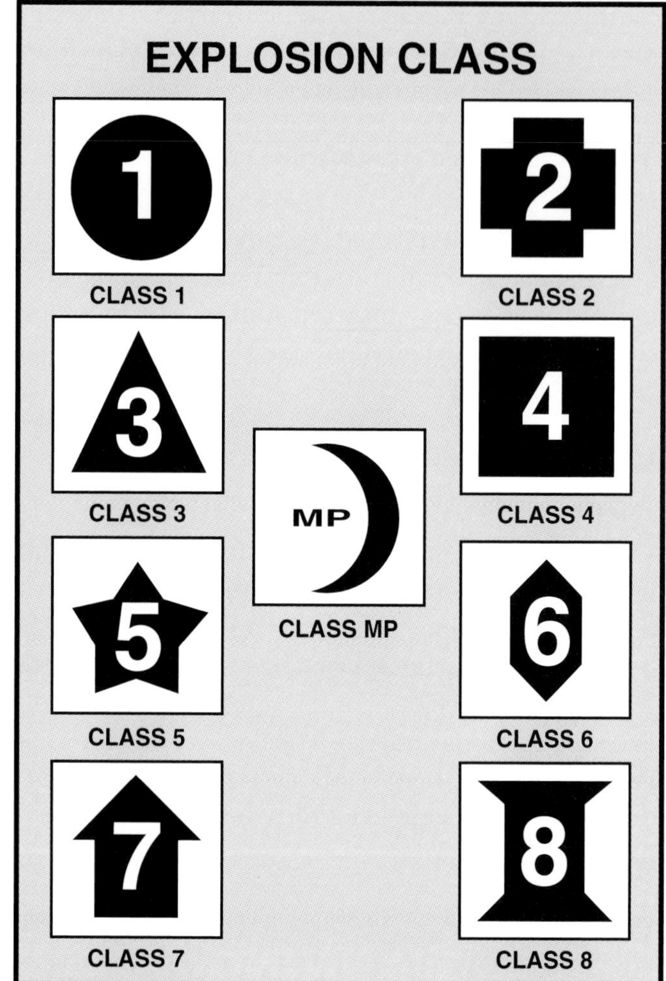

Figure 3.91 The Canadian Forces explosives marking system.

Pesticide Labels

[NFPA 472: 2-2.1.7(d)(f)];
[NFPA 472: 3-2.1.3.2 a through f]

The EPA regulates the manufacturing and labeling of pesticides. Each EPA label must contain one of the following signal words: DANGER/POISON, WARNING, or CAUTION. The words DANGER/POISON are used for highly toxic materials, WARNING for moderate toxicity, and CAUTION for chemicals with relatively low toxicity. The words EXTREMELY FLAMMABLE are also displayed on the label if the contents have a flash point below 80°F (27°C). An EPA registration number is listed on the label as well. The Chemical Transportation Emergency Center (CHEMTREC) can provide information about these chemicals if given the EPA registration number.

Materials originating in Canada carry a PCP Number, which is issued under the Pest Control

Products Act. The Canadian Transport Emergency Centre (CANUTEC) can provide information about these materials if given the PCP registration number.

Other information that may be found on these labels includes mode of entry into the body (inhalation, absorption, injection, or ingestion), requirements for storage and disposal, first aid information, and the antidote for poisoning if known.

Pipeline Identification

[NFPA 472: 2-2.1.7(e)]; [NFPA 472: 3-2.1.3.1 a through c]

Many types of materials, particularly of the petroleum variety, are transported across country in an extensive network of pipelines, most of which are buried in the ground. The DOT Office of Pipeline Safety regulates the pipelines that carry hazardous materials across state borders, navigable waterways, and federal lands. Pipeline companies maintain markers over each buried pipeline. These markers must be located at each railroad crossing, at each public road crossing, and in sufficient numbers along the rest of the pipeline to identify the pipe's location. Pipeline markers include the word WARNING and contain information describing the transported commodity and the name and telephone number of the carrier (Figure 3.92).

Figure 3.92 A common pipeline marker.

The EPA requires a warning label on any containers, transformers, or capacitors that contain polychlorinated biphenyls (PCBs). EPA's label can be used in conjunction with the appropriate DOT label (Figure 3.93).

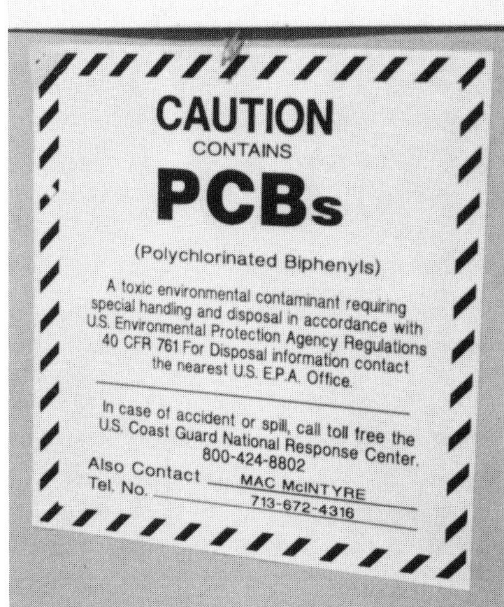

Figure 3.93 A typical PCB warning sign.

DETECTING AND MONITORING HAZARDOUS MATERIALS

[NFPA 472: 3-2.4.3]

The first responder must be able to detect the presence of a hazardous material at the scene of an emergency. Detection can usually be achieved by observing the leaking container from a safe distance outside the spill area. Examples of methods for detecting hazardous materials include observing container shapes, placards, labels, or physical signs of leaking containers such as vapor, frost, chemical reactions, etc. (Figure 3.94).

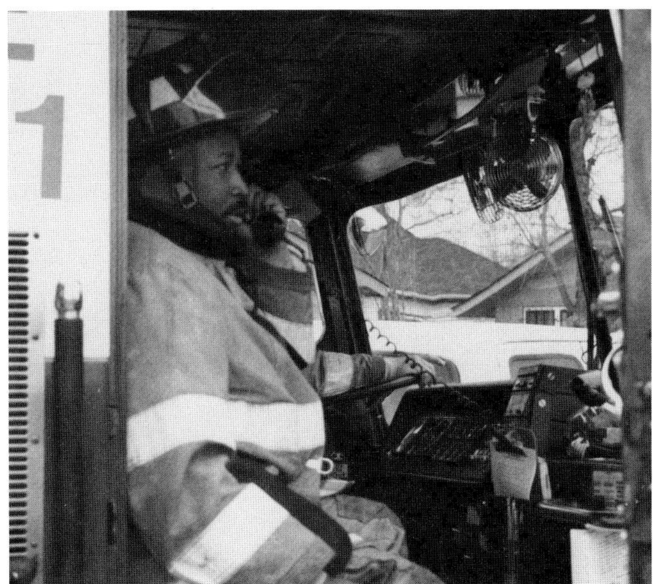

Figure 3.94 When first responders determine that hazardous materials are present, they should alert all other parties involved with the incident.

Detecting the presence of a hazardous material does not necessarily lead to identification. Monitoring instruments may be required to determine specific flammability, toxicity, or oxygen-deficient concentrations. The first responder is not expected to know how to operate the monitoring instruments and interpret the readings. However, the first responder must know where to obtain this equipment and how to contact the technicians who are required to operate it.

Simply stated, monitoring instruments are designed to determine specific concentrations of hazardous materials. The most common types of monitoring equipment available to first responders include the following instruments:

- Combustible-gas or explosive meters — These instruments are designed to measure the concentration of a flammable gas in air (Figure 3.95).
- Oxygen meters — These instruments measure the concentrations of oxygen in air and typically reveal both oxygen-enriched and oxygen-deficient atmospheres (Figure 3.96).
- Toxicity monitoring devices — These devices are designed to detect the presence and measure the concentration of specific chemicals. The type of instrument may range from simple colorimetric tubes to sophisticated instruments (Figure 3.97).

More than one instrument may be required to develop a complete picture of the hazards involved. These instruments are only of use when they are in the hands of trained haz mat specialists and technicians; however, first responders should be aware of their availability.

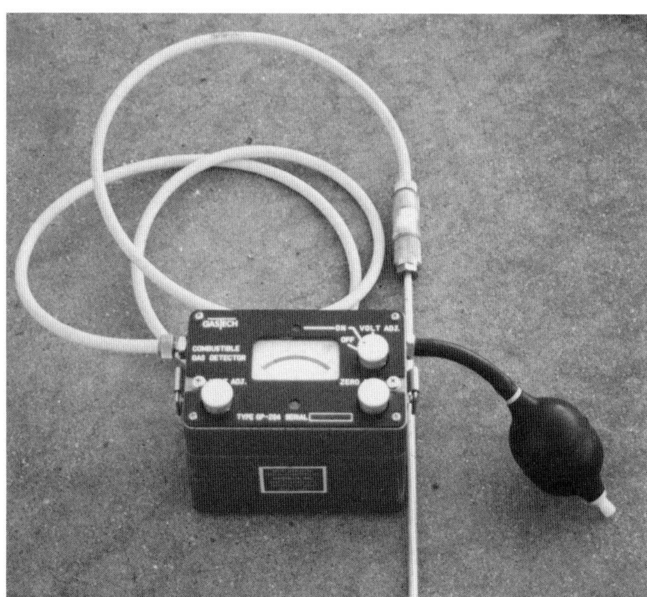

Figure 3.95 An explosimeter.

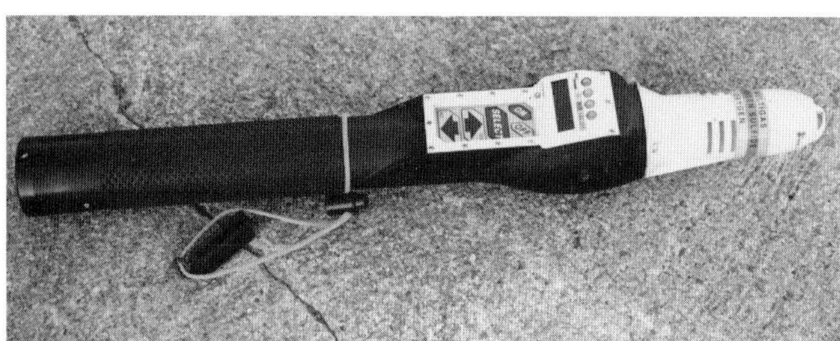

Figure 3.96 An oxygen meter.

Figure 3.97 A toxicity meter.

Chapter 4

HAZ MAT

Photo Courtesy of Scott D. Christiansen, Minot, N.D..

Hazard And Risk Assessment

LEARNING OBJECTIVES

This chapter provides information that will assist the reader in meeting the objectives from NFPA 472, *Standard for Professional Competence of Responders to Hazardous Materials Incidents* and 29 CFR 1910.120 that are listed below. The objective numbers are also noted directly in the text in the sections where they are addressed. Objectives in the list below that are denoted with an asterisk (*) are global in nature and are covered by reading the chapter in its entirety.

AWARENESS LEVEL

NFPA 472: 2-2.2 The first responder at the awareness level shall, given examples of facility and transportation situations involving hazardous materials, identify the hazardous material(s) in each situation by name, UN/NA identification number, and/or type placard applied.

OPERATIONAL LEVEL

- Knowledge of basic hazard and risk assessment techniques. [29 CFR 1910.120(q)(ii)(A)]

NFPA 472: 3-2.1.4 Identify and list the surrounding conditions that should be noted when surveying hazardous materials incidents.

NFPA 472: 3-2.3 The first responder at the operational level shall, given examples of facility and transportation hazardous materials incidents involving a single hazardous material, predict the likely behavior of the material and its container in each incident.

NFPA 472: 3-2.3.2 Identify three types of stress that could cause a container system to release its contents.

NFPA 472: 3-2.3.3 Identify five ways in which containers can breach.

NFPA 472: 3-2.3.4 Identify four ways in which containers can release their contents.

NFPA 472: 3-2.3.5 Identify at least four dispersion patterns that can be created upon release of a hazardous material.

NFPA 472: 3-2.3.6 Identify the three general time frames for predicting the length of time that exposures may be in contact with hazardous materials in an endangered area.

Chapter 4

Hazard And Risk Assessment

The first step in the successful mitigation of an incident is recognizing that a hazardous material is present. (**NOTE:** *Mitigation* is defined as those actions taken to lessen the harm or hostile nature of an incident.) However, the first responder's job does not stop here. The first responder must also be able to make reasonable determinations as to the amount or the level of hazard present and the risks associated with dealing with the incident.

[29 CFR 1910.120(q)(ii)(A)]

A number of different processes are available to help guide a first responder in the size-up and analysis of incidents (Table 4.1). Some processes are designed specifically for haz mat incidents; others are generic and may be applied to other types of incidents as well. A first responder should become very familiar with all of these helpful tools. What these methods have in common is that they are designed to identify specific areas of concern. This enables the user to develop the strategy and tactics to perform rescue, evacuation, probability determination, and a plan of attack to handle the incident. This chapter highlights some of the more common processes first responders may use.

TABLE 4.1
Mental Rules For Identifying Specific Areas Of Concern

Size-up	Decide
Rescue	Detect haz mat presence
Exposure	Estimate likely harm without intervention
Confinement	Choose response objectives
Extinguishment	Identify action options
Overhaul	Do best option
Ventilation	Evaluate progress
Salvage	

GEBMO
(General Emergency Behavior Model)

Stress (Identify the types of stress.)
Breach (Predict the type of breach.)
Release (Predict the type of release.)
Engulf (Predict the dispersion pattern.)
Contact (Predict the length of exposure.)
Harm (Predict the hazard causing the harm.)

IMMEDIATE CONCERN AND PRIMARY OBJECTIVE

One method of determining specific avenues of attack at chemical emergencies is to determine the immediate concern and the primary objective. The *immediate concern* includes specific tasks that are safety oriented. These tasks include activities such as isolating the incident and evacuating people from the area. Generally, the immediate concern is something that a first responder can execute. The *primary objective* includes specific goal-oriented tasks that will bring the incident to an end. In most cases, the primary objective will be something first responders can identify but cannot execute. Personnel with more advanced training and equipment will be required to carry out the primary objective.

Immediate Concern

Immediate concern tasks are preventive measures that can be performed by the first responder and can be carried out with minimal or no risk to the first responder. These tasks can be accomplished quickly (immediately upon arrival) and easily (require no special equipment). Performing these tasks will stabilize the scene, diminish or control the potential effects of the incident, and lower the level of personal anxiety as control of the incident is achieved. Executing immediate concern tasks increases life safety for the duration of the incident and prevents the situation from getting worse. The following are examples of immediate concern tasks:

- Isolating the area
- Denying access to the area (Figure 4.1)
- Evacuating or sheltering in place
- Diking and retaining the spill in a specific area for collection
- Diverting liquid and runoff water into an isolated location to control the speed of contamination
- Eliminating all ignition sources within an appropriate distance
- Cooling tanks involved in fire or exposed to heat by accepted methods to reduce the probability of rupture

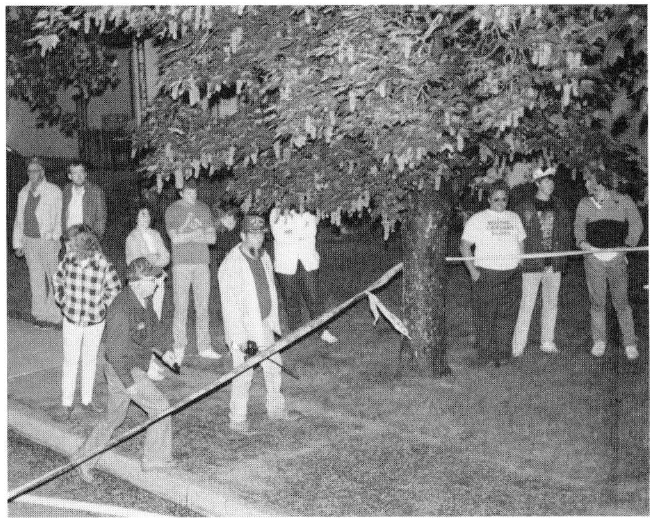

Figure 4.1 Cordoning off the area is an example of an immediate concern task.

In some cases, the first responder will be able to control the direction the incident is going to take. This may include predicting a possible explosion or deciding to let the fire burn and consume the hazardous material involved (Figure 4.2). Immediate concern tasks "buy time" for proceeding with other on-scene priorities, developing an overall plan of control, and implementing procedures for control.

Figure 4.2 In some cases, the most sound strategy is to allow the fire to burn out. *Courtesy of Joe Marino.*

Primary Objective

A *primary objective* is the operational goal at the incident. Execution of the tactical concepts to achieve the primary objective should bring the incident to a conclusion. First responders can identify a primary objective for a specific emergency. However, because of limitations to their training and equipment, first responders will generally need help from more highly trained personnel to achieve completion of the primary objective. Examples of primary objectives are as follows:

- Extinguishing fires and stabilizing the scene
- Controlling flammable and/or toxic gas clouds
- Stopping leaks by plugging and/or patching
- Diking and damming large-volume spills (Figure 4.3)

Completion of the primary objective is often difficult and time-consuming. Identifying the correct and desired objective may take only minutes, but planning to execute it safely can require considerable time. Putting the plan into motion is directly dependent upon the time it takes to assemble the personnel and equipment required. Other roadblocks are the unexpected delays encountered during the actual execution. How and when the primary objective is successfully achieved will depend on the following factors:

- Location and severity of the incident
- Properties of the involved material(s)

- Disintegration — The container suffers a general loss of integrity. Examples of disintegration are a glass bottle shattering and a grenade exploding.
- Runaway cracking — A crack develops in the container as a result of some type of damage and then continues to grow rapidly, breaking the container into two or more pieces. This type of breach is associated with closed containers such as drums or tank cars.
- Attachments (closures) open up — Attachments to the container, such as pressure-relief devices, discharge valves, or other related equipment, may open up or break off the container, leading to a total failure (Figure 4.10).
- Puncture — This breach typically occurs as a result of mechanical stress coming into contact with the container. Examples include forklifts puncturing drums and couplers puncturing a rail tank car (Figure 4.11).
- Split or tear — Examples of this type of breach include the failure of a welded seam on a tank or drum or a ripped seam on a bag of fertilizer.

Figure 4.10 Frequently a spout may be knocked off a barrel, resulting in a leak of the product within the barrel.

Figure 4.11 Careless forklift operators may drive the point of the fork through a barrel or other container.

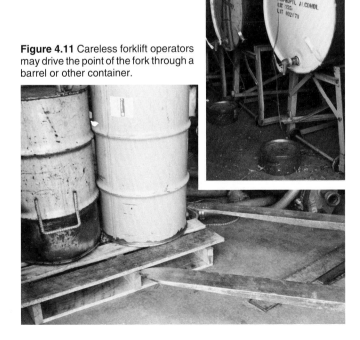

Containers stressed beyond their limits will breach and release their contents. This can include pressurized as well as nonpressurized containers. Each container will breach differently. The breach is dependent upon the type of container and the stress applied.

Release

[NFPA 472: 3-2.3.4]

The breaching of a container will most likely result in a release of the material. Depending on the situation, this release can occur quickly or over an extended period of time. Generally, when there is a great amount of stored chemical/mechanical energy, there will be a more rapid release of the material and a greater risk to the first responder. There are four ways in which containment systems can release their contents:

- Detonation — The instantaneous and explosive release of the stored chemical energy of the hazardous material. The results of this release include fragmentation, disintegration, or shattering of the container; extreme overpressure; and considerable heat release. The duration of a detonation can be measured in hundredths or thousandths of a second.
- Violent rupture — The immediate release of chemical or mechanical energy caused by runaway cracks. The results are ballistic behavior of the container and its contents and/or localized projection of container pieces/parts and hazardous material. Violent ruptures occur within a time frame of one second or less.
- Rapid relief — The fast release of a pressurized hazardous material through properly operating safety devices; through damaged valves, piping, or attachments; or through holes in the container (Figure 4.12). This action may occur in a period of several seconds to several minutes.
- Spill/leak — The slow release of a hazardous material under atmospheric or head pressure through holes, rips, tears, or usual openings/attachments. Spills and leaks can occur in a period of several minutes to several days.

Figure 4.12 A properly operating relief valve relieves the container of excess pressures.

The first responders must always be aware that there are two things that will endanger them during a release — energy and matter. High pressure may be the energy, while the hazardous material is the matter. These two things will have an effect on how quickly the breach will release the container contents.

Engulf

[NFPA 472: 3-2.3.5]

In the GEBMO model, the dispersion of material is referred to as *engulf*. Once the container has been compromised, the hazardous material will be distributed over the surrounding area according to the following factors:

- Its physical/chemical properties
- Prevailing weather conditions
- Local topography
- Duration of the release
- Control efforts of responders

The shape and size of the pattern also depends on how the material emerges from its container — whether the release is an instantaneous "puff," a continuous plume, or a sporadic fluctuation. The outline of the dispersing hazardous material, sometimes called its "footprint," can be described in a number of ways:

- Hemispheric — A semicircular or dome-shaped pattern of the airborne hazardous material that is still partially in contact with the ground or water (Figure 4.13).
- Cloud — A ball-shaped pattern of the airborne hazardous material where the material has collectively risen above the ground or water (Figure 4.14).
- Plume — An irregularly shaped pattern of the airborne hazardous material where wind and/or topography influence the downrange course from the point of release (Figure 4.15).
- Cone — A triangular-shaped pattern of the hazardous material with a point source at the breach and a wide base downrange (Figure 4.16).
- Stream — Surface-following pattern of liquid hazardous material affected by gravity and topographical contours (Figure 4.17).
- Pool — A flat and circle-shaped pattern (no wind, no contour) of the hazardous material on the surface of the ground or on water (Figure 4.18).
- Irregular — An irregular or indiscriminate deposit of the hazardous material (carried around by contaminated responders).

Figure 4.13 A hemispheric release.

Figure 4.14 A cloud of vapor.

Figure 4.15 A plume.

Figure 4.16 A cone has a wide base downrange.

Contact

[NFPA 472: 3-2.3.6]

Time frames are related to contact of surrounding exposures by the hazardous material. As a container is breached, it will release its materials and impinge upon people, the environment, and property. This contact may be harmful or may present no harm at all to the exposure. Contacts (impingements) are associated with four general time frames:

Figure 4.17 Gravity and topography affect a liquid stream.

Figure 4.18 Liquid releases pool in low lying areas.

- Immediate — milliseconds, seconds. Examples would be deflagration, explosion, or detonation.
- Short term — minutes/hours. An example would be a gas or vapor cloud.
- Medium term — days/weeks/months. An example would be a lingering pesticide.
- Long term — years/generations. An example would be a permanent radioactive source.

Harm

Harm is the injury or damage caused by exposure to the hazardous material. Three factors will determine the harm to the exposure:

- Timing of release — The speed at which the material is escaping; length of exposure to people, environment, and property.
- Size — The size of area covered by the release.
- Toxicity — The relative level of harm.

If it is not possible to make all predictions required to analyze the incident, then further assistance is needed. This assistance may be sought from the manufacturer, shipper and consignee, carrier, or other agencies involved with hazardous materials.

Chapter 5

HAZ MAT

Photo Courtesy of Scott D. Christiansen, Minot, N.D..

Personal Protective Equipment

LEARNING OBJECTIVES

This chapter provides information that will assist the reader in meeting the objectives from NFPA 472, *Standard for Professional Competence of Responders to Hazardous Materials Incidents* and 29 CFR 1910.120 that are listed below. The objective numbers are also noted directly in the text in the sections where they are addressed. Objectives in the list below that are denoted with an asterisk (*) are global in nature and are covered by reading the chapter in its entirety.

AWARENESS LEVEL

NFPA 472: 2-4.1.5.1* Given the current edition of the *Emergency Response Guidebook* and the name of a hazardous material, identify the recommended personal protective equipment for the particular incident from the following list of protective equipment:
(a) Street clothing and work uniforms;
(b) Structural fire fighters' protective clothing;
(c) Positive pressure self-contained breathing apparatus; and
(d) Chemical-protective clothing and equipment.

OPERATIONAL LEVEL

- Identify the appropriate personal protective equipment required for a given defensive option. [29 CFR 1910.120(q)(6)(ii)(B)]

NFPA 472: 3-3.3* The first responder at the operational level shall, given the name of the hazardous material involved and the anticipated type of exposure, determine whether available personal protective equipment is appropriate for implementing a defensive option.

NFPA 472: 3-3.3.1 Identify the appropriate respiratory protection required for a given defensive operation.

NFPA 472: 3-3.3.1.1 Identify the three types of respiratory protection and the advantages and limitations presented by the use of each at hazardous materials incidents.

NFPA 472: 3-3.3.1.2 Identify the required physical capabilities and limitations of personnel working in positive pressure self-contained breathing apparatus.

NFPA 472: 3-3.3.2 Identify the appropriate personal protective equipment required for a given defensive option.

NFPA 472: 3-3.3.2.2* Identify the purpose, advantages, and limitations of the following levels of protective clothing at hazardous materials incidents:

(a) Structural fire fighting clothing;

(b) High temperature-protective clothing; and

(c) Chemical-protective clothing:

1. Liquid splash-protective clothing; and
2. Vapor-protective clothing.

NFPA 472: 3-4.3 The first responder at the operational level shall demonstrate the ability to don, work in, and doff the personal protective equipment provided by the authority having jurisdiction.

NFPA 472: 3-4.3.5 Identify the physical capabilities required for and the limitations of personnel working in the personal protective equipment as provided by the authority having jurisdiction.

Chapter 5

Personal Protective Equipment

Personnel responding to a haz mat incident must be protected from the hazard by personal protective equipment. Personal protective equipment consists of self-contained breathing apparatus (SCBA) and either structural fire fighting, high-temperature, or chemical-protective clothing. Each of these ensembles and an SCBA protects the skin, eyes, face, hands, feet, body, head, and respiratory system. Structural fire fighting and high-temperature protective clothing offer very limited protection against chemical hazards, and those limitations must be recognized. Chemical-protective clothing offers protection against hazardous materials, but its use requires training above the first responder level. First responders, however, should be aware of chemical-protective clothing and understand that it, too, has limitations.

EPA CLASSIFICATION SYSTEM

[29 CFR 1910.120(q)(6)(ii)(B)] [NFPA 472: 3-3.3.2] [NFPA 472: 3-3.3.1]

The U.S. Environmental Protection Agency (EPA) has established the following levels of protective equipment: Level A, Level B, Level C, and Level D. These levels are also recognized by the National Institute of Occupational Safety and Health (NIOSH), the Occupational Safety and Health Administration (OSHA), and the United States Coast Guard (USCG).

Level A Protective Equipment

Level A provides the highest level of protection against vapors, gases, mists, and particles. The totally encapsulating suit not only envelops the wearer but also the SCBA (Figure 5.1). Level A protection is required when working directly with liquids, vapors, or gases that pose a severe threat of injury from any contact.

Level B Protective Equipment

Level B protection requires a garment, including SCBA, that provides protection against splashes from a hazardous chemical (Figure 5.2). This level of protection is worn when vapor-protective clothing (Level A) is not required. Wrists, ankles, facepiece and hood, and waist are secured to prevent any entry of splashed liquid. Depending on the chemical to be handled, specific types of gloves and boots are donned. These may or may not be attached to the garment.

Level C Protective Equipment

Level C protection differs from Level B in the area of equipment needed for respiratory protection. The same type of garment used for Level B protection is worn for Level C. Level C protection allows for the use of respiratory protection equipment other than SCBA. This protection includes any of the various types of air-purifying respirators. Emergency response personnel would not use this level of protection unless the specific material is known and can be measured.

Level D Protective Equipment

Level D protection requires no respiratory protection and provides minimal skin protection. This type of clothing consists of ordinary work clothes or uniforms (Figure 5.3). Level D protection is not adequate for first responders.

TYPES OF PERSONAL PROTECTIVE CLOTHING

[NFPA 472: 3-3.3][NFPA 472: 3-3.3.2.2][NFPA 472: 2-4-1.5.1][NFPA 472: 3-4.3]

There are several types of personal protective clothing. There is no single combination of protective equipment, including respiratory protection, that can protect against all hazards. NFPA 49,

Figure 5.1 Level A equipment is the highest level of protection that haz mat workers can wear. First responders at the awareness or operational level are not qualified to wear this level of protection.

Figure 5.2 Level B equipment provides protection against splashes from a hazardous chemical.

Figure 5.3 Level D protection is no more than a normal work uniform.

Hazardous Chemicals Data, defines two levels of protective clothing: structural fire fighting clothing and chemical-protective clothing. The following sections address these two types of clothing as well as high-temperature protective clothing.

Structural Fire Fighting Clothing

This clothing was designed for protection from heat, moisture, and the ordinary hazards associated with structural fire fighting. Structural fire fighting clothing includes a helmet, SCBA, turnout coat and pants, protective boots, protective hood, and gloves (Figure 5.4). In total, these pieces are referred to as *full protective clothing.*

Structural fire fighting clothing and SCBA provide minimal protection against hazardous materials. The multiple layers of the coat and pants may provide short-term exposure protection from some materials. However, the limitations of this level of protection are many and must be recognized by the first responder in order to avoid harmful exposures. This protective equipment is neither corrosive resistant nor vapor tight. Acids and bases can dissolve or deteriorate the outer layers, and gases can penetrate the garment. Gaps in protective clothing occur at the neck, wrists, waist, and the point where the pants and boots overlap. First responders must be aware of the composition of the materials used to manufacture their protective clothing and the hazardous materials that affect them.

Besides knowing what can deteriorate or destroy the protective clothing, the first responder must be alert for hazardous materials that permeate and remain in the protective equipment. Chemicals absorbed into the equipment can subject the wearer to repeated exposure or a later reaction with another chemical. The rubber or neoprene in boots, gloves, and SCBA masks can become permeated by chemicals and render them unsafe for use. It may be necessary to discard any equipment exposed to permeating types of chemicals. First responders should never clean turnout clothing at home, in departmental laundry facilities, at public laundries, or anywhere they might mix with other clothing. Turnout clothing should always be cleaned in a dedicated area, away from other clothing.

Chemical-Protective Clothing

Chemical-protective clothing is designed to protect against specific chemical hazards and provides overall body protection. The materials used in chemical-protective clothing offer good protection against some chemicals but not against all chemicals. The manufacturer of a particular suit must provide a list of chemicals with which the suit is compatible.

Chemical-protective clothing is available in both reusable and limited-use versions. Reusable garments can be decontaminated, tested, and put back into service. Limited-use garments, also known as disposable, can be reused if they are not contaminated or damaged. If contaminated or damaged, limited-use garments should be decontaminated and disposed of in an appropriate manner.

It should be noted that despite their ability to be used after decontamination, reusable chemical-protective clothing can retain some chemicals at the molecular level. This material can then leach out (inwards or outwards) days or even weeks later. If this leaching occurs, it could pose a hazard to the wearer.

Chemical-protective clothing should be used in conjunction with any other protective equipment required by the situation (Figure 5.5). The National Fire Protection Association has developed three standards for the performance of suits in the following categories: vapor-protective suits, liquid-splash-protective suits, and support-function protective garments.

Figure 5.4 A firefighter in full structural fire fighting clothing.

Figure 5.5 Chemical-protective clothing is worn by personnel who may be in contact with a hazardous chemical. *Courtesy of Joe Marino.*

VAPOR-PROTECTIVE SUITS

NFPA 1991, *Standard on Vapor-Protective Suits for Hazardous Chemical Emergencies*, was designed to protect responders against exposure to specified chemicals in vapor and liquid-splash environments. This standard sets the minimum criteria for suits designed to protect against chemical vapors, gases, and liquids. These suits are primarily for use as part of a Level A protective ensemble. They must be worn with SCBA.

Vapor-protective suits are made from a variety of special materials (Figure 5.6). Not all suits are compatible with all chemicals. The standard requires, as a minimum, that the suit be certified for the chemicals listed in Table 5.1, but the suit may be certified for other chemicals not listed.

In addition to chemical compatibility, these suits have other limitations. They do not allow body heat to escape and thus can contribute to heat stress. The wearer's mobility, vision, and communication are also impaired.

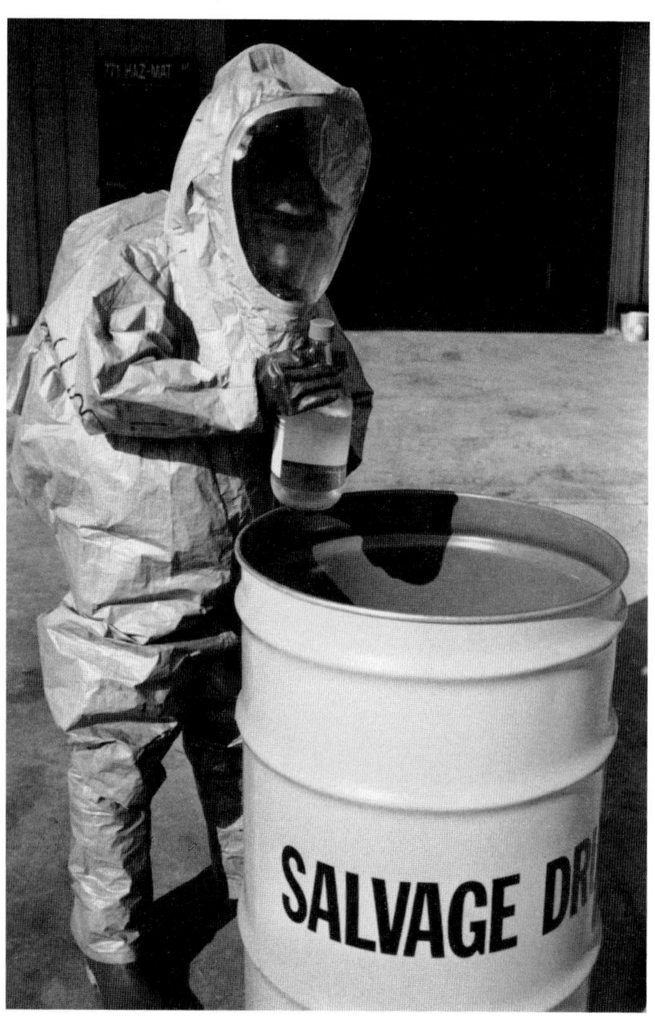

Figure 5.6 Vapor-protective suits are made from a variety of special materials.

TABLE 5.1 Minimum Chemical Certification For Vapor-Protective Suits
Acetone
Acetonitrile
Anhydrous ammonia
Carbon disulfide
Chlorine
Dichloromethane
Diethyl amine
Dimethyl formanide
Ethyl acetate
Hexane
Methanol
Nitrobenzene
Sodium hydroxide
Sulfuric acid
Tetrachloroethylene
Tetrahydrofuran
Toulene

WARNING:

First responders at the awareness and operational levels do not have sufficient training to operate in conditions requiring the use of vapor-protective suits.

LIQUID-SPLASH-PROTECTIVE SUITS

NFPA 1992, *Standard on Liquid Splash Protective Suits for Hazardous Chemical Emergencies*, sets the minimum design criteria for liquid-splash-protective suits. These suits are designed to protect users from chemical liquid splashes but not against chemical vapors or gases (Figure 5.7). They are made of the same types of materials used for vapor-protective suits. Depending on the situation, an SCBA, airline, or a full-face, air-purifying, canister-equipped respirator may be used with this suit. Liquid-splash-protective suits have many of the same limitations as the vapor-protective suit. These suits must be certified for the products listed in Table 5.2. They may also be certified for other products. First responders may wear this type of garment when participating in decontamination procedures that require a greater level of protection than the support-function garments offer.

Figure 5.7 Liquid-splash-protective suits are not designed to provide protection against vapors or gases.

TABLE 5.2 Minimum Chemical Certification For Liquid-Splash-Protective Suits
Acetone
Diethylamine
Ethyl acetate
Hexane
Sodium hydroxide
Sulfuric acid
Tetrahydrofuran
Toluene

SUPPORT-FUNCTION PROTECTIVE GARMENTS

NFPA 1993, *Standard on Support Function Protective Garments for Hazardous Chemical Operations*, establishes the minimum criteria for protective clothing to be worn by support personnel who are working outside the hot zone. (**NOTE:** The control zones will be discussed in Chapter 6.) Operations, such as decontamination, conducted outside the hot zone may require support personnel to wear protective clothing. Support-function suits may be fully encapsulating and worn with protective breathing equipment.

As with other types of encapsulating suits, support-function garments do not allow the body to give off excess heat. These suits also limit the mobility, vision, and communication of the responder. These suits cannot be used for protection against radiological, biological, or cryogenic materials; against immersion in liquids; or in flammable, explosive, or hazardous chemical vapor atmospheres. This type of suit is certified at a minimum for those chemicals listed in Table 5.3 but can be certified for other chemicals as well.

TABLE 5.3 Minimum Chemical Certification For Support-Function Protective Garments
Acetone
Diethylamine
Ethyl acetate
Hexane
Sodium hydroxide
Sulfuric acid
Tetrahydrofuran
Toluene

HIGH-TEMPERATURE CLOTHING

Another type of special protective clothing that first responders may encounter is high-temperature clothing. This is firefighter clothing designed for situations where heat levels exceed the capabilities of standard firefighter full protective clothing. There are three basic types of high-temperature clothing: approach suits, proximity suits, and fire entry suits. *Approach suits* are used for fire fighting operations that involve high levels of radiated heat (Figure 5.8). *Proximity suits* permit close approach to fires for the purpose of performing rescue, fire suppression, and property conservation activities (Figure 5.9). *Fire entry suits* are highly specialized garments that allow a person to work in total flame environments for short periods of time (Figure 5.10). (**NOTE:** Each suit has a specific use and is not interchangeable. These suits are not designed to protect the wearer against chemical hazards.)

There are several limitations to high-temperature protective clothing. As with chemical-protective clothing, high-temperature protective clothing contributes to heat stress by not allowing the body to give off excess heat. The suit is bulky

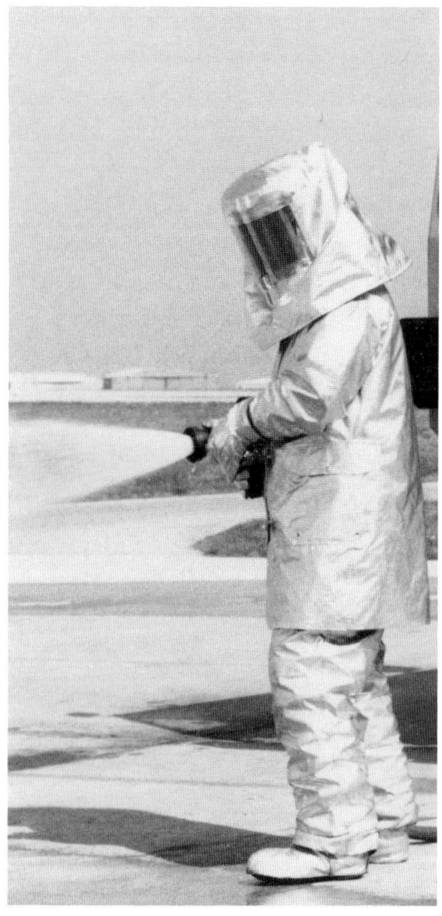

Figure 5.8 Approach suits deflect radiated heat away from the wearer.

Figure 5.9 Proximity suits permit the firefighter to make a close approach to the fire.

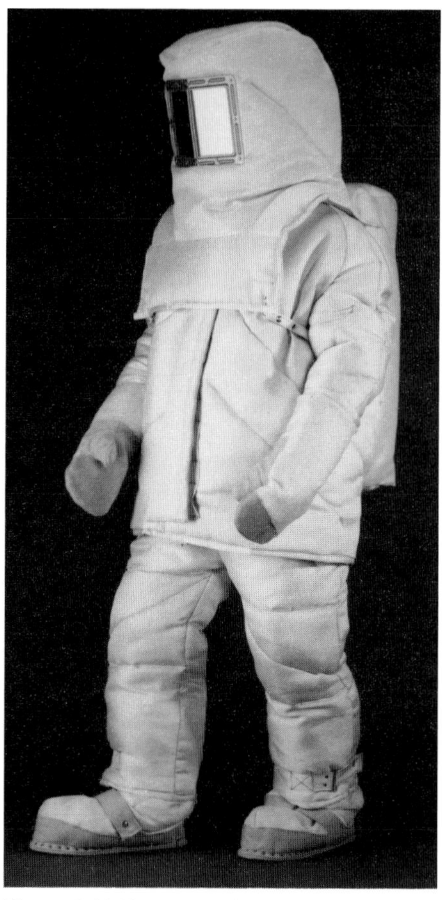

Figure 5.10 Fire entry suits allow the wearer to work in a total flame environment. *Courtesy of Fyrepel, Inc.*

and limits the vision, mobility, and communication of the individual wearing it. First responders require frequent and extensive training in the use of high-temperature clothing for efficient and safe use.

WARNING:
Only personnel who have been trained in the use and limitations of high-temperature protective suits should operate in them at an emergency scene.

PROTECTIVE BREATHING EQUIPMENT

[NFPA 472: 3-3.3.1.1]

Respiratory protection is of primary concern to the first responder because one of the major routes of exposure to hazardous substances is inhalation. Irritants, asphyxiants, poisons, toxins, and other health hazards discussed in Chapter 2 are respiratory hazards that first responders may encounter during an incident. Protective breathing equipment protects the body from inhaling these products. However, each type of respiratory protection equipment is limited in its capabilities. The four basic types of protective breathing equipment used by first responders are open-circuit SCBA, open-circuit airline breathing equipment, closed-circuit SCBA, and air-purifying respirators. (**NOTE:** Fire service first responders may only use SCBA. Airline breathing equipment and air-purifying respirators may be used only in industry and not by fire service first responders.)

Open-Circuit Self-Contained Breathing Apparatus (SCBA)

Open-circuit SCBA are the most commonly used protective breathing apparatus in the fire service (Figure 5.11). They are also commonly used in the private sector and in haz mat operations. The air supply in an open-circuit SCBA is compressed breathing air. Exhaled air is vented to the outside atmosphere. The air supply duration of an open-circuit SCBA varies depending on the design of the unit, the fitness of the wearer, and the amount of physical exertion required of the wearer. Most SCBA allow at least 15 to 20 minutes of heavy work to be performed. Some units may provide up to 45 minutes of actual duration under heavy work conditions. (**NOTE:** In haz mat incidents, decontamination time must be included in the overall time a first responder can operate in the SCBA.)

Before July 1, 1983, two types of open-circuit SCBA were available: demand and positive pressure. Since that date, demand units are no longer in compliance with OSHA requirements and should not be used. NFPA and ANSI standards also require that only positive-pressure breathing apparatus be used in the fire service. (**NOTE:** Any organizations still using demand apparatus should be aware that these units can and must be converted to positive-pressure units as soon as possible.)

The main reason for the change to positive pressure is the greater protection factor afforded by the positive-pressure units. Positive-pressure SCBA maintains a slightly increased pressure (above atmospheric) in the user's facepiece. This positive pressure significantly increases protection against contaminants entering the facepiece (Figure 5.12).

Numerous companies manufacture open-circuit SCBA, each with different design features or mechanical construction. Certain parts, such as cylinders and backpacks, are interchangeable; however, such substitution voids National Institute for Occupational Safety and Health (NIOSH) and Mine Safety and Health Administration (MSHA) certification and is not a recommended practice. Substituting different parts may also nullify warranties and leave the department or first responder liable for any injuries incurred.

The following are the four basic SCBA component assemblies:

- Backpack and harness assembly
- Air cylinder assembly — includes cylinder, valve, and pressure gauge
- Regulator assembly — includes high-pressure hose and low-pressure alarm
- Facepiece assembly — includes low-pressure hose (breathing tube) and exhalation valve on SCBA with harness-mounted regulators and head harnesses on all types

Open-Circuit Airline Breathing Equipment

Incidents involving hazardous materials often require a longer air supply than can be obtained from standard open-circuit SCBA. In these situations, an airline attached to one or several large air cylinders can be connected to an open-circuit facepiece, regulator, and egress cylinder (Figure 5.13). Airline equipment enables the first responder to travel up to 300 feet (90 m) from the *regulated* air supply source (Figure 5.14). This type of respiratory protection enables the first responder to work for several hours without the encumbrance of a backpack. If greater mobility is needed, the first responder can also wear a standard SCBA with an

Figure 5.11 Open-circuit SCBA is the most common type used by firefighters. *Courtesy of Bob Esposito.*

Figure 5.12 Modern SCBA maintains a positive pressure on the facepiece to keep contaminants from entering.

Figure 5.13 Airline equipment is useful during extended operations. *Courtesy of MSA.*

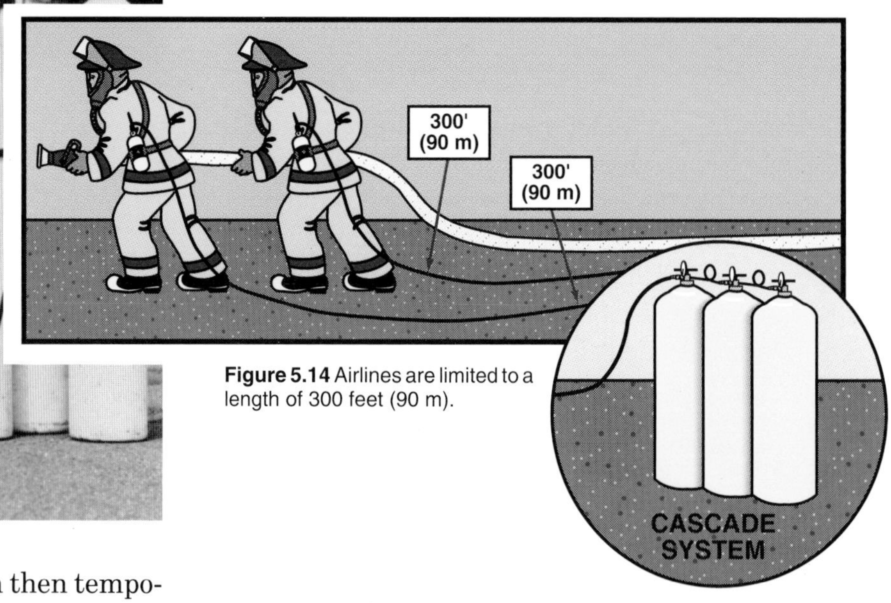

Figure 5.14 Airlines are limited to a length of 300 feet (90 m).

airline option. The first responder can then temporarily disconnect from the airline supply, using the SCBA to provide breathing air, and perform necessary tasks beyond the range of the airline equipment. Airline equipment is not commonly used by fire service personnel but is more commonly used in industry and by special haz mat cleanup companies.

Any airline respirator must provide enough breathing air for the wearer to escape the atmosphere in the event the airline is severed. This requirement is usually accomplished by attaching a very small breathing cylinder, called an escape cylinder, to the airline unit (Figure 5.15). The five-minute escape cylinder must not be used for untethered work. To perform untethered work (detached from the airline), a 30- or 60-minute SCBA that can be augmented by an airline must be used.

Closed-Circuit Self-Contained Breathing Apparatus (SCBA)

Closed-circuit SCBA are sometimes used for haz mat incidents because they have air supply durations that exceed open-circuit SCBA (Figure 5.16). Closed-circuit SCBA are available with durations of 30 minutes to 4 hours and usually weigh less than open-circuit SCBA. They weigh less because a smaller cylinder containing pure oxygen is

Figure 5.15 Airline apparatus should be equipped with an emergency escape cylinder that can be used if the airline becomes inoperable. *Courtesy of MSA.*

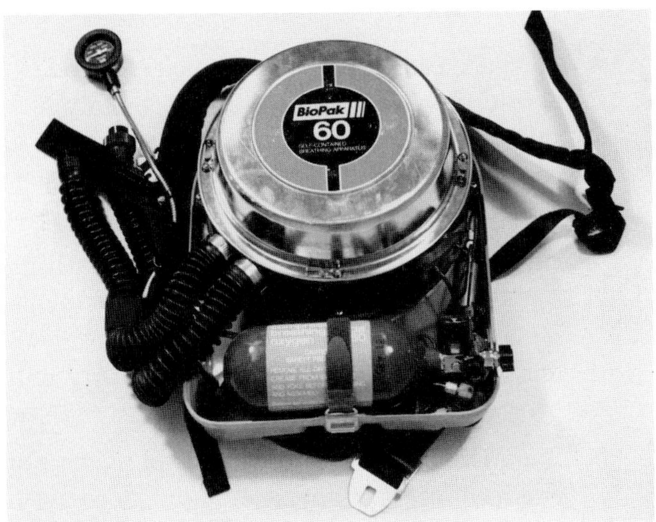

Figure 5.16 Closed-circuit SCBA have longer durations than do open-circuit SCBA.

used. An open-circuit SCBA uses a large compressed-air cylinder. A closed-circuit SCBA "recycles" exhaled air by mixing pure oxygen into it, thus allowing the wearer to rebreathe the air. The operation of these units is considerably more complicated than open-circuit SCBA, and wearers require special training.

Air-Purifying Respirators

Air-purifying respirators use ambient air that is purified through a filter before inhalation. Air-purifying respirators enhance the mobility of the first responder because they do not require an air cylinder or backpack-type assembly.

> **WARNING:**
>
> **Do not wear air-purifying respirators during emergency operations. Wear them only in controlled atmospheres where the hazards present are completely understood and at least 19.5 percent oxygen is present. SCBA must be worn during emergency operations.**

Three categories of air-purifying respirators are particulate-filtering, vapor- and gas-removing, and powered air-purifying.

PARTICULATE-FILTERING RESPIRATORS

These respirators are used for protection against dusts and/or mists. A filter constructed of fibrous material is used to trap the particulate as it is inhaled (Figure 5.17). Different filters are designed for specific chemicals.

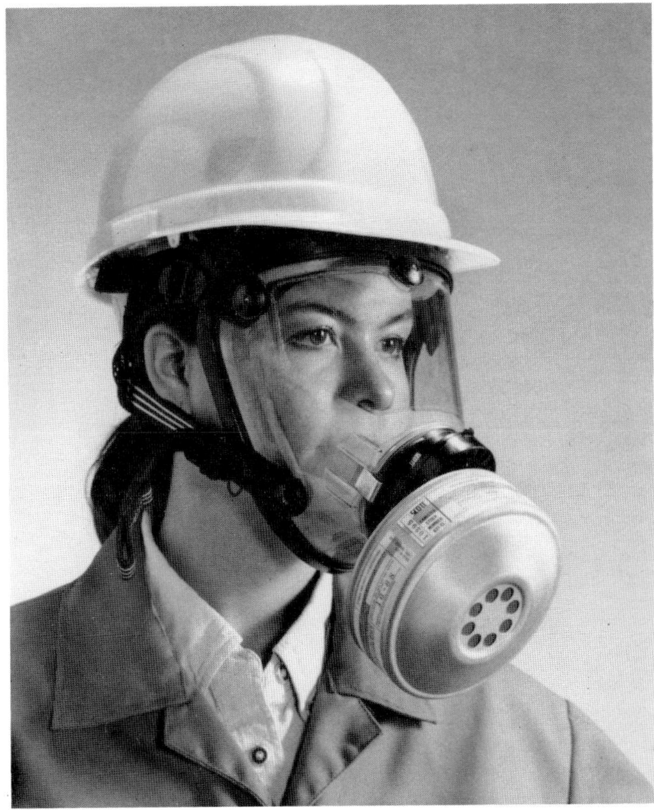

Figure 5.17 Particulate-filtering respirators protect the wearer from inhaling vapors and dusts. *Courtesy of Scott Aviation.*

VAPOR- AND GAS-REMOVING RESPIRATORS

These respirators are available for protection against specific gases and vapors such as ammonia gas and mercury vapor. They are also available for classes of gases and vapors such as acid gases and organic vapors (Figure 5.18). These respirators work by removing the contaminant through interaction of its molecules with a granular, porous material commonly called a *sorbent*. The general method of removal is called *sorption*. Some respirators use catalysts that react with the contaminant to produce a less toxic gas or vapor. These respirators are divided into three classes: chemical cartridge, gas masks, and particulate-vapor and gas-removing air-purifying respirators.

POWERED AIR-PURIFYING RESPIRATORS

The powered air-purifying respirator uses a blower to pass contaminated air through a product that removes the contaminants and supplies the

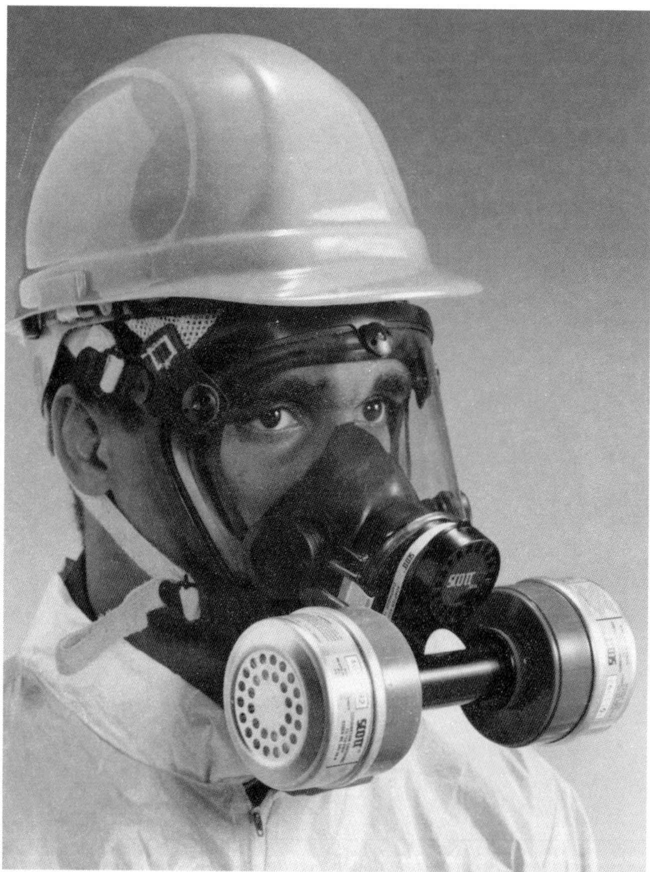

Figure 5.18 Vapor- and gas-removing respirators are useful against specific types of hazardous materials. *Courtesy of Scott Aviation.*

purified air to the facepiece. There are several types of powered air-purifying respirators. Some units are supplied with a small blower and are battery operated. The smaller size allows the user to wear it on his or her belt. Other units have a stationary blower, usually mounted on a vehicle, connected by a long, flexible tube to the respirator facepiece.

Protective Breathing Equipment Limitations

[NFPA 472: 3-3.3.1.2]

To operate effectively, the first responder must be aware of the limitations of protective breathing equipment. These include limitations of the wearer, the equipment, and the air supply (both open- and closed-circuit types).

LIMITATIONS OF WEARER

[NFPA 472: 3-4.3.5]

Several factors affect the first responder's ability to use SCBA effectively. These factors include physical, medical, and mental limitations.

Physical

- Physical condition — The wearer must be in sound physical condition in order to maximize the work that can be performed and to stretch the air supply (both open- and closed-circuit types) as far as possible.
- Agility — Wearing a protective breathing apparatus with an air cylinder or backpack restricts the wearer's movements and affects his or her balance. Good agility will overcome these obstacles.
- Facial features — The shape and contour of the face affect the wearer's ability to get a good facepiece-to-face seal (Figure 5.19).

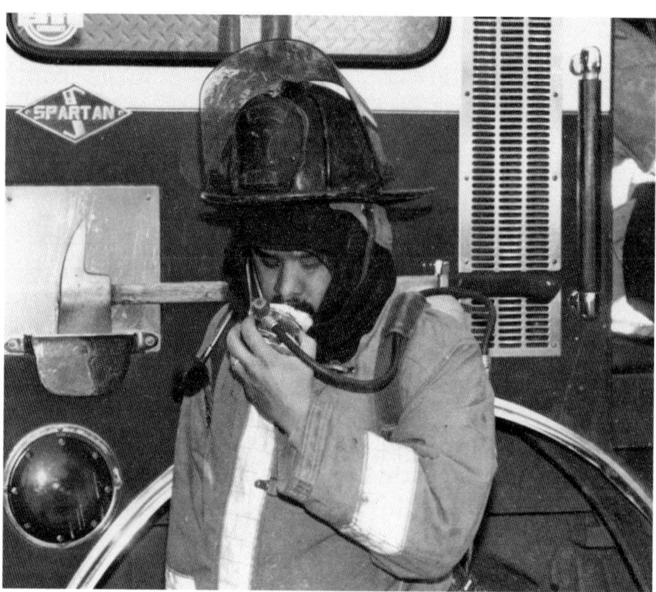

Figure 5.19 The shape of the wearer's face affects the ability to get a good facepiece-to-face seal.

Medical

- Neurological functioning — Good motor coordination is necessary for operating in protective breathing equipment. First responders must be of sound mind to handle emergency situations that may arise.
- Muscular/skeletal condition — The first responder must have the physical strength and size required to wear the protective equipment and to perform necessary tasks.
- Cardiovascular conditioning — Poor cardiovascular conditioning can result in heart attacks, strokes, or other related problems during strenuous activity.

- Respiratory functioning — Proper respiratory functioning will maximize the wearer's operation time in an SCBA.

OSHA 29 CFR 1910.134 states that no one should be assigned a task requiring use of respirators unless the individual is physically able to do the work while wearing the respirator. The wearer's medical history should be reviewed and a physical given to determine the capabilities of the individual. Responders are at risk when wearing a respirator if they have any of the following conditions:

- Asthma
- Emphysema
- Chronic lung disease
- Psychological problems or symptoms including claustrophobia
- Physical deformities or abnormalities of the face
- Medication usage
- Intolerance to increased heart rate, which can be produced by heat stress

The medical doctor who completes the physical exam to determine fitness for duty for respirator use has the authority to deem the individual fit or unfit (Figure 5.20).

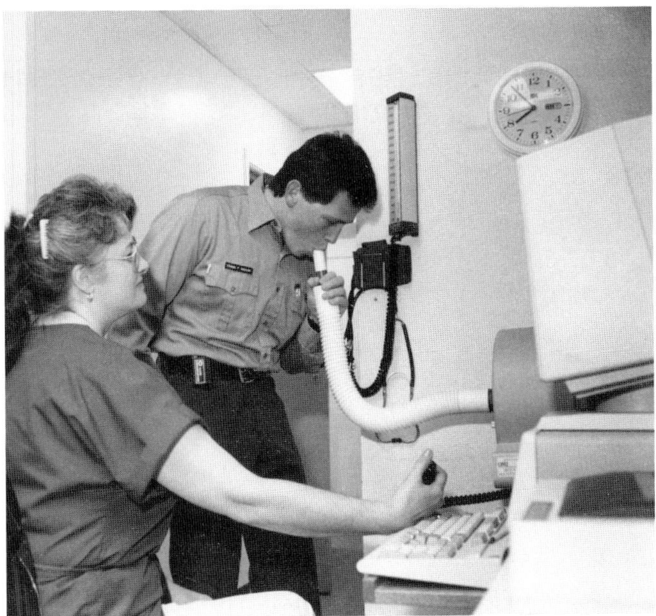

Figure 5.20 A doctor determines if individuals are fit to wear protective breathing apparatus.

Mental

- Training in equipment use — The first responder must be knowledgeable in every aspect of protective breathing apparatus use.
- Self-confidence — The first responder's belief in his or her ability has an extremely positive overall effect on the actions that are performed.
- Emotional stability — The ability to maintain control in an excited or high-stress environment reduces the chances of a serious mistake being made.

LIMITATIONS OF EQUIPMENT

In addition to being concerned about the limitations of the wearer, first responders must also be cognizant of the limitations of the equipment and the air supply. These limitations are as follows:

- Limited visibility — The facepiece reduces peripheral vision, and facepiece fogging can reduce overall vision.
- Decreased ability to communicate — The facepiece hinders voice communication.
- Increased weight — Depending on the model, the protective breathing equipment adds 25 to 35 pounds (11 kg to 16 kg) of weight to the first responder.
- Decreased mobility — The increase in weight and the splinting effect of the harness straps reduce the wearer's mobility (Figure 5.21).
- Inadequate oxygen levels — Air-purifying respirators cannot be worn in IDLH or oxygen-deficient atmospheres.
- Chemical specific — Air-purifying respirators can only be used to protect against certain chemicals. The specific type of cartridge depends on the chemical the wearer is exposed to.

LIMITATIONS OF AIR SUPPLY

Open- and closed-circuit SCBA will have maximum air supply durations that will limit the amount of time a first responder has to perform the tasks at hand. The following are some of the limitations placed upon the air supply:

Figure 5.21 Breathing apparatus harness straps restrict the wearer's ability to move.

- Physical condition of user — If the wearer is in poor physical condition, the air supply will be expended faster.
- Degree of physical exertion — The harder the wearer exerts himself or herself, the faster the air supply is expended (Figure 5.22).
- Emotional stability — A person who becomes excited will increase his or her respirations and use air faster.

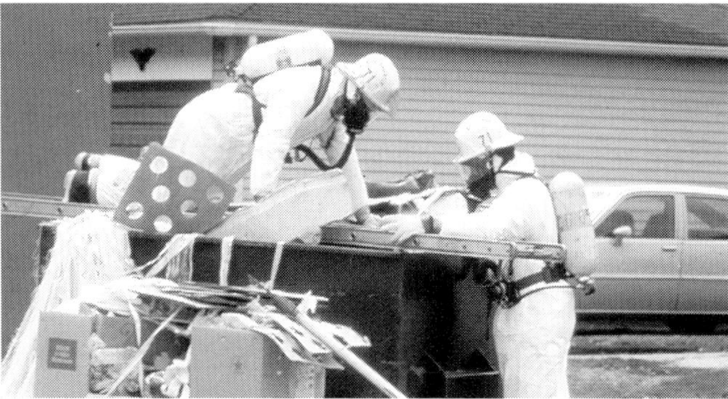

Figure 5.22 First responders engaged in heavy work expend their air supply more quickly than those who are not working as strenuously. *Courtesy of Bob Esposito.*

- Condition of apparatus — Minor leaks and poor adjustment of regulators result in excess air loss.
- Cylinder pressure before use — If the cylinder is not filled to capacity, the amount of working time will be reduced proportionately.
- Training and experience — Poorly trained and inexperienced personnel will use air at a fast rate.

Use And Maintenance

Responders need to know how to don, use, and doff respiratory protective equipment. Because each respirator may vary in the manufacturer's instructions for these procedures, the technical data sheets sent with the respirator need to be consulted. The first responder should not wait until a haz mat incident to learn to put on a respirator.

After each use, the respirator needs to be cleaned, sanitized, and put back into working order. All respiratory protection equipment needs to be placed on a regular inspection list. See the manufacturer's technical data for servicing instructions.

For more information on protective breathing equipment donning, doffing, use, and servicing, see the IFSTA **Essentials of Fire Fighting** or **Self-Contained Breathing Apparatus** manuals.

SELECTION AND USE OF NEEDED EQUIPMENT FOR SPECIFIC INCIDENTS

There are many sources that can be consulted for determining which type of equipment to use and what level to use. Responders need to become familiar with the references available that recommend appropriate equipment.

The first-arriving responder may not have all resources available upon arrival, but in the U.S. he or she can consult the *Emergency Response Guidebook* to determine the minimum level of protection required. This information can be found on the action guide pages in the health hazards emergency action section. For Canadians, the *Initial Emergency Response Guide* (IERG) contains similar information under the protective clothing section in the action guides.

Chapter 6

Photo Courtesy of Scott D. Christiansen, Minot, N.D.

HAZ MAT

Command, Safety, And Scene Control

LEARNING OBJECTIVES

This chapter provides information that will assist the reader in meeting the objectives from NFPA 472, *Standard for Professional Competence of Responders to Hazardous Materials Incidents* that are listed below. The objective numbers are also noted directly in the text in the sections where they are addressed. Objectives in the list below that are denoted with an asterisk (*) are global in nature and are covered by reading the chapter in its entirety.

AWARENESS LEVEL

NFPA 472: 2-4.1* First responders at the awareness level shall, given examples of facility and transportation hazardous materials incidents, identify the actions to be taken to protect themselves and others and to control access to the scene using the local emergency response plan, the organization's standard operating procedures, or the current edition of the *Emergency Response Guidebook.*

NFPA 472: 2-4.1.1 Identify the location of both the local emergency response plan and the organization's standard operating procedures.

NFPA 472: 2-4.1.2 Given a copy of the current edition of the DOT *Emergency Response Guidebook*, describe the difference between the protective action distances in the orange-bordered guide pages and the green-bordered pages in the document.

NFPA 472: 2-4.1.3 Given the local emergency response plan or the organization's standard operating procedures, identify the role of the first responder at the awareness level during a hazardous materials incident.

NFPA 472: 2-4.1.4 Given the local emergency response plan or the organization's standard operating procedures, identify the basic precautions to be taken to protect himself/herself and others in a hazardous materials incident.

NFPA 472: 2-4.1.4.1 Identify the precautions necessary when providing emergency medical care to victims of hazardous materials incidents.

NFPA 472: 2-4.1.4.2 Identify typical ignition sources found at the scenes of hazardous materials incidents.

NFPA 472: 2-4.1.5.2 Given the current edition of the *Emergency Response Guidebook,* identify the definitions for each of the following protective actions:

(a) Isolate hazard area and deny entry;
(b) Evacuate; and
(c) In-place protection.

NFPA 472: 2-4.1.5.3 Given the current edition of the DOT *Emergency Response Guidebook*, identify the shapes of recommended initial isolation and protective action zones.

NFPA 472: 2-4.1.5.4 Given the current edition of the DOT *Emergency Response Guidebook*, describe the difference between small and large spills as found in the table of isolation distances.

NFPA 472: 2-4.1.5.5 Given the current edition of the DOT *Emergency Response Guidebook*, identify the circumstances under which the following distances are used at a hazardous materials incident:

(a) Table of initial isolation and protective action distances; and
(b) Isolation distances in the numbered guides.

NFPA 472: 2-4.1.6 Identify the techniques used to isolate the hazard area and deny entry to unauthorized persons at hazardous materials incidents.

NFPA 472: 2-4.2 The first responder at the awareness level shall, given either a facility or transportation scenario of hazardous materials incidents, identify the appropriate notifications to be made and how to make them, consistent with the local emergency response plan or the organization's standard operating procedures.

NFPA 472: 2-4.2.1 Identify the initial notification procedures for hazardous materials incidents in the local emergency response plan or the organization's standard operating procedures.

OPERATIONAL LEVEL

NFPA 472: 3-4.1 The first responder at the operational level shall, given scenarios for facility and/or transportation hazardous materials incidents, identify how to establish and enforce scene control including control zones, emergency decontamination, and communications.

NFPA 472: 3-4.1.1 Identify the procedures for establishing scene control through control zones.

NFPA 472: 3-4.1.1.1 Identify the criteria for determining the locations of the control zones at hazardous materials incidents.

NFPA 472: 3-4.1.2 Identify the basic techniques for the following protective actions at hazardous materials incidents:
(a) Evacuation; and
(b) In-place protection.

NFPA 472: 3-4.1.5 Identify the items to be considered in a safety briefing prior to allowing personnel to work on a hazardous materials incident.

NFPA 472: 3-4.2 The first responder at the operational level shall, given simulated facility and/or transportation hazardous materials incidents, initiate the incident management system (IMS) specified in the local emergency response plan and the organization's standard operating procedures.

NFPA 472: 3-4.2.1 Identify the role of the first responder at the operational level during hazardous materials incidents as specified in the local emergency response plan and the organization's standard operating procedures.

NFPA 472: 3-4.2.2 Identify the levels of hazardous materials incidents as defined in the local emergency response plan.

NFPA 472: 3-4.2.3 Identify the purpose, need, benefits, and elements of an incident management system (IMS) at hazardous materials incidents.

NFPA 472: 3-4.2.4 Identify the considerations for determining the location of the command post for a hazardous materials incident.

NFPA 472: 3-4.2.5 Identify the procedures for requesting additional resources at a hazardous materials incident.

NFPA 472: 3-4.2.6 Identify the responsibilities of the safety officer.

NFPA 472: 3-4.3.1 Identify the importance of the buddy system in implementing the planned defensive options.

NFPA 472: 3-4.3.2 Identify the importance of the back-up personnel in implementing the planned defensive options.

NFPA 472: 3-4.3.3 Identify the safety precautions to be observed when approaching and working at hazardous materials incidents.

NFPA 472: 3-5.1.2 Describe the circumstances under which it would be prudent to pull back from a hazardous materials incident.

NFPA 472: 3-5.2.2 Identify the methods for immediate notification of the incident commander and other response personnel about critical emergency conditions at the incident.

Chapter 6

Command, Safety, And Scene Control

The initial recognition and identification of hazardous materials are critical phases of haz mat emergencies. However, if the first-arriving responder fails to use this information to exercise adequate protective actions, then lives, property, and incident control may be lost. It is the responsibility of every first responder to ensure the safety of his or her peers and to facilitate rapid control of the incident. These responsibilities are most easily accomplished when the first responder has the benefit of a standardized method or plan for assessing risk and determining the best course of action.

Title III of the Superfund Amendments and Reauthorization Act (commonly referred to as "SARA Title III") of 1986 required jurisdictions in the United States to develop a local emergency response plan (LERP) for haz mat incidents. The LERP was to be developed by the local emergency planning committee (LEPC), which is composed of representatives of the emergency response agencies. Also included on the LEPC are representatives of the following local and state agencies: government, emergency management, and transportation. There are also representatives from the health community, the media, environmental groups, and industry. The local emergency planning committee must address the following areas:

- Haz mat facilities and transportation routes (Figure 6.1)
- Methods and procedures for handling haz mat incidents
- Methods to warn people at risk
- Haz mat equipment and information resources

Figure 6.1 The LEPC must address hazards such as chemical plants.

- Evacuation plans
- Training of first responders
- Schedule for exercising the LERP

There is no legislation in Canada that is comparable to SARA Title III. However, the local response agencies are required to develop plans that are similar to those used in the United States. The Canadian plans are referred to as emergency measures organization (EMO) plans. These plans basically cover the same information that the LERPs in the United States cover. (**NOTE:** For consistency in presentation, the term *local emergency response plan* (LERP) is used throughout this chapter. Canadian first responders should remember that this is the same as an EMO plan.)

[NFPA 472: 2-4.1.1]

First responders should be familiar with the plan and with the location of the plan. Copies of the plan must be made available to all first responders and their agencies. First responders should thoroughly understand their organization's standard operating procedures (SOPs). Because written copies of the actual plan itself are too bulky to be easily used on a response, first responders should carry a checklist of the initial actions to be taken.

This chapter covers areas that are of concern before and immediately after agencies learn of a developing haz mat emergency. It also focuses on available resources that assist the first responder in making the initial decisions to carry out the local emergency response plan.

EMERGENCY INFORMATION MANAGEMENT

The collection and use of information regarding a particular incident typically begin when the call for help is received; this initial receipt of information is called *external communication*. Additional information regarding the incident is received when emergency personnel arrive on the scene; this is called *internal communication*. The following sections look at these two types of communication.

External Communication Of Hazardous Materials Incidents

To some extent, external communications really begin with the collection of information regarding hazardous materials and operations during pre-incident planning. The availability of this information, in addition to that which is received when an emergency is actually reported, can make a significant difference in the outcome of the incident.

The telephone is the most commonly used medium for reporting an emergency. Haz mat incidents compose a very small portion of the overall calls (often referred to as *complaints*) handled by dispatchers. For this reason, dispatchers require special training to ensure that they recognize the significance of the call and act accordingly. Dispatchers must be trained to obtain as much information as possible regarding a reported incident (Figure 6.2). They should ask questions of the caller that will help determine whether a hazardous material may be involved and, if so, what the material is. It is best if the dispatcher has a prepared list of questions to ask the caller. This list of questions will help determine the nature of the incident. If possible, the following information should be gathered by the dispatcher:

Figure 6.2 Well-trained dispatchers obtain as much information as possible from the reporting party.

- Location of the incident with a cross-street reference, including primary and alternate access points
- Name, phone number, and location of the caller reporting the incident
- Identity of the substance involved (spelled out if possible)
- Approximate quantity of the material, the type of container, and the condition of the container
- Prevailing weather conditions at the scene, including wind direction and speed, precipitation, and temperature if possible
- Number and proximity of persons or properties threatened
- Brief description of the events leading to the incident and any obvious threatening effects
- Summary of control actions taken or underway
- Type of assistance needed
- Arrangements for recontacting the reporting party

With this information and the response determined by SOPs, dispatchers start the operation by dispatching appropriate entities. Additional equipment and personnel can be dispatched upon request of the incident commander.

In addition to initial training, dispatchers should be given periodic refresher training on how to obtain the correct information from a reporting party. The dispatcher must also be cognizant of the LERP, the department's SOPs, and special procedures for making haz mat incident notifications and information requests. **Dispatchers should be included in responder training sessions or exercises.**

Internal Communication Of Hazardous Materials Incidents

[NFPA 472: 2-4.2]; [NFPA 472: 3-4.2.5]

Internal communication begins when emergency personnel arrive at the scene. If either the dispatcher or the first responder realizes that a haz mat incident is in progress, he or she should initiate the organization's SOP for this type of incident. Depending on the capabilities of local emergency forces and the requirements of the LERP, this initial response may be no more than that normally dispatched to conduct an investigation of unknown trouble. Where resources are available, specialized personnel and equipment may be sent immediately. Additionally, the dispatcher must be prepared to change the response whenever additional information warrants such a change. The first responder executes the preliminary activities contained in the departmental SOP, facility plan, or those activities recommended by references such as the *Emergency Response Guidebook (ERG*/U.S.) or the *Initial Emergency Response Guide (IERG*/ Canada). However, the dispatcher's responsibilities may include the following procedures:

- Establishing internal or external clear-line communications with technical advisors
- Notifying mutual aid agencies
- Activating other prescribed departmental procedures
- Advising next-in-line supervisors and chief officers of the incident

Dispatchers must relay all available information to the responding units, and each unit should feed back reports to the dispatcher (Figure 6.3). Dispatchers must not filter, edit, delete, or change the information that they receive. They must pass on all information as promptly and accurately as possible. The amount of information requested and transmitted between command, tactical, and support units during these incidents may seriously strain the abilities of single-channel, single-operator dispatching systems. Ideally, dispatchers relaying information between units should have no other assignments in order to avoid confusion and loss of information. In addition, it is desirable for all the units operating at the haz mat incident to be switched to a radio frequency that is separate from radio frequencies used for normal daily operations.

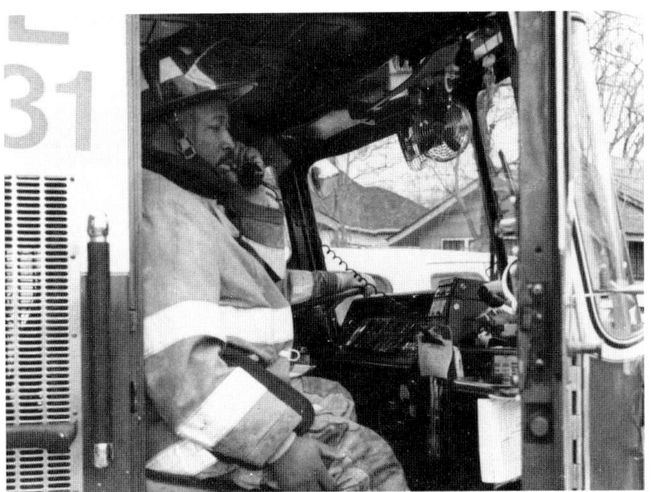

Figure 6.3 There must be a good line of communication between the dispatcher and the personnel in the field.

Whenever possible, on-scene command personnel should talk directly to technical advisors (Figure 6.4). However, when direct communications with technical advisors are not possible, the dis-

Figure 6.4 Technical advisors provide invaluable information to on-scene personnel.

patcher must perform the function of liaison between the technical advisor and on-scene personnel. Dispatchers should gather as much information about the nature of the incident as possible before contacting technical sources. CHEMTREC and CANUTEC are examples of highly useful services with which dispatchers may be required to communicate. While these organizations primarily provide immediate information on material properties, their hazards, and suggested control techniques, they also serve an equally important role as a communications link with other technical support resources. These organizations should be contacted only for their intended purpose: to provide information in the early stages of the emergency.

[NFPA 472: 3-5.2.2]

The dispatchers and on-scene personnel must communicate all pertinent information regarding the incident to the incident commander (IC). The incident commander should state clearly all directives to on-scene personnel and confirm with personnel that the orders are understood.

MISSION OF INCIDENT OPERATIONS

The first-arriving responder may be a firefighter, police officer, emergency medical technician, public works employee, plant operator, or member of an industrial emergency response brigade. It is critical that each worker, regardless of his or her position, is prepared to carry out the mission for the awareness or operational levels of response, depending on the level to which he or she has been trained. These levels are clearly explained in OSHA 1910.120 and NFPA 472. The following sections list the scope of responsibilities for each of these missions.

Awareness Level

[NFPA 472: 2-4.1.3]

The first responder's mission at the awareness level is to implement the jurisdiction's SOPs. The following are the standard procedures that the first responder will perform:

- Recognize the presence of hazardous materials.
- Call for the appropriate help to mitigate the incident.
- Secure the area of the emergency and prevent anyone else from entering (Figure 6.5).
- Survey the incident from a safe distance to determine the identity of the materials involved (Figure 6.6).
- Determine the appropriate actions to be taken as recommended by the *ERG* or *IERG*.

No action must be taken that will place the responder or others in a position of danger or in contact with the material.

Figure 6.5 Awareness level responders should be able to secure the scene.

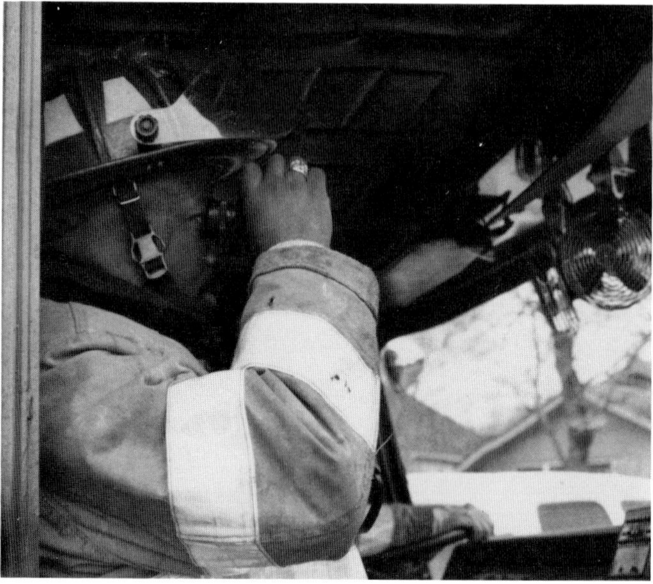

Figure 6.6 Binoculars are helpful in making an accurate scene assessment from a safe distance.

Operational Level

[NFPA 472: 3-4.2.1]

The first responder at the operational level must perform the awareness level tasks. In addition, he or she must be able to define the problem, design a defense, and direct execution of the incident action plan.

DEFINE THE PROBLEM

In order to define the problem, the first responder must analyze the incident to determine the extent of the problem and to predict the possible outcomes. The first responder accomplishes these tasks by performing the following functions:

- Surveying the condition of containers
- Estimating the nature and extent of release
- Observing present conditions
- Gathering and sharing material information with all involved parties (Figure 6.7)
- Predicting the incident's future course
- Estimating likely harm to responders, civilians, and the environment

Figure 6.7 Gather as much information as possible from parties on the scene.

DESIGN A DEFENSE

The first responder at the operational level must determine the initial defensive actions necessary to control the haz mat incident. These initial actions must fall within the first responder's level of training and the equipment he or she has available. In order to design a response, the first responder must perform the following functions:

- Establish defensive goals.
- Identify defensive tactical options to meet the goals.
- Ensure the appropriateness of the training and personal protective equipment of on-scene personnel for the actions to be taken.
- Prepare for emergency decontamination of contaminated responders or civilians (Figure 6.8).

Figure 6.8 Preparing for emergency decontamination is one part of designing a response plan.

DIRECT EXECUTION OF THE INCIDENT ACTION PLAN

Once the plan is formulated, the first responder at the operational level must be prepared to have limited participation in the implementation of the plan. The level of training will limit the actions of the first responder to the following tasks:

- Establishing protective zones
- Activating the incident management system
- Using personal protective equipment (Figure 6.9)
- Carrying out defensive activities
- Evaluating and reporting incident progress
- Performing emergency decontamination procedures

Figure 6.9 First responders should don protective equipment to assist in the execution of the incident action plan.

INCIDENT MANAGEMENT SYSTEMS

[NFPA 472: 3-4.2.3]

Perhaps the most crucial aspect of controlling any incident, including haz mat incidents, is the implementation and use of an incident management system (Figure 6.10). An incident management system is a management framework used to organize emergency incidents. Whether the system is developed locally or adopted from a nationally recognized model system, the command structure should satisfy the following requirements:

- *Common Terminology* — Names of organizational elements, resources, and facilities should be consistent.
- *Modular Organization* — The management system should be built from the top down, with branches/sections added as needed according to the size and complexity of the incident.
- *Integrated Communication* — All of the agencies involved in the incident must be able to communicate with each other.
- *Unified Command Structure* — All of the individuals or agencies that have jurisdictional responsibility should be represented within the command structure (Figure 6.11).

The need for an integrated and cooperative incident management system cannot be overemphasized. For more information on the model IMS systems listed, refer to the following Fire Protection Publications manuals:

Figure 6.10 All incidents, be they haz mat emergencies or routine incidents such as this vehicle extrication, should be managed under the guidance of an incident management system. *Courtesy of Bob Esposito.*

Figure 6.11 All agencies on the scene should be represented at the command post.

- *Incident Command System*
- *National Fire Service Incident Management System*

ESTABLISHING INCIDENT COMMAND

[NFPA 472: 3-4.2]

Regardless of which incident management system is used, one directive is clear: Either the first person on the scene or the ranking individual of the first company on the scene should assume command of the incident. This command is maintained by that individual until a higher ranking or more extensively trained responder arrives on the scene and assumes command following a briefing (Figure 6.12). As more responders arrive on the scene, the incident commander can assign them to oversee specific areas of the IMS structure.

All incident management systems have a myriad of roles to be performed. Two very important roles in any system are those of the incident commander and the safety officer. The following sections highlight the major responsibilities of each position.

Figure 6.12 Command should be transferred after the higher ranking officer has been briefed completely on the activities that have occurred to that point.

Incident Commander (IC)

The incident commander assumes command of an incident and briefs other responders after the overall situation is reviewed. The IC is responsible for the development, implementation, and documentation of the incident action plan. No aggressive plan should be undertaken unless sufficient information is available to make logical decisions and the safe coordination of operations can be accomplished. The IC must make it known to the dispatcher and to other responders when command is assumed or transferred.

[OSHA 1910.120 (q)(3)]

The incident commander is responsible for the following functions:

- Establishing the site safety plan
- Implementing a site security and controlling plan to limit the number of personnel operating in the control zones
- Designating a safety officer
- Identifying the materials or conditions involved in the incident
- Implementing appropriate emergency operations
- Ensuring that appropriate personal protective equipment is worn
- Establishing a decontamination plan and operation

The incident commander does not have to actually perform or supervise each of these functions but may choose to delegate them to others.

Safety Officer

[NFPA 472: 3-4.2.6]; [OSHA 1910.120 (q)(3)(vii & viii)]

The safety officer is responsible for monitoring and identifying hazardous and unsafe situations and developing measures for ensuring personnel safety (Figure 6.13). Although the safety officer may exercise emergency authority to stop or prevent unsafe acts when immediate action is required, he or she will generally choose to correct them through the regular line of authority. Many of these unsafe acts or conditions are addressed in the incident action plan. The safety officer is required to perform the following duties:

- Maintain communications with the IC.
- Identify hazardous situations at the incident scene.
- Participate in incident planning.
- Review incident action plans for safety issues.
- Identify and immediately correct potentially unsafe situations if necessary.

Figure 6.13 The safety officer monitors the scene for unsafe conditions. *Courtesy of Ron Jeffers.*

HAZARD ASSESSMENT

[NFPA 472: 3-4.3.3]

Hazard assessment, sometimes referred to as size-up, is the mental process of considering all available factors that will immediately affect the incident during the course of the operation. The information gained from the hazard assessment is used to determine the strategy and tactics that will be applied to the incident.

Hazard assessment is a continual evaluation. It starts with pre-incident planning, extends through the receipt of an alarm, and continues throughout the course of the incident. When the first incident commander arrives on the scene, he or she conducts an extensive hazard assessment. During the course of the operation, the incident commander will periodically conduct a hazard assessment to ensure that progress is being made.

Much information needed for hazard assessment can be obtained at the time of alarm:

- Nature of the call
- Location of the call (Figure 6.14)
- Equipment responding
- Time of day
- Weather

Figure 6.14 The location of the call can have a dramatic impact on the scope of the activities required to handle the incident. An incident in the vicinity of this nursing home might pose serious evacuation problems.

Additional information relating to size-up may be considered by responders while still en route:

- Evaluating response route
- Reviewing plans and sketches (Figure 6.15)
- Noting arrival time of other responding units
- Noting exposure types and distances
- Reviewing hydrant and water supply conditions
- Considering access to the scene
- Making preliminary plans for apparatus placement at the scene
- Securing any additional information from the dispatcher
- Deciding if and what additional units are needed

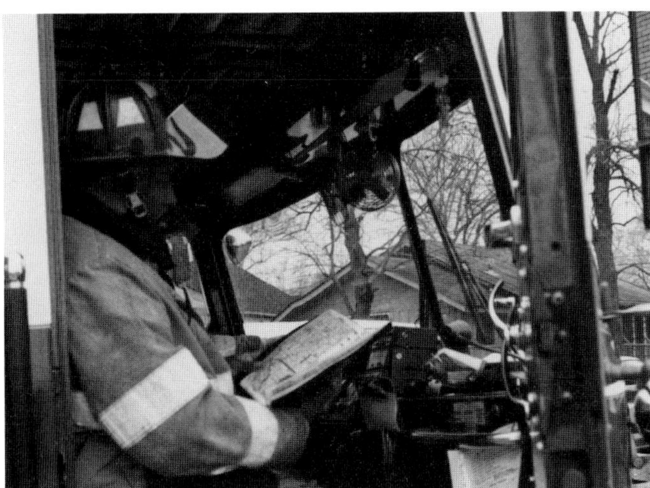

Figure 6.15 The first responder reviews plans and maps while en route to the scene.

None of these factors, however, should interfere with the safe operation of the apparatus to the emergency scene.

Once on the scene, the final pieces of the hazard assessment are added to the information made available before arrival (Figure 6.16). With this combined information, a formal plan of action may be decided upon and implemented. The following conditions on the scene must be evaluated:

- Unusual signs (smoke, fire, explosions, leaking material, vapor clouds, etc.)
- Life hazards
- Material(s) involved
- Path of fire or material travel
- Actions already taken by people on the scene

Figure 6.16 The size-up is completed when the first responder is actually on the scene.

[NFPA 472: 2-4.2.1]

Based on the assessment, the incident commander develops a plan of action. Overall, the plan must focus on three strategic goals. In order of priority, these goals are the following:

1. Life safety
2. Environmental protection
3. Property conservation

The IC evaluates the situation with these priorities in mind and determines the potential for successfully accomplishing them. The incident commander's first priority is the safety of the responders. If there is no immediate threat either to responders or to civilians, the next consideration is that of protecting the environment from further damage. When the first two priorities are satisfied, the conservation of property can be addressed.

In some cases, the accomplishment of a lower priority may, in fact, be the prescription for satisfying a higher priority. For example, confinement of material to prevent further environmental damage might result in controlling a life safety risk.

Modes Of Operation

When the immediate priority has been established, the IC will determine the mode of operation to be followed. The safety of the first responders is the uppermost consideration in selecting a mode of operation. The three modes of operation are nonintervention, defensive, and offensive.

NONINTERVENTION OPERATIONS

[NFPA 472: 3-5.1.2]

Nonintervention operations are those operations in which the responders take no direct actions on the actual problem (Figure 6.17). Not taking any action is the only safe strategy in many types of incidents and is the best strategy in certain types of incidents. An example of nonintervention is when a pressure vessel exposed to fire cannot be adequately cooled. In such incidents, responders should withdraw to a safe distance. The nonintervention mode is selected when one or more of the following circumstances exist:

- The facility or LERP calls for it based on a pre-incident evaluation of the hazards present at the site.

Figure 6.17 Nonintervention operations are used when attacking the incident poses more harm than good. *Courtesy of Mike Sanphy, Westbrook (ME) Fire Department.*

- The situation is clearly beyond the capabilities of responders.
- Explosions are imminent.
- Serious container damage threatens a massive release.

When operating in the nonintervention mode, responders will take the following actions:

- Withdraw to a safe distance.
- Report scene conditions to dispatch.
- Establish scene control.
- Initiate the incident management system.
- Initiate evacuation where needed.
- Call for additional resources.

DEFENSIVE OPERATIONS

Defensive operations are those in which the responders seek to confine the emergency to a given area, without directly contacting the materials causing the emergency. This mode of operation is the upper limit of risk that first responders may take at the operational level. The defensive mode is selected when one of the following circumstances exists:

- The facility or LERP calls for it based on a pre-incident evaluation of the hazards present at the site.
- The responders have the training and equipment necessary to confine the incident to the area of origin.

When operating in the defensive mode, responders will take the following actions:

- Report scene conditions to dispatch.
- Establish scene control.
- Initiate the incident management system.
- Establish and indicate zone boundaries.
- Commence evacuation where needed.
- Control material spread by diverting it to a safe location.
- Construct dikes or dams to confine the materials (Figure 6.18).
- Control ignition sources.
- Call for additional resources.

Figure 6.18 Constructing a dike is a defensive tactic.

OFFENSIVE OPERATIONS

Offensive operations are those where responders take aggressive, direct action on the material, container, or process equipment involved in the incident. These operations may result in contact with the material and therefore require responders to wear appropriate chemical-protective clothing and respiratory protection (Figure 6.19). Offensive operations are beyond the scope of responsibilities for first responders and are conducted by more highly trained haz mat personnel.

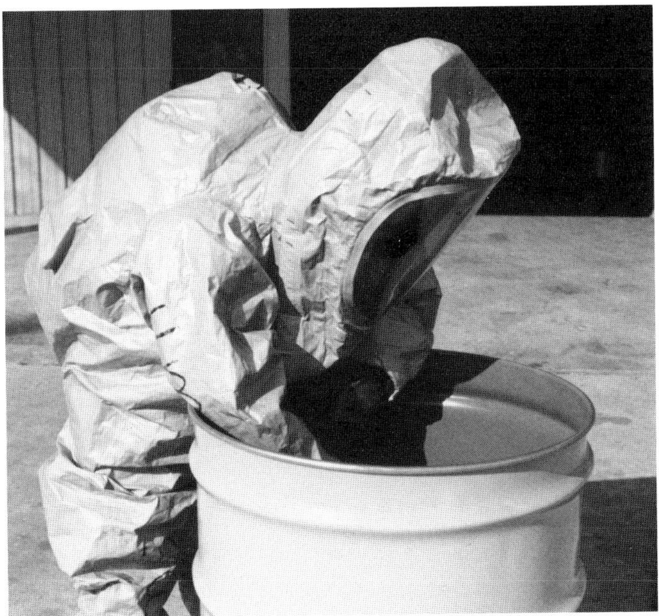

Figure 6.19 Haz mat specialists or technicians usually undertake offensive operations.

Determining Levels Of Hazardous Materials Incidents

[NFPA 472: 3-4.2.2]

First responders must understand that haz mat incidents are different from other types of emergencies. They must also understand who needs to get involved in the incident and why. This understanding comes through the knowledge of the organization's SOPs and the LERP. These documents dictate how resources, depending on the level of risk, will be devoted to incidents.

Most models used for determining the level of a haz mat incident define three levels of response. The levels graduate from Level I (least serious) to Level III (most serious). By defining the levels of response, an increasing level of involvement and necessary resources can be identified based on the gravity of the incident. The criteria for one model used to determine the level of a haz mat incident are as follows:

- Extent of municipal, county, state, and federal involvement (or potential involvement)
- Level of technical expertise required at the scene
- Extent of evacuation of civilians
- Extent of injuries or deaths

LEVEL I INCIDENT

A Level I incident is the least serious and the easiest to handle. It may pose a serious threat to life or property, although this is not usually the case. This type of incident is within the capabilities of the fire department or other first responders having jurisdiction. Evacuation, if required, will be limited to the immediate area of the incident.

The following are examples of Level I incidents:

- Small amount of gasoline or diesel fuel spilled from an automobile (Figure 6.20).
- Leak from domestic natural gas line on the consumer side of the meter.
- Broken containers of "consumer commodity" such as paint, thinners, bleach, swimming pool chemicals, and fertilizers. The owner or proprietor is normally responsible for cleanup and disposal.

Figure 6.20 An example of a Level I incident is a small amount of fuel spilled at a vehicle accident.

LEVEL II INCIDENT

A Level II incident is one that is beyond the capabilities of the first responders on the scene and may be beyond the capabilities of the first response agency having jurisdiction. Level II incidents require the services of a formal haz mat response team. A properly trained and equipped response team could be expected to respond in the following manner:

- Use chemical protective clothing.
- Dike and confine within the contaminated areas.
- Perform plugging and patching (Figure 6.21).

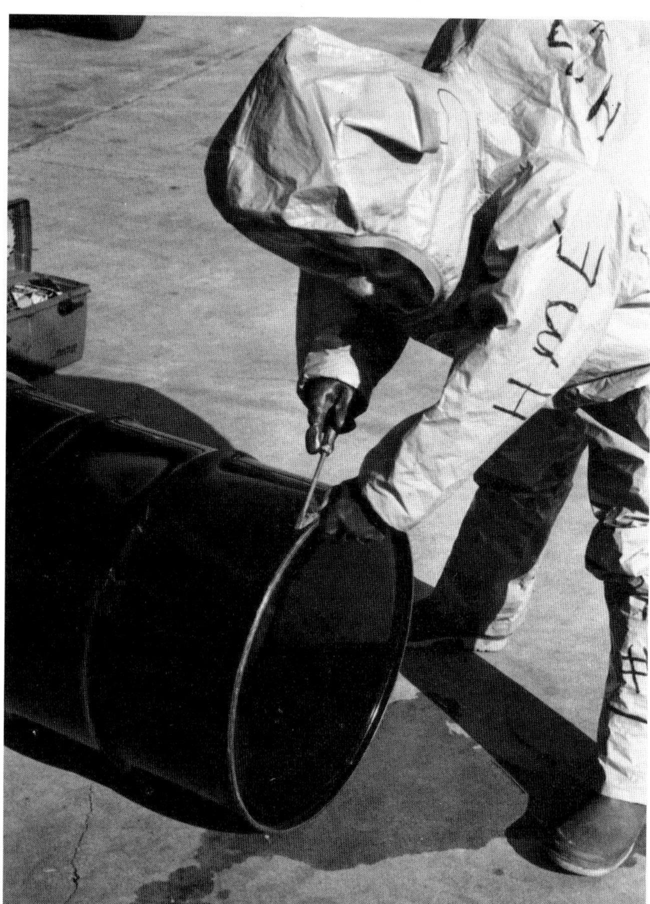

Figure 6.21 Plugging and patching are performed at Level II incidents.

- Sample and test unknown substances (Figure 6.22).
- Perform at various levels of decontamination.

The following are examples of Level II incidents:

- Spill or leak requiring large-scale evacuation
- Any major accident, spillage, or overflow of flammable liquids (Figure 6.23)
- Spill or leak of unfamiliar or unknown chemicals
- Accident involving extremely hazardous substances (Figure 6.24)
- Rupture of an underground pipeline
- Fire that is posing a BLEVE threat

Figure 6.24 Level II incidents are large-scale incidents involving extremely hazardous substances at fixed facilities. *Courtesy of Harvey Eisner.*

Figure 6.22 First responders may be required to perform air sampling at Level II incidents. *Courtesy of Joe Marino.*

Figure 6.23 A leaking, overturned tanker is an example of a Level II incident. *Courtesy of the Chicago Fire Department.*

LEVEL III INCIDENT

A Level III incident is the most serious of all haz mat incidents. These incidents require resources from state agencies, federal agencies, and/or private industry. A large-scale evacuation may be required. Most likely, the incident will not be concluded by any one agency. Successful handling of the incident will require a collective effort:

- Specialists from industry and governmental agencies (Figure 6.25)
- Sophisticated sampling and monitoring equipment
- Specialized leak and spill control techniques
- Decontamination on a large scale

The following are examples of Level III incidents:

- Those that require an evacuation extending across jurisdictional boundaries
- Incidents beyond the capabilities of the local haz mat response team
- Incidents that activate, in part or in whole, the federal response system

In addition to this system of determining the levels of hazardous materials incidents, the National Fire Protection Association has defined three levels of response in NFPA 471, *Recommended Practice for Responding to Hazardous Materials Incidents* (Table 6.1). This table is primarily used to determine the level of incident in planning a response or training exercise. The application of this information is basically the same as that for the three levels previously described.

Safety Considerations

[NFPA 472: 3-4.1.5]; [NFPA 472: 3-4.3.2]; [NFPA 472: 3-4.3.1]; [NFPA 472: 2-4.1.4]

There are a number of safety-related concerns involving personnel, procedures, and precautions that the incident commander should consider before committing to action at haz mat incidents:

- Are the responders working as members of a team?
- Have all the responders been adequately briefed on the incident action plan and the hazards of the situation? Each responder should be aware of the following aspects of the response:
  - The immediate goal
  - Who performs which task
  - Operation completion time
  - How to call for help
  - The escape route
  - The material's effects
  - Signs/symptoms of exposure
- Can reconnaissance be made visually? (Figure 6.26)
- Can approach be made from upwind/uphill?
- Can contact with the material be avoided?
- Can the vapor cloud, mist, dust, or smoke spread?
- Is the risk worth the benefit?

In addition to the safety responsibilities of the incident commander and/or the safety officer, each

Figure 6.25 Help from governmental agencies, such as the Coast Guard, may be required on Level III incidents. *Courtesy of Ron Jeffers.*

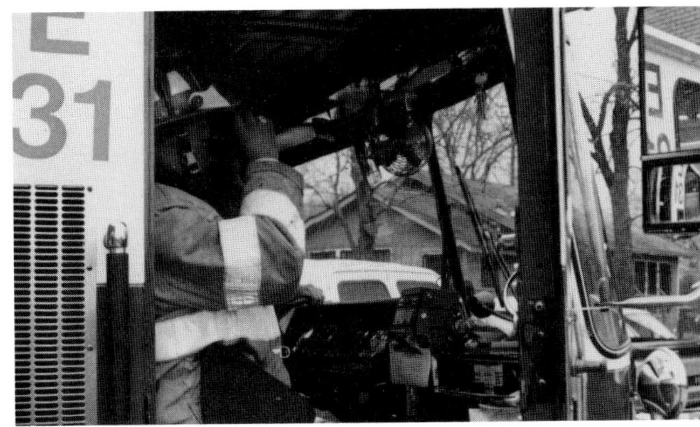

Figure 6.26 When possible, perform reconnaissance visually.

TABLE 6.1
Planning Guide For Determining Incident Levels, Response, And Training

Incident Conditions	Incident Level One	Incident Level Two	Incident Level Three
Product identifications	Placard not required, NFPA 0 or 1 all categories, all ORM A, B, C, and D.	DOT placarded, NFPA 2 for any categories, PCBs without fire, EPA regulated waste.	Poison gas, explosives, organic peroxide, flammable, solid, materials dangerous when wet, chlorine, fluorine, anhydrous ammonia, radioactive materials, NFPA 3 and 4 for any categories including special hazards, PCBs and fire DOT inhalation hazard, EPA extremely hazardous substances, and cryogenics.
Container size	Small (e.g., pail, drums, cylinders except 1-ton, packages, bags).	Medium (e.g., 1-ton cylinder, portable containers, nurse tanks, multiple small packages).	Large (e.g., tank cars, tank trucks, stationary tanks, hopper cars/trucks, multiple medium containers).
Fire/Explosion potential	Low	Medium	High
Leak severity	No release or small release contained or confined with readily available resources.	Release may not be controllable without special resources.	Release may not be controllable even with special resources.
Life safety	No life-threatening situation from materials involved.	Localized area, limited evacuation area.	Large area, mass evacuation area.
Environmental impact (potential)	Minimal	Moderate	Severe
Container integrity	Not damaged.	Damaged but able to contain the contents to allow handling or transfer of product.	Damaged to such an extent that catastrophic rupture is possible.

responder must be able to recognize a number of threats to his or her own safety. To accomplish this, the first responder must respect the material, note the surroundings, observe the container, and eliminate the ignition sources if the material is flammable.

RESPECT THE MATERIAL

The first responder must not minimize the potential threat posed by the material involved in the incident. To maximize safety, the first responder should observe the following precautions:

- Avoid contact with mists, vapors, dusts, and smoke.
- Maintain a safe distance, and stay outside the hot zone.
- Use available shielding, and stay with emergency vehicle if necessary.
- Anticipate change such as delayed reactions by the material and weather conditions.

NOTE THE SURROUNDINGS

The first responder must be cognizant of the area surrounding the incident and what effect it will play in the formation of an incident action plan. The following factors may have a bearing on the formation of the plan:

- Weather — Will wind shift (speed/direction), precipitation/humidity, air temperature, and/or sunlight affect the level of hazard present?

- Topography — Will the slope, altitude, and/ or aspect have a bearing on the potential for the material to spread?
- Water — Does the incident affect — or have the potential to affect — oceans, lakes, rivers, streams, ponds, puddles, estuaries, flood-control channels, storm and sewer drains, cisterns, reservoirs, or catch basins (Figure 6.27)?

Figure 6.27 Incidents that affect waterways can be particularly demanding.

- Occupancies — Does the area of the incident include public occupancies such as stadiums, churches, schools, hospitals, jails, and malls? If it does not currently affect those occupancies, does it have the potential to spread to such places?
- Community transportation systems — Will the course of the incident affect depots, terminals, stations, and stops in the public transportation system?
- Utilities — Does the incident involve — or have the potential to involve — overhead and underground electrical transmission equipment, telephone and cable systems, natural gas, sanitary sewers, steam and chilled-water piping, cooling towers, or utilidors?
- Zero energy state — Has every power (electric, pneumatic, hydraulic, mechanical) or energy (chemical, thermal, gravitational) source that can produce unexpected movement of machines or containers been locked off, locked out, de-energized, adequately secured, or otherwise accounted for?

OBSERVE THE CONTAINER

First responders should continually watch the container or problem area from a safe distance to observe changing conditions. Any change in the status of an incident should be communicated immediately to the incident commander. The incident commander may revert to nonintervention operations if the incident deteriorates. First responders must be ready to pull back immediately if they observe adverse changes. The first responder should observe the container for the following conditions:

- Container integrity — Does the container show signs of failure?
- Safety devices — Is the container equipped with vents or relief devices (Figure 6.28)? Are the vents or relief devices operating properly?
- Leaking — Are seals, gaskets, and connections still intact?
- Stability — Is the container likely to move?

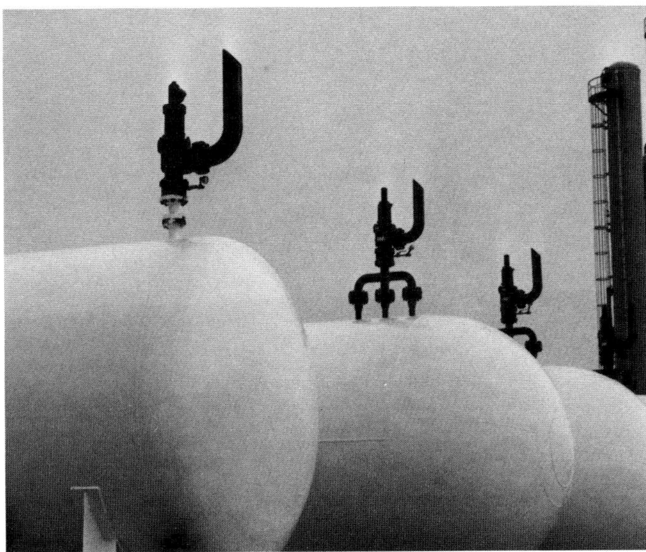

Figure 6.28 Look to make sure that the container is equipped with a relief device.

If there is a rapid deterioration of the incident, there must be some method to signal a withdrawal. Common signaling methods are the sounding of air horns or sirens or an emergency radio broadcast. The use of sirens or air horns should be in a distinctive manner such as a long blast or several short blasts. All responders should be briefed on these signals before operating at a scene and should be familiar with these signals at the incident.

ELIMINATE IGNITION SOURCES

[NFPA 472: 2-4.1.4.2]

To maximize scene safety, potential ignition sources must be eliminated if the involved material is flammable or explosive. There are several potential ignition sources:

- Internal combustion engines
- Electric motors, switches, and controllers
- Lighting equipment
- Fuel-powered equipment
- Open or pilot flames
- Electrostatic and frictional sparks
- Heated metal surfaces
- Smoking materials
- Fuses, flares, torpedoes, and lanterns
- Radios, hand lights, pagers, and PASS devices (Figure 6.29)

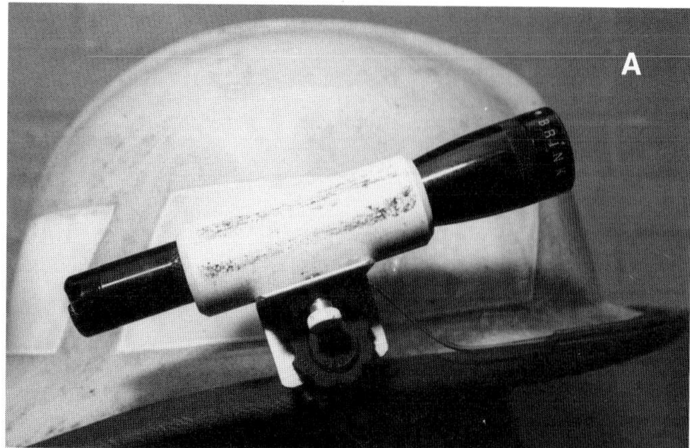

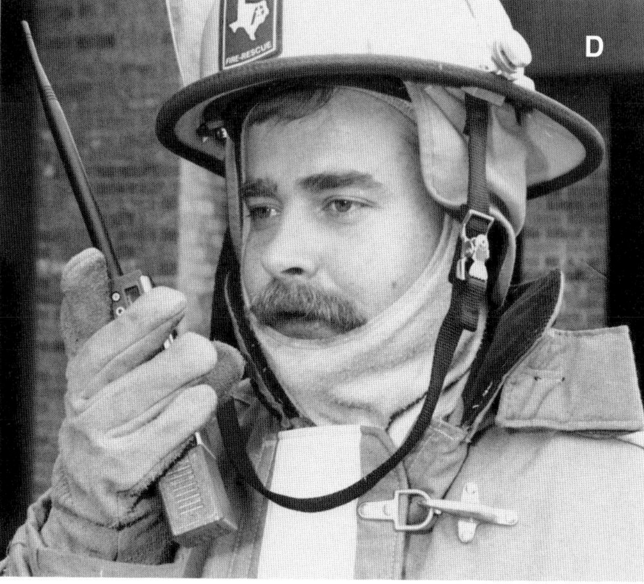

Figure 6.29 Flashlights (a), PASS devices (b), pagers (c), and portable radios (d) are all examples of devices commonly carried by firefighters that can be ignition sources if not operating properly.

ESTABLISHING SCENE CONTROL

[NFPA 472: 2-4.1]; [NFPA 472: 3-4.1]

In order for a haz mat incident to be concluded smoothly, it is essential to deal with it appropriately at the onset. Identification of the involved hazardous material, which was covered earlier in this manual, is the first step. The second step is to establish control of the scene. Control of the scene begins by isolating the site, removing people who are within the isolation area, and denying entry of unauthorized persons. The process continues, if needed, with evacuation or protection-in-place of people located within the "protective action" distance. The process of controlling the scene concludes with the establishment of scene control zones. The following sections detail the process of handling each function of scene control.

Establishing The Initial Isolation Distance

[NFPA 472: 3-4.1.1.1]; [NFPA 472: 2-4.1.5.2]; [NFPA 472: 2-4.1.5.3]; [NFPA 472: 2-4.1.5.5]; [NFPA 472: 2-4.1.2]

The process for establishing the initial isolation distance is covered in the *ERG*. The *ERG* contains a Protective Action Distances section in the back part of the book (green-bordered pages). In order to use this information, the first responder must have already identified the material involved in the incident and have looked it up in either the yellow- or blue-bordered pages of the *ERG*. Those sections contain some chemicals that are highlighted in color and others that are not. The chemicals that are highlighted are those chemicals that were selected for inclusion in the *Table of Initial Isolation and Protective Action Distances* because of their poison/inhalation hazards. (**NOTE:** If the materials are on fire or have been leaking for longer than 30 minutes, this table does not apply. Seek more detailed information on the involved material on the appropriate orange-bordered page in the *ERG*.)

The Canadian *IERG* does not contain initial isolation distances. However, the *IERG* does contain initial evacuation distances. These distances let the first responder know how large an area to evacuate if the material is involved in a fire. This information is found in the yellow guide pages of the *IERG*. In addition, the yellow guide pages contain information on immediate isolation distances for public safety if the material is not on fire.

[NFPA 472: 2-4.1.5.4]

In addition to determining the material involved in the incident, the first responder must also determine the amount of material involved in the incident. The *Table of Initial Isolation and Protective Action Distances* gives parameters for establishing isolation and protective action distances. These distances are based on whether the spill is small or large. According to the *ERG*, a small spill is one which involves a single, small package (i.e., up to a 55-gallon drum), small cylinder, or a small leak from a large package. A large spill is one which involves a spill from a large package, or multiple spills from many small packages. The smallest isolation distance given for any chemical is 500 feet (Figure 6.30).

When the first responder has determined the initial isolation distance for the particular chemical, all people within that distance of the spill should be directed to move in a crosswind direction away from the spill until they are outside of the isolation distance (Figure 6.31).

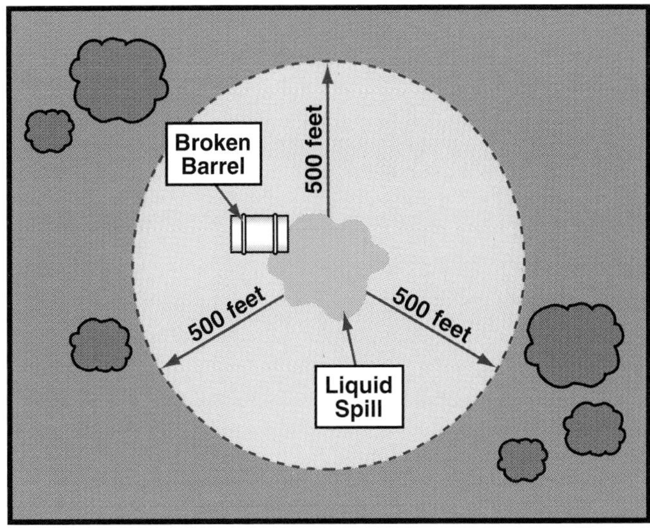

Figure 6.30 The smallest isolation distance for any chemical is 500 feet.

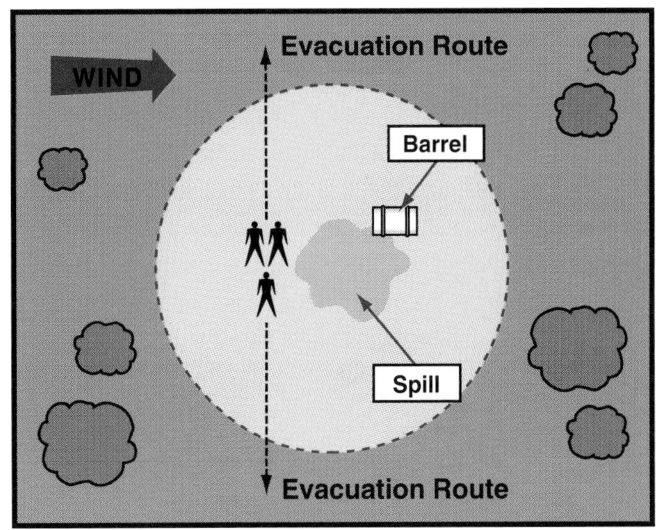

Figure 6.31 When possible, evacuate in a crosswind direction.

INITIAL ISOLATION ZONE

The initial isolation zone is the area created by circling the spill using the minimum isolation distance. No one should be allowed to enter the initial isolation zone once it has been established.

[NFPA 472: 2-4.1.6]

Access to the initial isolation zone can be established and controlled in a number of ways:

- Station a responder at approaches and refuse entry.
- Activate local alarm devices.
- Reroute traffic away from the scene.
- Put up physical barriers — tape, rope, and/or barricades (Figure 6.32).

Figure 6.32 Boundary tape is one way of marking the initial isolation zone. *Courtesy of Sally McCann.*

- Transmit warnings over public address systems.
- Broadcast an alert via the media.
- Stage responders an adequate distance away.

Evacuees should be told the nature of the emergency, the route upon which they are to proceed, the location of the assembly area, and the approximate amount of time that they will be inconvenienced. People who are directed to leave an area should be sent to a given location for assembly. Once the evacuees have assembled, perform the following procedures:

- Decontaminate evacuees (if necessary).
- Record evacuees' identities.
- Perform triage or give treatment (**NOTE:** Performing triage/treatment may also reduce false claims of injury later).

These evacuees may also have valuable information on the cause of the emergency, materials or containers involved, and the condition/location of remaining disabled victims.

Establishing The Protective Action Distance

[NFPA 472: 3-4.1.2 a,b]

The *protective action zone* is the area immediately adjacent to and downwind from the initial isolation zone. This area is in imminent danger of being contaminated by airborne vapors within 30 minutes of material release (Figure 6.33). The boundaries of this area are given in the *ERG* green pages in units of tenths of miles. The protective actions are those steps taken to preserve the health and safety of both emergency workers and the public. The options for action are evacuation, protection-in-place, or a combination of both. The IC will select the best option, or combination of options, based on factors that include but are not limited to the following:

- Material considerations
  - — Toxicity
  - — Quantity
  - — Rate of release
  - — Possibility of control
  - — Direction of spread
- Environmental conditions
  - — Wind direction
  - — Wind velocity
  - — Temperature
  - — Humidity
  - — Precipitation
  - — Topography

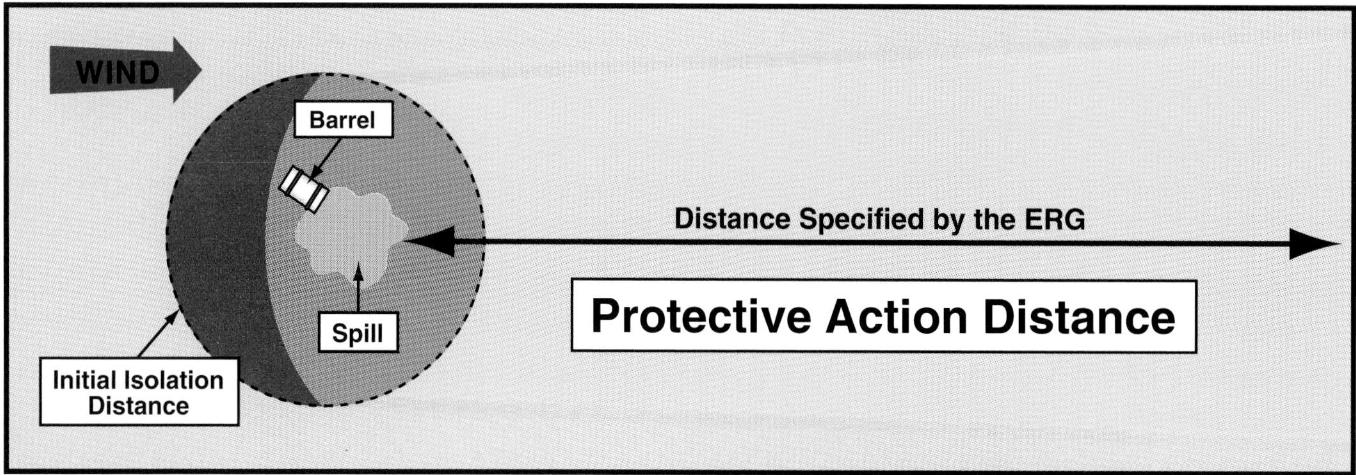

Figure 6.33 The protective action zone is immediately adjacent to and downwind from the initial isolation zone.

- Population at risk
  - Population density
  - Proximity
  - Warning/notification systems
  - Method of transport
  - Ability to control
  - Special needs

EVACUATION

Evacuate means to move all people from a threatened area to a safer place. To perform an evacuation, there must be enough time for people to be warned, to get ready, and to leave the area. Generally, if there is enough time for evacuation, it is the best protective action. First responders should begin evacuating people nearby, downwind, or crosswind of the incident within the distance recommended by the *ERG/IERG*. Even after people move the recommended distances, they are not completely safe from harm. The evacuees should not be permitted to congregate at these "safe" distances. They should be sent by a specific route to a designated place upwind of the incident so that they will not have to be moved again, even if the wind shifts.

The number of responders needed to perform an evacuation varies with the size of the area and the number of people to be evacuated. Evacuation can be a labor-intensive operation. Enough personnel resources should be assigned to an incident to carry out the evacuation.

PROTECTION-IN-PLACE

Protection-in-place means to direct people to go quickly inside a building and to remain inside until the danger passes. It may be determined that protection-in-place is the preferred option. The decision to protect-in-place may be guided by the following factors:

- The population is unable to initiate evacuation because of health care, detention, or educational occupancies (Figure 6.34).
- The material is spreading too rapidly.
- The material is too toxic to risk exposure.
- Vapors are heavier than air, and the people are in a high-rise structure.

When protecting people inside, direct them to close all doors and windows and to shut off all heating, ventilating, and air-conditioning (HVAC) systems. Protection-in-place may not be the best option if the vapors or gas is explosive, if it will take a long time for the vapors or gas to clear the area, or if the building cannot be closed tightly. Vehicles are not as effective as buildings for protection-in-place, but they can offer temporary protection if the windows are closed and the ventilation system is turned off.

Figure 6.34 A secure facility, such as this high-rise county jail, may be a candidate for protection-in-place.

In either case, evacuation or protection-in-place, the public needs to be informed as early as possible and needs to receive additional instructions and information throughout the course of the emergency.

By adding an "s" to the IFSTA acronym, the following list can be used to remember the order of action needed when establishing scene control:

- **I**dentify the material name and ID number (highlighted in the yellow or blue *ERG* pages or green or orange *IERG* pages).
- **F**ind corresponding name and ID number in green *ERG* pages. (**F**ind the appropriate yellow guide page in the *IERG*.)
- **S**ize the spill (by container and amount).
- **T**ake the distance from the *Table of Initial Isolation and Protective Action Distance* in the *ERG*.
- **A**pply the appropriate protective action.
- **S**eek additional information in the *ERG* or *IERG* 2-digit guide pages.

Scene Control Zones

[NFPA 472: 3-4.1.1]; [NFPA 472: 3-4.1.1.1]

Control zones are necessary to provide the rigid scene control needed at haz mat incidents. The zones prevent sightseers and other unauthorized persons from interfering with first responders, help to regulate movement of first responders within the zones, and minimize contamination. Control zones are not necessarily static and can be adjusted as the incident changes. Zones divide the levels of hazard of an incident, and what a zone is called generally depicts this level. NFPA 472 refers to the zones as the hot zone, warm zone, and cold zone (Figure 6.35).

HOT ZONE

The *hot zone* is an area surrounding the incident that has been contaminated by the released material. This area will be exposed to the gases, vapors, mists, dusts, or runoff of the material. It is generally the same as the isolation distance and could include the protective action area. The zone extends far enough to prevent people outside the zone from suffering ill effects from the released material. Work performed inside the hot zone is generally limited to haz mat technicians.

WARNING:

First responders at the awareness level do not operate in the hot zone. Responders at the operational level must have proper training and appropriate personal protective equipment to support work being done inside the hot zone.

Some agencies may refer to the hot zone by one of the following descriptions:

- Restricted Zone
- Exclusion Zone
- Red Zone

WARM ZONE

The *warm zone* is an area abutting the hot zone and extending to the cold zone. It is considered safe for workers to enter briefly without special protective clothing, unless assigned a task requiring increased protection. The warm zone is used to support workers in the hot zone and to decontaminate personnel and equipment exiting the hot zone. Decontamination usually takes place within a cor-

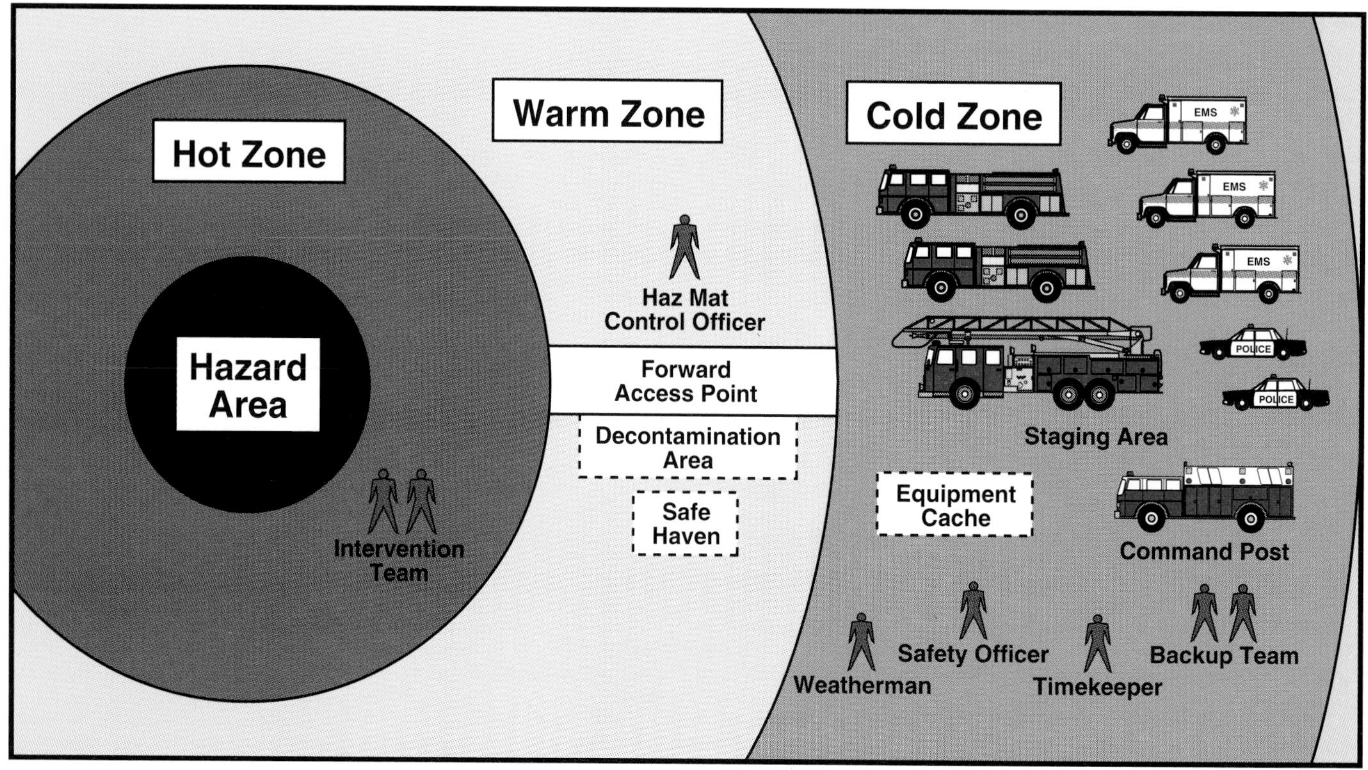

Figure 6.35 The various zones that surround the haz mat scene.

ridor (decon corridor) located in the warm zone. Other terms for the warm zone include the following names:

- Contamination Reduction Zone
- Limited Access Zone
- Yellow Zone

COLD ZONE

The *cold zone* encompasses the warm zone and is used to carry out all other support functions of the incident. Workers in the cold zone are not required to wear personal protective clothing because the zone is considered safe. The command post, the staging area, and the triage/treatment area are located within the cold zone. Other names for the cold zone include the following terms:

- Support Zone
- Green Zone

Command Post

[NFPA 472: 3-4.2.4]

Establishing a command post to which information flows and from which orders are issued is vital to a smooth operation. The incident commander must be accessible, either directly or indirectly, and command posts ensure this accessibility. A command post can be a predetermined location at a facility, a conveniently located building, or a radio-equipped vehicle located in the cold zone (Figure 6.36). Ideally, the command post should be located where the incident commander can observe the scene, although such a location is not absolutely necessary. The location of the command post should be relayed to the fire dispatcher and responding fire companies. Command posts should be readily identifiable. One common identifier is a green flashing light (Figure 6.37). Other methods include pennants, signs, and flags.

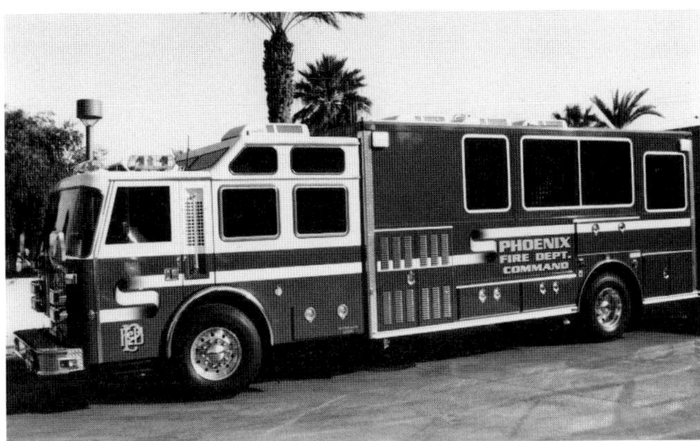

Figure 6.36 The command vehicle is located in the cold zone.

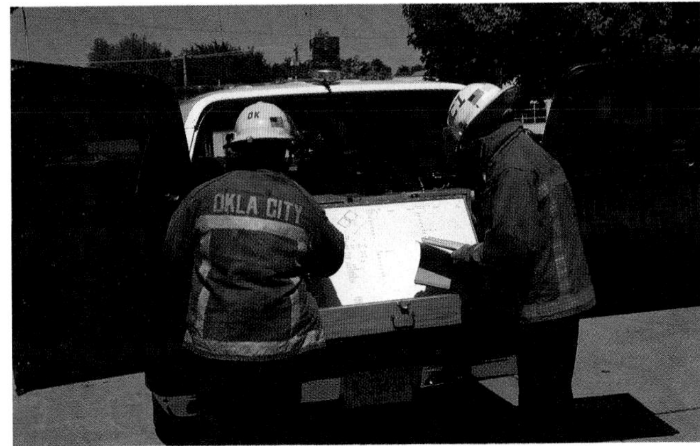

Figure 6.37 Some jurisdictions mark the command post with a green light.

Staging Area

Personnel and equipment awaiting assignment to the incident are held in the staging area. This keeps the responders and their equipment out of the way until needed. It minimizes confusion at the scene. The staging area should be located at an isolated spot in the cold zone where occupants cannot interfere with ongoing operations.

Triage/Treatment Area

[NFPA 472: 2-4.1.4.1]

Victims of the incident are brought to the triage/treatment area for medical assessment (triage) and stabilization (treatment). When responders are involved in transporting victims from decon to the triage/treatment area, they should protect themselves in case a contaminated victim has not been thoroughly decontaminated. Triage and treatment of victims should be delegated to those first responders who are trained and equipped to provide medical aid. Requirements for these personnel are contained in NFPA 473, *Standard for Competencies for EMS Personnel Responding to Hazardous Materials Incidents.*

Chapter 7

HAZ MAT

Photo Courtesy of Scott D. Christiansen, Minot, N.D..

Tactical Priorities And Defensive Control Strategies

LEARNING OBJECTIVES

This chapter provides information that will assist the reader in meeting the objectives from NFPA 472, *Standard for Professional Competence of Responders to Hazardous Materials Incidents* and 29 CFR 1910.120 that are listed below. The objective numbers are also noted directly in the text in the sections where they are addressed. Objectives in the list below that are denoted with an asterisk (*) are global in nature and are covered by reading the chapter in its entirety.

OPERATIONAL LEVEL

- First responders shall perform basic control and confinement operations within the capabilities of the resources and personal protective equipment available with their unit. [29 CFR 1910.120(q)(ii)(D)]*
- First responders shall perform basic containment operations within the capabilities of the resources and personal protective equipment available with their unit. [29 CFR 1910.120(q)(ii)(D)]

NFPA 472: 3-2.1 The first responder at the operational level shall, given examples of both facility and transportation situations involving hazardous materials, survey the hazardous materials incident to identify the containers and materials involved, whether hazardous materials have been released, and the surrounding conditions.

NFPA 472: 3-3.1* The first responder at the operational level shall, given simulated facility and transportation hazardous materials problems, describe the first responder's response objectives for each problem.

NFPA 472: 3-3.1.1 Identify the steps for determining the number of exposures that could be saved by the first responder with the resources provided by the authority having jurisdiction and operating in a defensive fashion, given an analysis of a hazardous materials problem and the exposures already lost.

NFPA 472: 3-3.1.2 Describe the steps for determining defensive response objectives given an analysis of a hazardous materials incident.

NFPA 472: 3-3.2 The first responder at the operational level shall, given simulated facility and transportation hazardous materials problems, identify the defensive options for each response objective.

NFPA 472: 3-3.2.1 Identify the defensive options to accomplish a given response objective.

NFPA 472: 3-3.2.2 Identify the purpose for, and the procedures, equipment, and safety precautions used with, each of the following control techniques:

(a) Absorption;
(b) Dike, dam, diversion, retention;
(c) Dilution;
(d) Vapor dispersion; and
(e) Vapor suppression.

NFPA 472: 3-4.4.1 Using the type of fire fighting foam or vapor suppressing agent and foam equipment furnished by the authority having jurisdiction, demonstrate the proper application of the fire fighting foams(s) or vapor suppressing agent(s) on a spill or fire involving hazardous materials.

NFPA 472: 3-4.4.1.1 Identify the characteristics and applicability of the following foams:

(a) Protein;
(b) Fluoroprotein;
(c) Special purpose:
   1. Polar solvent alcohol-resistant concentrates;
   2. Hazardous materials concentrates.
(d) Aqueous film-forming foam (AFFF); and
(e) High expansion.

NFPA 472: 3-5.1 The first responder at the operational level shall, given simulated facility and/or transportation hazardous materials incidents, evaluate the status of the defensive actions taken in accomplishing the response objectives.

NFPA 472: 3-5.1.1 Identify the considerations for evaluating whether defensive options are effective in accomplishing the objectives.

Chapter 7

Tactical Priorities And Defensive Control Strategies

In order for haz mat incidents to be handled properly, first responders must have a working knowledge of the order of tactical priorities to be performed for any incident. These tactical priorities place the greatest amount of emphasis on life safety. When life safety is ensured, the remaining priorities are directed at protecting the environment and property.

This chapter covers the tactical priorities that can be applied to all haz mat incidents and the various defensive control strategies that are needed to achieve each tactical priority. The last part of this chapter discusses the types and application of fire fighting foams commonly used to assist in the control of haz mat incidents.

TACTICAL PRIORITIES

[NFPA 472: 3-2.1]

When the incident commander has established strategic objectives, the officer in charge of each assigned work group should develop tactical objectives to accomplish the strategic objectives. These tactical objectives should reflect the conditions at the scene, the material involved, and the responder's capabilities. Responders will initially be concerned with the following two issues. These issues determine tactical objectives that are necessary to successfully achieve the strategic objectives.

- The identity of the material and its potential threat to life, the environment, and property
- How much material has escaped from the container(s), if any, and the current condition of the container(s)

Material Identity. Materials can be placed into three categories:

- Material is known and poses a substantial threat.
- Material is known and poses no immediate threat.
- Material is unknown.

The more first responders know about the material, the easier it is for them to make decisions. The identification methods described earlier in this manual, coupled with information from reference materials, pre-incident plans, or technical advisors, are the bases for deciding the relative hazards and the potential threat of the material. There will be a need for conservative tactics if the threat of the material is high and not much is known about the material.

When little is known about the material, the first responder should presume the worst-case scenario. The first responder should consider that the material may be potentially toxic, it may burn fiercely or explode, and it may harm the environment. Acting on these presumptions, the first responder must take the following actions:

- Control all ignition sources.
- Protect the material from excess heat, shock, or contamination.
- Confine material runoff as quickly as possible.
- Avoid contact with the material.

When the properties and hazards of the material are better known, strategies can be tailored to the data available.

Container Integrity. The second issue of concern, container integrity, also affects the selection of tactics. Stress on the container may result from mechanical, chemical, or thermal exposure. Stress on the container may have one of four consequences:

- No apparent container damage
- Container damaged with no material release (Figure 7.1)
- Container damaged with material release and no fire (Figure 7.2)
- Container damaged with material release and fire (Figure 7.3)

[NFPA 472: 3-3.1.2]

After evaluating the identity of the material and the container integrity, the incident commander must select and carry out the most appropriate tactics for achieving the incident strategies. The first responder should consider the following scenario:

> The driver is trapped in the cab of a tank truck carrying gasoline. The properties of the material are well known, the container is intact, and there is no fire. The only immediate threat to life, environment, and property involves the trapped driver. The proper strategy is rescue (first considered priority). The tactics for achieving this objective are stabilizing the vehicle, assessing and treating the victim's injuries, and using tools and techniques to free the driver without damaging the container (Figure 7.4). The incident commander provides the support first responders need if the circumstances change. If fire is present, the strategy is still rescue, but the tactics focus on fire control as a way to effect the rescue.

[NFPA 472: 3-3.1]

When the material is identified and container integrity ensured, there is a set order in which tactical priorities must be considered and carried out for every incident. The following is the order in which the tactical priorities must be performed:

Figure 7.1 Some containers may be damaged but do not release their contents. *Courtesy of Gary Girod.*

Figure 7.2 In some cases, the product may be released but not be on fire. *Courtesy of Gary Girod.*

Figure 7.3 Some incidents involve both product release and fire. *Courtesy of Harvey Eisner.*

1. Rescue
2. Exposure protection
3. Fire extinguishment
4. Confinement
5. Containment
6. Recovery

Figure 7.4 Rescue is always the first tactical priority. *Courtesy of Bob Esposito.*

For every incident, the incident commander must consider the tactical priorities in the order presented. If one or more of these priorities do not pertain to the specific incident, that priority may be skipped and the next one addressed. (For example, if the incident does not involve a rescue situation, then the next priority, exposure protection, becomes the most important concern.) Each tactical priority is addressed in detail in the following sections.

Rescue

[NFPA 472: 3-4.4.4]

Rescue is the first tactical priority to consider for every incident. Incidents in which human lives are immediately threatened by fire or chemical contamination present the most difficult decisions for first responders. Initially, the first responder must establish the actual need for a rescue attempt. If the rescue is too time consuming or too dangerous for emergency personnel, the proper decision may be to protect the victims in place. Another possibility may be to move victims to an area that is less dangerous before moving them to an area that offers complete safety. In any event, the emergency personnel's safety should be the incident commander's first consideration before any rescues are attempted (Figure 7.5). The following factors affect the ability of personnel to perform a rescue:

- Nature of the hazardous material and incident severity
- Availability of appropriate personal protective equipment
- Number of victims and their condition
- Time needed (including a safety margin) to complete rescue
- Tools, equipment, and other devices needed to effect rescue

Before the incident commander assigns personnel to the incident, he or she must consider the possibility that the incident may deteriorate suddenly. Personnel concentrating on a rescue may not notice important, potentially deadly changes in incident status; therefore, a safety officer must observe evolving conditions. Before attempting a rescue, the incident commander must establish an escape plan and a signal to initiate the plan. When there are several victims, the first responder should rescue first the victims that can be the most readily saved.

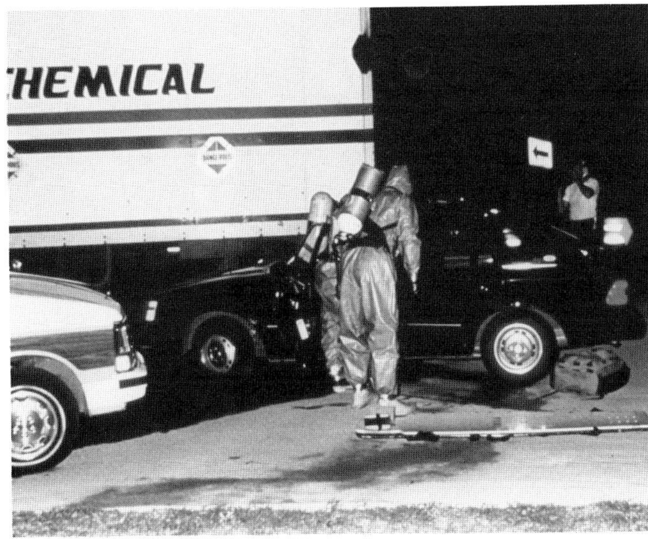

Figure 7.5 Emergency personnel must take appropriate precautions to ensure that they do not become contaminated while performing a rescue. *Courtesy of the Chicago Fire Department.*

First responders should make a primary search over the area as quickly and thoroughly as possible when conditions permit. Incident commanders establishing strategies, officers supervising tactical operations, and personnel working at the scene must not take unnecessary risks that would jeopardize the safety of the first responder.

Exposure Protection

[NFPA 472: 3-3.1.1]

Exposure protection, the second tactical priority to consider, involves the following tactics:

- Protecting the people
- Protecting the environment
- Protecting property not yet directly involved in but threatened by an expanding incident (Figure 7.6)

Figure 7.6 First responders should protect property that is not yet directly involved in the incident. *Courtesy of Harvey Eisner.*

Protecting the people. The most important exposure consideration is the threat to life. Life safety must be preserved not only during the initial stages of an incident but over the complete cycle of the incident. Ideally, the best way to protect lives that may be exposed to hazardous materials is to evacuate the people to a place of safety. However, evacuation is not always possible. In some cases, it will be necessary to protect people without actually moving them from the proximity of the incident. Methods of protecting people are covered in detail in Chapter 6 of this manual.

Protecting the environment. Environmental damage is the next most important concern. The air, surface water, wildlife, water table, and land surrounding an incident may be seriously affected by released materials. The nonbiodegradable nature of many materials means that the consequences of contamination may take years for the full effect to be realized, or the result of contamination may require large sums of money to repair. All released materials should be confined and held until their impact on the environment can be determined.

Protecting property. The property risk is similar to that created by fires except that the threatening material may not be readily evident. Gases, mists, and vapors can contaminate exposed goods with no visible signs. It may be more effective to contain leaking materials than to dilute or divert them. Action must be tailored to the material, its properties, and any reactions to the proposed protective medium. Incident commanders have appropriately "written off" property when operations were potentially risky. Lives or the environment must not be unduly compromised to save property.

Fire Extinguishment

If the haz mat incident involves a fire, the third tactical priority may be fire extinguishment. In some cases, such as fire around a leaking gasoline tanker, it may be necessary to extinguish the fire before containing the material. In other cases, such as a pressurized natural gas fire, it is the containment of the gas (turning it off) that will actually extinguish the fire.

The fire department's practice has always been to attack fires aggressively. However, this may not be the best approach for a haz mat first responder when a fire involves a hazardous material. The incident commander should always carefully evaluate the situation before committing personnel to fire suppression. When the incident commander considers it safe and appropriate to begin fire suppression, sufficient fire suppression capabilities must be available to attack the fire and to protect each exposure.

First responders applying fire streams must avoid making the situation worse. Fire streams can dramatically increase the size and intensity of a fire. Whether using foam or water, the size of a flammable liquid spill will increase when the agent is applied. Increased size adds to the runoff from the spill. When agents are applied to burning tanks, the tanks could overflow and threaten adjacent containers. Runoff from streams applied to hazardous materials should be confined until the runoff can be analyzed.

When personnel or environmental risks are determined to be too great, the practice of letting hazardous materials burn is an appropriate option.

This process is particularly useful when dealing with pesticides and flammable liquids. Personnel should let the fire burn freely to destroy as much of the material as possible, but they should be aware of the possibility of downwind contamination from the smoke. Protecting exposures during burn-off can also be a problem because of the time it takes for fuel consumption. First responders should extinguish burning spills and leaks completely only when the flow of materials has been or can be immediately controlled (Figure 7.7). When shutoffs cannot be accomplished immediately, hoselines and other portable equipment can be used to decrease the intensity of the fire while permitting controlled burning at the leak site.

Figure 7.7 Pressurized gas fires should not be extinguished until the flow of gas can be stopped. *Courtesy of the Chicago Fire Department.*

Confinement

The fourth tactical priority at an incident is confinement. *Confinement* is the process of controlling the flow of a haz mat spill and capturing it at some specified location. Confinement is primarily a defensive action. The objectives of confinement operations are to capture the moving materials as soon as possible in the most convenient location and to recover the materials with the smallest amount of exposure to people, the environment, and other property. Materials may be confined by building dams or dikes near the source, catching the material in another container, or directing (diverting) the flow to a remote location for collection (Figure 7.8). Generally, the following necessary tools are carried on fire apparatus:

- Shovels for building earthen dams
- Salvage covers for making catch basins
- Charged hoselines for creating diversion channels

Before using the equipment to confine spilled materials, incident commanders should seek advice from technical sources to determine if the spilled materials will affect the equipment. Large or rapidly spreading spills may require the use of heavy construction-type equipment, floating confinement booms, or special sewer and storm drain plugs.

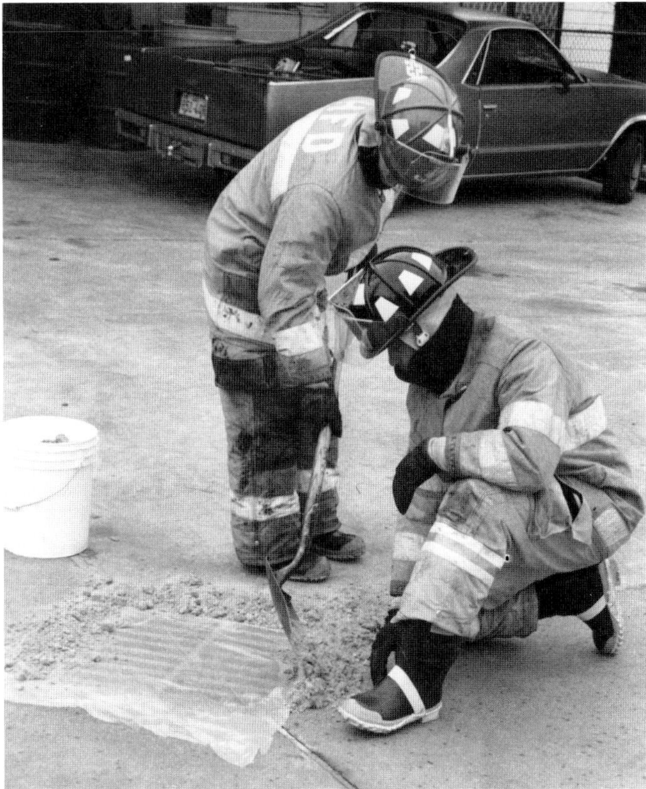

Figure 7.8 Building a dike is one method of confining a material.

Confinement is not restricted to controlling liquids. Dusts, vapors, and gases can also be confined. A protective covering of a fine spray, earth, plastic sheets, or a salvage cover can keep dusts from blowing about at incidents. Foams can be used on liquids to reduce the release of vapors (Figure 7.9). Gases can be directed by strategically placed streams, or they can be absorbed by water. Reference sources should be checked for the proper procedure for confining gases. Confinement efforts will be dictated by the material type, rate of release, speed of spread, number of personnel available, tools and equipment needed, weather, and topography.

Figure 7.9 A foam blanket on a liquid spill is one way of confining the vapors from the spill.

Containment

The fifth tactical priority at an incident is containment. *Containment* is the act of stopping the further release of a material from its container. Typically, first responders do not perform containment activities. Containment is an offensive action that is performed at the technician level. However, first responders should be aware of the following containment conditions that may have adverse effects on the incident:

- The condition of the container
- The properties of the material
- The rate of release
- The incident assessment

The first responder should ask the following questions regarding containment conditions:

- *The condition of the container*. Can the container withstand the stress of the operation? Will changes in material behavior, the weather, or other nearby operations further compromise the container?
- *The properties of the material*. Will containing the material after a partial release be undesirable, especially if contamination has occurred or a chemical reaction has started? Will the operation cause first responders to come in contact with the material, and if so, what risk will that contact present? Is the material changing physical states?
- *The rate of release*. Can the material be contained before the vessel is empty, or will the leaking stop soon because of the position of the breach (rupture)? Are sufficient resources available for the size of the breach?
- *The incident assessment*. Will the leaking materials ignite, explode, or react violently before or during the operation? Will containment devices and the container hold until recovery operations end?

Of course, the location of the incident, the type and number of exposures, the number and capabilities of responding forces, and a complex variety of other concerns will bear on the decision to contain haz mat leaks.

Confinement of the spilled material is often performed initially to protect exposures and then leak containment is started. Both processes may be started simultaneously if personnel and equipment are available. Most containment efforts should include setting up a confinement area in the event that the patch or plug fails.

Recovery

Recovery, which includes hazard removal and cleanup, is the final priority of an incident. Every action taken by personnel during the incident affects recovery in some manner. If recovery is not considered throughout the incident, the time needed to achieve full recovery can be greatly extended, and the cost of cleanup can increase significantly. The tactics chosen should result in a minimum effort to return a scene to normal.

Sometimes first responders will be needed during cleanup to provide cooling streams, to maintain scene control, or to assist with salvage. However, in most cases, first responders will not directly participate in recovery operations. During recovery it is important that first responders do not relax attention to detail and to safety. In many cases, the potential for tragic accidents still exists even after an incident has been "stabilized."

DEFENSIVE CONTROL ACTIONS

[NFPA 472: 3-5.1]; [NFPA 472: 3-5.1.1]; [NFPA 472: 3-3.2 & 472: 3-3.2.1]; [29 CFR 1910.120(q)(ii)(D)]

Defensive control measures are those measures used to contain and/or confine a material. Defensive control measures may be taken by first responders who are properly trained, who are outfitted with appropriate personal protective equipment, and who have the equipment to carry out these measures.

CAUTION: These actions are undertaken only if the first responder is reasonably ensured that he or she will not come in contact with or be exposed to the material.

The following defensive control actions will be discussed in the following sections:

- Absorption
- Confinement
- Dilution
- Vapor dispersion
- Vapor suppression

Absorption

[NFPA 472: 3-3.2.2(a)]

Absorption is a physical and/or chemical event occurring during contact between materials that have an attraction for each other. This event results in one material being retained in the other. The bulk of the material being absorbed enters the cell structure of the absorbing medium. An example of absorption is soaking an axe head in water to swell the handle. Diatomaceous earth, sawdust, and ground corn cobs are common absorbents (Figure 7.10). The absorbent is spread directly onto the hazardous material or in a location where the material is expected to flow.

Confinement

[NFPA 472: 3-3.2.2(b)]

Diking, damming, diverting, and retaining are ways to confine the hazardous material. These actions are taken to control the flow of liquid hazardous materials away from the point of discharge. The responder can use available earthen materials, or materials carried on their response vehicle, to construct curbs that direct or divert the flow away from gutters, drains, storm sewers, flood control channels, and outfalls (Figure 7.11). In some cases, it may be desirable to direct the flow into locations, such as gutters and drains, to capture and retain the material for later pickup and disposal. Dams may be built that permit surface water or runoff to pass over the dam while holding back the hazardous material. Again, any construction materials that contact the spilled material must be properly disposed of.

Figure 7.10 Absorbents come in a variety of forms.

Figure 7.11 Divert hazardous materials from drain openings.

Dilution

[NFPA 472: 3-3.2.2(c)]

Dilution is the application of water to a water-soluble material to reduce the hazard. However, dilution of liquid materials has few practical applications at a haz mat incident. The amount of water needed to reach an effective dilution increases overall volume and creates a runoff problem. This is especially true of slightly water-soluble liquids.

Vapor Dispersion

[NFPA 472: 3-3.2.2(d)]

Vapor dispersion is the action taken to direct or influence the course of airborne hazardous materials. Pressurized streams of water from handlines or unmanned master streams may be used to help disperse the vapor (Figure 7.12). These streams are used to create turbulence, which increases the rate of mixing with air and reduces the concentration of the hazardous material. After using handlines for vapor dispersion, it is necessary for first responders to confine and analyze the runoff for contamination.

Figure 7.12 In some cases, vapors may be dispersed using fog streams. *Courtesy of Joe Marino.*

Vapor Suppression

[NFPA 472: 3-3.2.2(e)]; [NFPA 472: 3-4.4.1]

Vapor suppression is the action taken to reduce the emission of vapors at a haz mat spill. Fire fighting foams are effective on spills of flammable and combustible liquids if the foam is compatible with the material (Figure 7.13). Water miscible (capable of being mixed) materials, such as alcohols, esters, and ketones, destroy regular fire fighting foams and require an alcohol-resistant foam. In general, the required application rate for applying foam to an unignited liquid spill will be substantially less than that required to extinguish a fire.

Figure 7.13 Foam is useful for vapor suppression operations.

WARNING

First responders performing vapor suppression tactics must constantly be aware of the threat of ignition of the spilled material. First responders must stay out of vapors at all times while applying foam.

Foam concentrates vary in their foam quality and, therefore, their effectiveness for suppressing vapors. Foam quality is measured in terms of its 25 percent drainage time and its expansion ratio. *Drainage time* is the time required for one-fourth of the total liquid solution to drain from the foam. *Expansion ratio* is the volume of finished foam that results from a unit volume of foam solution. Long drainage time results in a long-lasting foam blanket. The greater the expansion ratio, the thicker the foam blanket that can be developed in a given period of time. Air-aspirating nozzles produce a larger expansion ratio than do water fog nozzles.

First responders must exercise care when applying any of the foams onto a spill. All foams (except fluoroprotein types) should not be plunged directly into the spill but should be applied onto the ground at the edge of the spill and rolled gently onto the material (Figure 7.14). If the spill surrounds some type of obstacle, the foam can be banked off the obstacle (Figure 7.15). When possible, first responders should use air-aspirating nozzles rather than water fog nozzles for vapor suppression.

Figure 7.14 One method of applying foam is to roll it onto the spill.

Figure 7.15 In some cases, foam may be banked off an object and allowed to flow down onto the spill.

All Class B foams, except the special foams made for acid and alkaline spills, may be used for both fire fighting and vapor suppression. These special foams for acid or alkaline fuels were limited to mitigating vapor production and were ineffective for suppressing fire. However, these foams may still be found in use today but are no longer produced or sold.

Selection of the proper foam for vapor suppression is important. Because foam is composed principally of water, it should not be used to cover water-reactive materials. Some fuels destroy foam bubbles, so a foam must be selected that is compatible with the liquid.

Other points to consider when using foam for vapor suppression are as follows:

- Water destroys and washes away foam blankets; water streams should not be used in conjunction with the application of foam.
- A material must be below its boiling point; foam cannot seal vapors of boiling liquids.
- If you are unable to see the film that precedes the foam blanket, such as with AFFF, the foam blanket may be unreliable. Reapply aerated foam periodically until the spill has been covered.

First responders must be trained in the techniques of vapor suppression. Training for extinguishment of flammable liquid fires does not necessarily qualify a first responder to mitigate vapors produced by haz mat spills. More information on foam is contained in the next section of this chapter.

FOAM CONTROL ACTIONS

Extinguishment of flammable and combustible liquid haz mat fires requires properly applied foam in adequate quantities. All first responders must be knowledgeable in the principles of foam, the different types of foam concentrate, and how foam is applied to materials.

Hydrocarbon fuels, such as crude oil, fuel oil, gasoline, benzene, naphtha, jet fuel, and kerosene, are petroleum based and float on water. Standard fire fighting foam is effective as an extinguishing agent and vapor suppressant because it can float on the surface of hydrocarbon fuels.

Polar solvent fuels, such as alcohol, acetone, lacquer thinner, ketones, and esters, are flammable liquids that are miscible in water — much like a positive magnetic pole attracts a negative pole. Fire fighting foam can be effective on these fuels but only in special alcohol-resistant (polymeric) formulations.

Foam extinguishes and/or prevents fire in several ways (Figure 7.16):

- *Smothering* prevents air and flammable vapors from combining.
- *Separating* intervenes between the fuel and the fire.
- *Cooling* lowers the temperature of the fuel and adjacent surfaces.
- *Suppressing* prevents the release of flammable vapors.

In general, foam works by forming a blanket on the surface of a burning fuel. The foam blanket excludes oxygen and stops the burning process.

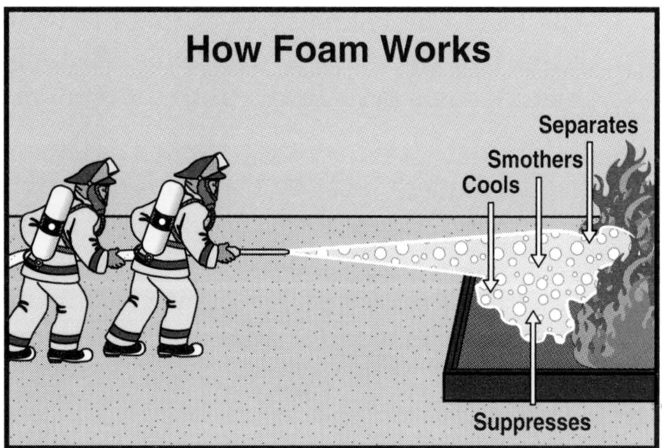

Figure 7.16 Foam cools, smothers, separates, and suppresses vapors.

How Foam Is Generated

Before discussing types of foams and the foam-making process, it is important to understand the following terms:

- Expansion — The ratio of final foam volume to original foam solution volume
- Foam Concentrate — The raw foam liquid before the introduction of water and air, usually stored in a 5-gallon (20 L) pail, 55-gallon (220 L) drum, or an apparatus storage tank (Figure 7.17)
- Foam Proportioner — The device that introduces the correct amount of foam concentrate into the water stream to make the foam solution (Figure 7.18)
- Foam Solution — A homogeneous mixture of foam concentrate and water before the introduction of air

Figure 7.17 Foam is most commonly found in 5-gallon (20 L) pails.

Figure 7.18 A typical foam proportioner.

- Finished Foam — The completed product after the foam solution reaches the nozzle and air is introduced into the solution (aeration)

There are two stages in the formation of foam. First, water is mixed with foam liquid concentrate to form a *foam solution.* This is the proportioning stage of foam production. Second, the foam solution passes through the hoseline to the foam nozzle (Figure 7.19). The foam nozzle aerates the foam solution to form *finished foam*, herein simply referred to as foam.

All foams in use today are of the mechanical type; that is, they must be proportioned (mixed with water) and aerated (mixed with air) before they can be used. Four elements are necessary to produce high-quality fire fighting foam:

- Foam concentrate
- Water

Figure 7.19 The foam nozzle mixes air with the foam concentrate to produce an effective foam blanket.

- Air
- Mechanical agitation (aeration) (Figure 7.20)

All these elements must be present and blended in the correct ratios. Removing any element results in either no foam or a poor-quality foam.

Aeration produces an adequate amount of foam bubbles to form an effective foam blanket. Proper aeration also produces uniform-sized bubbles, which form a longer-lasting blanket. A good foam blanket is required to maintain an effective cover over the flammable liquid.

Line proportioners and foam nozzles (also called foam makers) are engineered to work together. Using a line proportioner that is not hydraulically matched to the foam nozzle (even if the two are made by the same manufacturer) can result in unsatisfactory foam or no foam at all. There are numerous appliances for making and applying foam. A number of these are discussed later in this chapter.

Foam Concentrate

Fire fighting foam is 94 to 99½ percent water. Foams in use today are designed to be used at ½, 1, 3, or 6 percent concentrations. Class A foams are used at ½ percent or 1 percent concentrations. In general, foams designed for hydrocarbon fires are used at 1 to 6 percent concentrations. Polar solvent fuels require 3 or 6 percent concentrates, depending on the particular brand being used. Medium- and high-expansion foams are typically used at 1½, 2, or 3 percent concentrations.

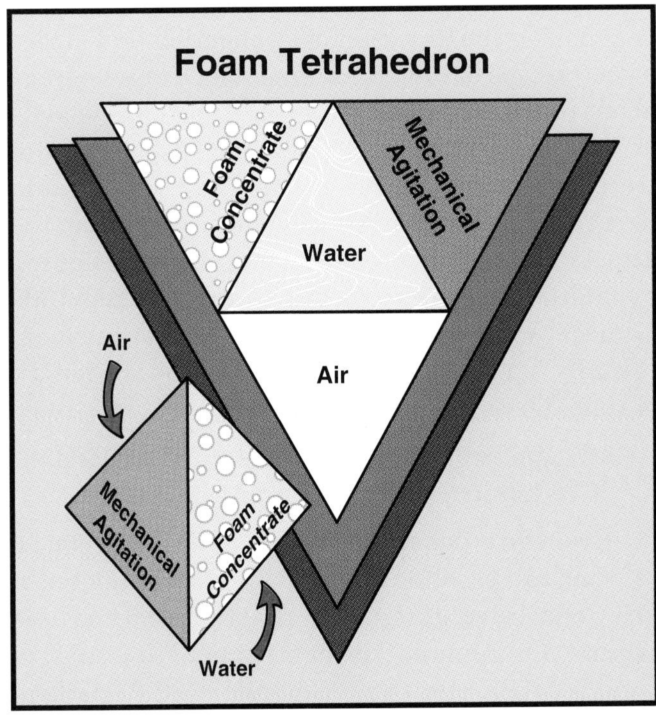

Figure 7.20 The foam tetrahedron.

NOTE: To ensure maximum effectiveness, use foam concentrates *only* at the specific percentage for which they are intended to be proportioned. This percentage rate is part of their Underwriters Laboratories rating and is clearly marked on the outside of every foam container.

Foams designed solely for hydrocarbon fires will not extinguish polar solvent fires regardless of the concentration at which they are used. However, foams that are designed for polar solvents may be used on hydrocarbon fires.

Depending on their purpose, foams are designed for low, medium, or high expansion. Low-expansion foams, such as aqueous film forming foam (AFFF) or film forming fluoroprotein foam (FFFP), have a small air/solution ratio generally in the area of 7:1 to 20:1. They are used primarily to extinguish fires in liquid fuels. Low-expansion foams are also used for vapor suppression on unignited spills. Low-expansion foams are most effective when the temperature of the fuel liquid does not exceed 212°F (100°C). If the fuel temperature exceeds this figure, much higher flow rates of foam will be required to reduce the temperature to below 212°F (100°C).

Medium-expansion foams typically have expansion ratios between 20:1 and 200:1. High-expansion foams generally have expansion ratios of 200:1 to about 1,000:1. Both foams are especially useful as space-filling agents in such hard-to-reach spaces as basements, mine shafts, and other subterranean areas (Figure 7.21). Steam caused by the vaporization of the foam in heated areas displaces gas and smoke, thus cooling the environment and extinguishing confined-space fires.

Application Rates

All AFFFs and FFFPs require an application rate of 0.10 gpm foam solution per square foot (4.1 L/min/square meter) for *ignited* hydrocarbon fuel. Fires involving polar solvent fuels generally re-

Figure 7.21 High-expansion foam may be used to fight hard-to-reach basement fires. Proper vertical ventilation tactics must be used in conjunction with this type of attack.

quire a higher rate of application and an alcohol-resistant foam concentrate. As a general rule, apply at least 0.20 gpm per square foot (8.2 L/min/square meter) to polar solvent fires. Apply protein and fluoroprotein foams at a rate of 0.16 gpm per square foot (6.5 L/min/square meter) on *ignited* hydrocarbon fuels. Remember that these figures are only general rules. Each foam will have its own recommended flow rates, and the user must be familiar with these specific figures.

Many fire departments have the capabilities needed to supply the application rate for extinguishing a hydrocarbon fire in a small storage tank, 35 feet (10.7 m) in diameter. A 35-foot (10.7 m) tank requires an application rate of about 160 gpm (600 L/min) of foam solution. Two 1½-inch (38 mm) handlines will supply the flow. A 60-minute supply of concentrate will be required.

Many departments have the capability of delivering the necessary flow rate to extinguish a fire. Frequently, however, they do not have enough concentrate available to sustain the attack until the fire is extinguished. There is also the logistical problem of providing the foam rapidly enough to keep up with the consumption of the foam. These factors usually limit the average fire department from effectively handling larger containers of burning fuels.

It is essential that enough concentrate is available before starting the application of foam on a flammable liquid fire. The quantity of foam needed depends on the duration of application and varies with the type of fuel. Recommended times average about 60 minutes. For the 35-foot tank previously described, 288 gallons (1 152 L) of foam concentrate, or 58 standard five-gallon (20 L) pails, should be available at the scene before beginning the attack. The amount of concentrate needed is figured by multiplying the application rate (160 gpm [600 L/min]) by the concentrate percentage of the foam (usually 3 percent for hydrocarbons) by the length of time (60 minutes).

160 gpm x 0.03 x 60 minutes = 288 gallons
[600 L/min x 0.03 x 60 minutes = 1 090 L]

Every fire department may be called upon to respond to a gasoline or diesel transport tank truck fire. Foam calculations show that hydrocarbon fires require a minimum flow of 60 gpm (240 L/min) of foam solution and twenty 5-gallon (20 L) pails of foam. A review of one fire department's experience

Figure 7.27 The foam nozzle and the eductor must be compatible.

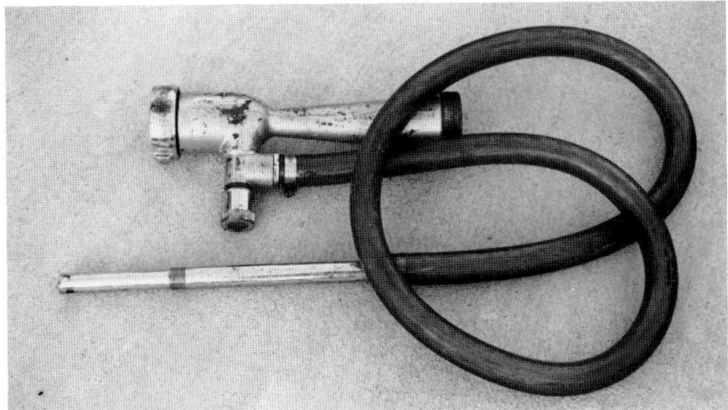

Figure 7.28 A typical in-line eductor.

Figure 7.29 Foam master streams either may be mounted on the apparatus or may be portable.

LINE EDUCTORS

The line eductor is the simplest proportioning device to use and the least expensive proportioning device to buy. It has no moving parts in the waterway, which makes it durable and dependable. The line eductor may be attached to the hoseline, or it may be part of the nozzle. Two different types of line eductors are the in-line eductor and the self-educting master stream nozzle (Figures 7.28 and 7.29).

Both types of eductors use the Venturi principle to draft foam concentrate into the water stream. As water at high pressure passes over a reduced opening, it creates a low-pressure area near the outlet side of the eductor. This low-pressure area creates a suction effect. The eductor pickup tube is connected to the eductor at this low-pressure point (Figure 7.30). The pickup tube submerged in the foam concentrate draws concentrate into the water stream, creating a foam water solution. Basic in-line eductors are designed to be operated at a fixed pressure and flow, using hose of a fixed size and length. Variable pressure in-line eductors are also available. These eductors can be operated at a variety of pressures, which allows for adjustments in hose size and length and flow rates.

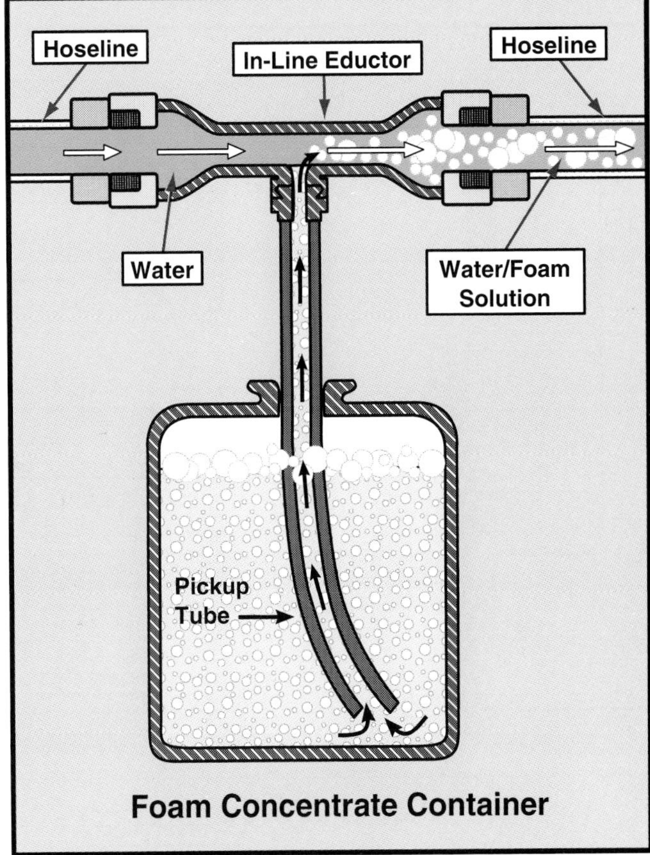

Figure 7.30 As water under pressure passes over the opening, foam concentrate is drawn into the tube, creating a foam solution.

BALANCED-PRESSURE AND AROUND-THE-PUMP PROPORTIONERS

Balanced-pressure proportioners and around-the-pump proportioners are both systems that are built into the apparatus fire pump. Apparatus equipped with a balanced-pressure proportioner have a foam concentrate line connected to each discharge outlet (Figure 7.31). This line is supplied by a foam concentrate pump separate from the main fire pump. The foam concentrate pump draws the concentrate from a fixed tank on the apparatus. This pump is designed to supply foam concentrate to a desired outlet at the same rate at which the fire pump is supplying water to that discharge.

The around-the-pump proportioning system consists of a small return line from the discharge side of the pump back to the intake side of the pump (Figure 7.32). An in-line eductor is positioned on the pump bypass. These units are rated for a specific flow and should be used at this rate, although they do have some flexibility. For example, a unit designed to flow 500 gpm (2 000 L/min) at a 6 percent concentration will flow 1,000 gpm (4 000 L/min) at a 3 percent rate.

Foam Nozzles

Nozzles designed to add air to the foam solution are called foam nozzles. There are many types that can be used, including standard water stream nozzles. The following sections highlight some of the more common foam nozzles.

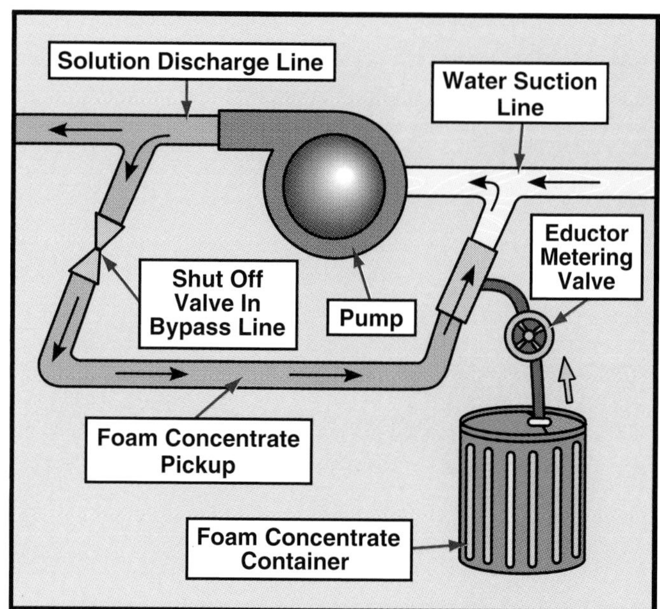

Figure 7.32 An around-the-pump proportioning system.

AIR-ASPIRATING FOAM NOZZLE

The most effective appliance for the generation of low-expansion foam is the air-aspirating foam

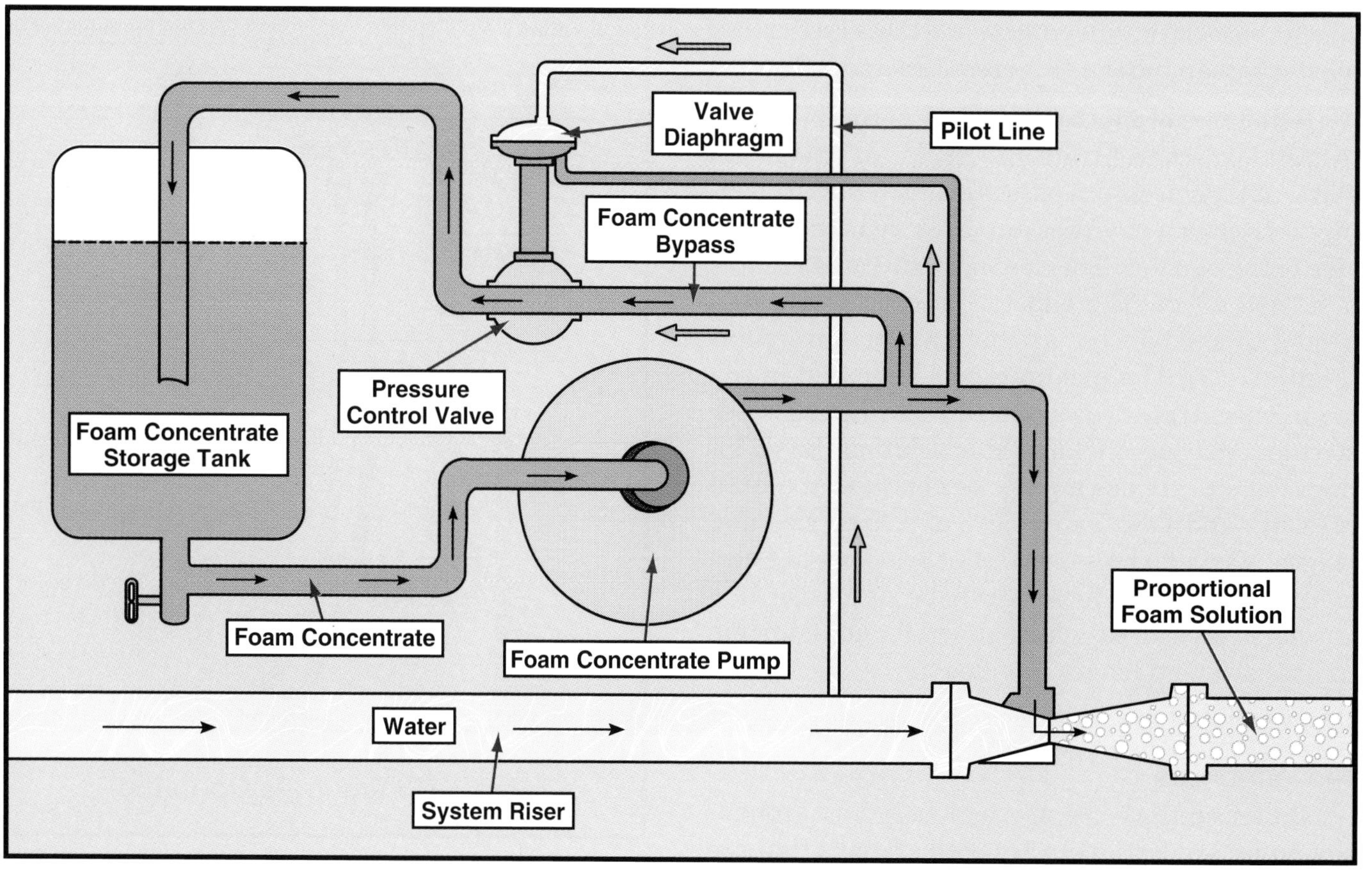

Figure 7.31 A balanced-pressure proportioning system.

nozzle. The foam nozzle is especially designed to provide the aeration required to make the highest quality foam possible. These nozzles may be hand-held (30 to 250 gpm [125 L/min to 1 000 L/min]) or monitor-mounted (250 gpm [1 000 L/min] or more) units (Figure 7.33).

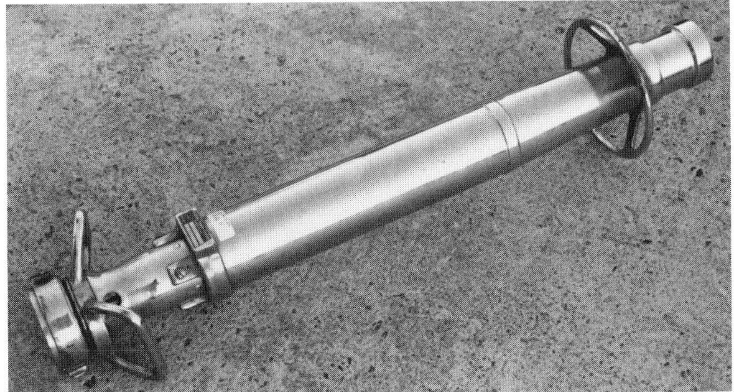

Figure 7.33 A typical hand-held foam nozzle.

Figure 7.34 A standard fog nozzle can be used with foam.

STANDARD FIXED-FLOW FOG NOZZLE

The standard fixed-flow fog nozzle is used with foam solution to produce a low-expansion, short-lasting foam (Figure 7.34). This nozzle breaks the foam solution into tiny droplets and uses the agitation of water droplets moving through air to achieve its foaming action. Its best application is when used with regular AFFF concentrate because its filming characteristic does not require a high-quality foam to be effective. These nozzles cannot be used with the following types of foam:

- Protein
- Fluoroprotein
- Alcohol-resistant FFFP
- Alcohol-resistant AFFF

AUTOMATIC NOZZLES

An automatic nozzle operates with an eductor in the same way that a fixed-flow fog nozzle operates, providing that the eductor is operated at the inlet pressure for which it was designed and providing that the nozzle is fully open. An automatic nozzle may cause problems if the eductor is operated at a lower pressure than that recommended by the manufacturer or if the nozzle is gated down.

HIGH-EXPANSION FOAM GENERATORS

High-expansion foam generators produce a high-air-content, semi-stable foam. The air content ranges from 200 parts air to 1 part foam solution (200:1) to 1,000 parts air to 1 part foam solution (1,000:1). The two basic types of high-expansion foam generators are the mechanical blower and the water-aspirating types. The water-aspirating type is similar to the other foam-producing nozzles except that it is much larger and longer. The back of the nozzle is open to allow airflow. The foam solution is pumped through the nozzle in a fine spray that mixes with air to form a moderate-expansion foam. The end of the nozzle has a screen, or series of screens, that breaks up the foam and further mixes it with air (Figure 7.35). These nozzles typically produce a lower-air-volume foam than do mechanical blower generators.

Figure 7.35 A high-expansion foam nozzle.

Mechanical blower generators resemble smoke ejectors in appearance (Figures 7.36 a and b). They operate along the same principle as the water-aspirating nozzle except the air is forced through the foam spray instead of being pulled through by the water movement. These devices produce a higher-air-content foam and are typically associated with total-flooding applications.

Figure 7.36a High-expansion foam blowers produce the best results. *Courtesy of Walter Kidde.*

Figure 7.36b Some departments have special high-expansion foam apparatus. *Courtesy of Ron Bogardus.*

Assembling A Foam Fire Stream System

To provide a foam fire stream, the first responder or apparatus driver/operator must be able to assemble correctly the components of the system. The following procedure describes the steps for placing a foam line in service using an in-line proportioner:

Step 1: Select the proper foam concentrate for the burning fuel involved and make sure that enough of it is available to extinguish the fire and to reapply as required.

Step 2: Check the eductor and nozzle to make sure that they are hydraulically compatible (rated for same flow).

Step 3: Check to see that the foam concentration listed on the foam container matches the eductor percentage rating. If the eductor is adjustable, set it to the proper concentration setting (Figure 7.37).

Step 4: Attach the eductor to a hose capable of efficiently flowing the rated capacity of the eductor and the nozzle (Figure 7.38).

- — Avoid kinks in the hose.
- — If the eductor is attached directly to a pump discharge outlet, make sure that the ball valve gates are completely open. In addition, avoid connections to discharge elbows. This is important because any condition that causes water turbulence will adversely affect the operation of the eductor.

Figure 7.37 Adjust the eductor to match the concentrate.

Step 5: Attach the attack hoseline and desired nozzle to the discharge end of the eductor (Figure 7.39). The length of the hose should not exceed the manufacturer's recommendations.

Step 6: Open enough pails of foam concentrate to handle the task. Place them at the eductor

Figure 7.38 Attach the supply hose to the intake of the eductor.

Figure 7.39 Attach the attack hose to the discharge of the eductor.

so that the operation can be carried out without interruption in the flow of concentrate (Figure 7.40).

Step 7: Place the eductor suction hose into the concentrate (Figure 7.41). Make sure that the bottom of the concentrate is no more than 6 feet (2 m) below the eductor (Figure 7.42).

Step 8: Increase the water supply pressure to that required for the eductor. (**NOTE:** Consult the manufacturer's recommendations for your specific eductor.) Foam should now be flowing.

Figure 7.40 Open enough pails of foam to do the job.

Figure 7.41 Insert the eductor into the foam pail.

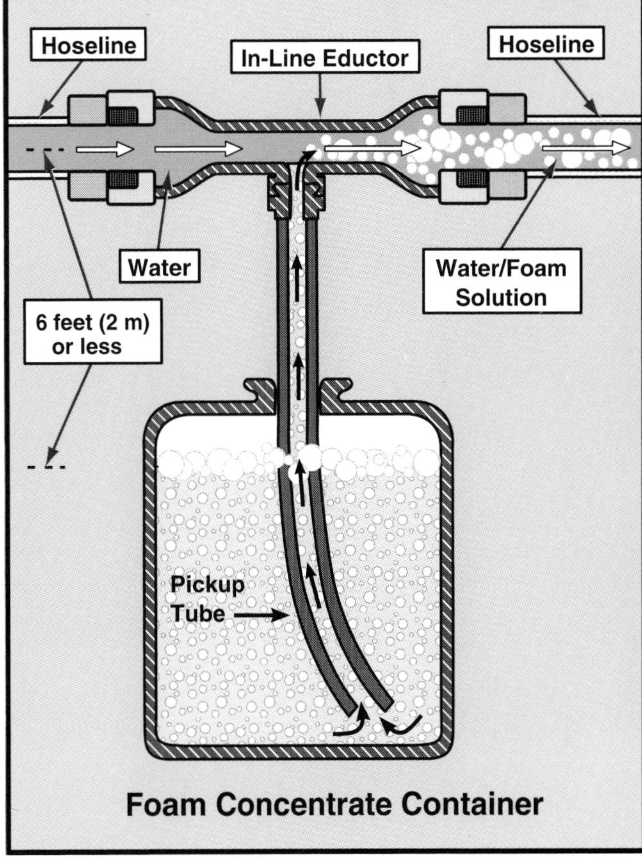

Figure 7.42 The eductor must be no more than 6 feet (2 m) above the foam concentrate.

Troubleshooting Foam Operations

There are a number of reasons for failure to generate foam or for generating poor-quality foam. The most common reasons for failure are as follows:

- Failure to match eductor and nozzle flow, resulting in no pickup of foam concentrate
- Air leaks at fittings that cause loss of suction
- Improper cleaning of proportioning equipment that results in clogged foam passages
- Partially closed nozzles that result in a higher nozzle pressure
- Too long a hose lay on the discharge side of the eductor
- Kinked hose
- Nozzle too far above eductor (results in excessive elevation pressure)
- Mixing different types of foam concentrate in the same tank

When using other types of foam proportioning equipment, such as balanced-pressure proportioners or around-the-pump proportioners, the apparatus driver/operator should be well trained in the proper operation of that particular equipment. The equipment manufacturer should be consulted for specific operating instructions.

Chapter 8

HAZ MAT

Photo Courtesy of Scott D. Christiansen, Minot, N.D.

Incident Control Strategies And Tactics

LEARNING OBJECTIVES

This chapter provides information that will assist the reader in meeting the objectives from NFPA 472, *Standard for Professional Competence of Responders to Hazardous Materials Incidents* that are listed below. Objectives in the list below that are denoted with an asterisk (*) are global in nature and are covered by reading the chapter in its entirety.

Operational Level

NFPA 472: 3-4.4* The first responder at the operational level shall, given a plan of action for a hazardous materials incident within his or her capabilities, demonstrate the ability to perform the defensive control actions set out in the plan.

gases are referred to as *pressure vessels*. The pressure in these vessels may range from 40 psi (276 kPa) to 4,000 psi (27 580 kPa). There are many sizes and shapes of pressure vessels such as one-ton chlorine cylinders, breathing air cylinders on self-contained breathing apparatus, fire extinguishers, and cylinders used in cascade systems (Figures 8.14 a through d).

Figure 8.14a Pressure cylinders are found in all locations.

Figure 8.14b Chlorine is often delivered to water treatment plants in one-ton pressure vessels.

Figure 8.14c A large pressure vessel containing ammonia.

Figure 8.14d Welding cylinders are pressure vessels.

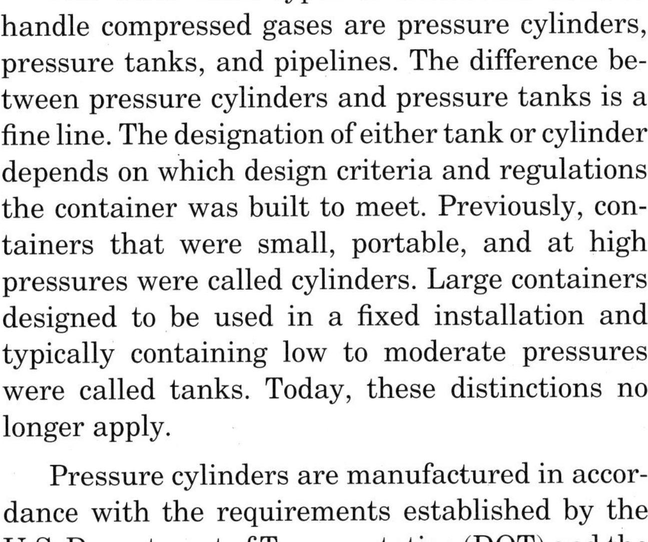

The three basic types of containers used to handle compressed gases are pressure cylinders, pressure tanks, and pipelines. The difference between pressure cylinders and pressure tanks is a fine line. The designation of either tank or cylinder depends on which design criteria and regulations the container was built to meet. Previously, containers that were small, portable, and at high pressures were called cylinders. Large containers designed to be used in a fixed installation and typically containing low to moderate pressures were called tanks. Today, these distinctions no longer apply.

Pressure cylinders are manufactured in accordance with the requirements established by the U.S. Department of Transportation (DOT) and the Canadian Transport Commission (CTC). (**NOTE:** The CTC's requirements are the same as those of the DOT.) A wide variety of gases are transported in cylinders, and these cylinders are considered the workhorses of the industry. Cylinders range in size from the very small, hand-held types to those with a maximum capacity of a 1,000 pounds (454 kg) (Figure 8.15). Cylinders are usually transported in an upright position (Figure 8.16).

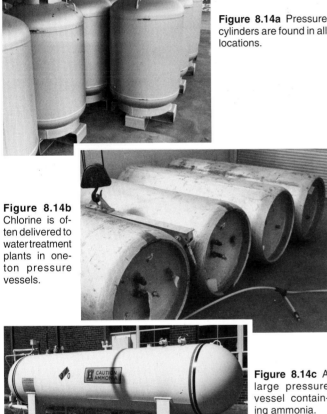

Figure 8.15 A typical cylinder.

Figure 8.16 Cylinders and other small pressure vessels are transported in the upright position.

Figure 8.17 A typical fixed-installation pressure tank.

Tanks are constructed to comply with requirements set forth by the American Petroleum Institute (API) or the American Society of Mechanical Engineers (ASME). Tanks are most commonly found in fixed installations but may also be found on motor vehicles and railcars (Figure 8.17). Tanks used in transportation are subject to additional criteria beyond API and ASME requirements.

All cylinders and tanks made of metal are steel, except for disposable and lift-truck types that can be made of aluminum. All pressure cylinders and tanks, with a few exceptions, must have some type of pressure-relieving device. Many disposable cylinders and some containers for poison gases, such as methyl bromide and hydrogen cyanide, have no pressure-relieving device.

Pipelines carrying compressed gases are found in many forms. The most common type is the municipal natural gas distribution system (Figure 8.18). Other industrial gases, such as oxygen, anhydrous ammonia, and hydrogen, may be piped within or between facilities. The U.S. DOT regulates all U.S. pipelines except for those on the consumer's property.

Emergencies Involving Gases

Immediately upon their arrival to the incident, first responders should pursue answers to the following questions:

- What gas is involved?
- What is the type and size of container involved?
- Is there mechanical damage to the container?
- Is there a leak?

Figure 8.18 Municipal natural gas systems are found in most communities.

- Is there fire?
- Is there flame impingement on the container?
- What is the availability of water?
- Can the fuel supply valve be shut off safely?

Transportation emergencies that involve gases may include large highway tankers and rail tank cars (Figure 8.19). Accidents involving these vehicles can be very serious because of the quantity of gas involved and the locations in which the accidents occur. The first responders must realize that their abilities to deal directly with this type of incident are very limited. If there is a leak and/or fire, the first responder should call more highly trained personnel for assistance.

Leaks in municipal natural gas distribution systems are also commonly encountered by emergency personnel. Most of the safety procedures described for transportation emergencies can also be applied to these incidents. In most cases, the following are the best procedures for first responders to follow in any gas emergency:

- Execute any feasible rescues.
- Rely on the *ERG/IERG* for suggestions on evacuation distances.
- Determine wind direction and initiate evacuation downwind.
- Set up unmanned portable master stream nozzles to cool tanks and exposures (if there is a fire) and then have personnel withdraw to a safe distance.
- Do not allow anyone in or near the area until the arrival of specialists who have both the necessary technical knowledge and the resources to handle the emergency.

The following sections describe the emergencies involving gases and the immediate concern and primary objective that should be taken for each.

Figure 8.19 Emergencies involving compressed gases test the resources of any agency. *Courtesy of Harvey Eisner.*

NONFLAMMABLE GAS LEAKS NOT INVOLVING FIRE

The immediate concern is to protect life safety and exposures. The primary objective is to shut off the flow of gas. The size of the container dictates how a first responder reacts. Obviously, a small container can be quickly isolated. If it is leaking, it will not be a major problem unless a poison is involved. With proper protective clothing and training, the first responder may attempt to either roll the cylinder or move it to an upright position to change a liquid leak to a gas leak. If a gas is toxic, such as chlorine or methyl bromide, a larger isolation area should be identified.

NONFLAMMABLE GAS LEAKS INVOLVING FIRE OR FLAME IMPINGEMENT

The immediate concern is to protect exposed tanks by cooling. The primary objective is to shut off the flow of gas.

FLAMMABLE GAS LEAKS NOT INVOLVING FIRE

The immediate concern is to prevent ignition. The primary objective is to shut off the flow of gas. When working with a flammable gas leak, the first responder should remove all ignition sources. If the gas is a liquefied petroleum gas (LPG), downwind ignition sources up to one-half mile away should be considered because the vapors are heavier than air and can flash back great distances. The area should be isolated until the incident is completely stabilized.

The first responder should locate the valves to see if the leak can be controlled. On residential natural gas systems, the valve is the petcock on the gas meter (Figure 8.20). The first responder should turn the petcock 90 degrees to stop the flow (Figure 8.21). If valves, valve stems, pressure-relief valves,

Figure 8.20 A typical residential gas meter.

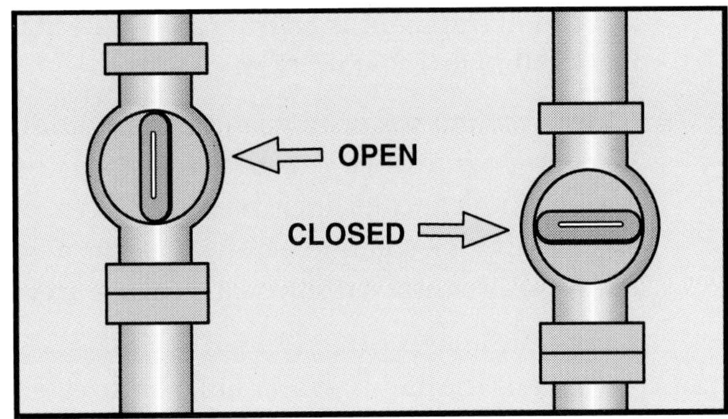

Figure 8.21 This figure shows the petcock in both the open and closed positions.

or fusible plugs have been damaged or if the cylinder has been punctured, the first responder should isolate the cylinder, move all civilians and personnel back, and control all ignition sources.

FLAMMABLE GAS LEAKS INVOLVING FIRE OR FLAME IMPINGEMENT

The immediate concern is to protect exposed tanks by cooling. The primary objective is to shut off the flow of gas. Fire situations involving flammable gases are extremely serious. Ideally, the objective is to direct large quantities of water onto all sides of the tank (or tanks) as quickly as possible. Water is used at a pressurized container fire for cooling the tank and reducing the internal vapor pressure — not for fire extinguishment.

WARNING

Do not attack this type of emergency without a continuous water supply. This can be the beginning of a serious mistake if you fail to plan ahead.

Lessons learned from historical review of LPG catastrophes provided the following facts:

- Large-capacity containers may fail violently within 10 to 20 minutes of direct-flame impingement. First responders must remember to factor both the response time and the possible set-up time for equipment when determining a course of action.
- BLEVEs of large-capacity containers typically create nonsurvivable fire conditions within 500 feet (150 m) and create severe tank shell fragmentation from 2,500 to 4,000 feet (750 m to 1 200 m) (Figure 8.22).
- Unless cooling water in adequate and uninterrupted quantities can be applied on the exposed tank or container, rescue and evacuation activities should be quickly performed followed by a total withdrawal from the area.

A flame impinging directly onto a tank shell quickly weakens the metal. Water must be directed onto the area being hit by the flame, especially if it is in the vapor space. A minimum of two 250 gpm

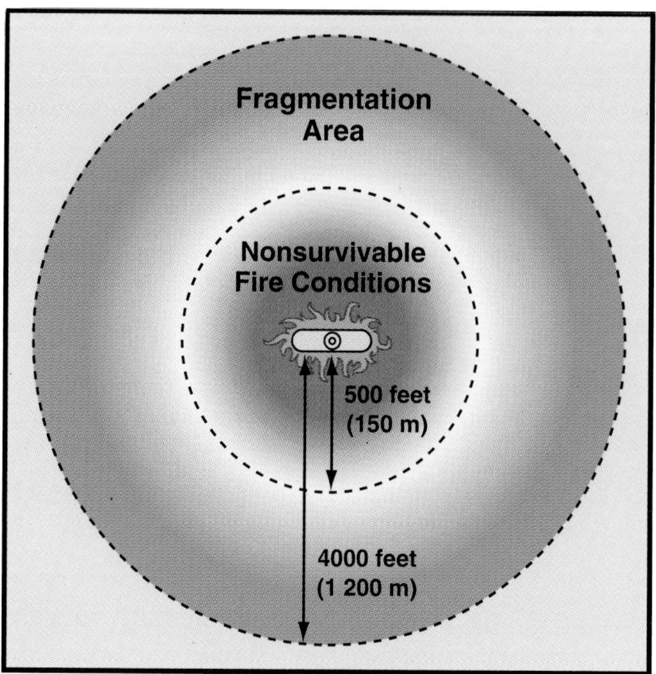

Figure 8.22 The danger area around a BLEVE can be quite large.

(1 000 L/min) streams must be played on each large highway tanker or rail tank car at the point of flame impingement (Figure 8.23). An obvious sheen of water should be seen rolling down the tank shell to ensure that the water is not being converted to steam before it cools the tank. Streams must be directed from each side to maximize total coverage of the tank shell. The first responder should concentrate on the upper vapor space and let the water flow down the sides.

Unfortunately, there are no absolute clues as to when tank rupture will occur. However, there are some warning signals that indicate a situation is becoming worse:

- Operation of the pressure-relief device indicates that pressure is building up in the tank. Actions must be taken to reverse this trend.

WARNING

Never extinguish flames that come from a pressure-relief device. This may allow flammable vapors to build in the area and reignite violently.

- The pitch of sound from the pressure-relief valve increases (becomes sharper) as the gas

Figure 8.23 Play water directly onto the impinged tank.

exits at a greater velocity, indicating a continuing increase in pressure.

- The torch, volume, and pitch of sound from the pressure-relief valve continue to increase because of a greater volume of gas rushing out, indicating an increase in internal boiling and vapor production. If the fire or torch coming from the pressure in the tank is also increasing, on-scene cooling techniques are obviously not adequate (Figure 8.24).
- A pinging, popping, or snapping sound occurs, indicating the metal has been softened by high heat and that it is stretching.
- Visible steam is coming from the tank surface. If this occurs when water is applied, the tank shell is over 212°F (100°C) and more water is needed. Insufficient cooling may also be noticed if there are dry spots on the tank surface.
- An impinging flame, usually in an isolated location, causes discoloration of the shell. The color of the tank turns from gray to an off-white, and small pieces of paint and metal flake off.
- A bulge or bubble indicates a serious localized heating of the shell in the vapor space. The metal is softening and beginning to deform because of inadequate cooling of the shell.

POISONOUS GAS LEAKS NOT INVOLVING FIRE

The immediate concern is to protect life safety and exposures. The primary objective is to shut off the flow of gas. If a poisonous gas is leaking, evacuation should proceed in accordance with the distances suggested in the *ERG/IERG*. First responders should not attempt to plug these leaks.

POISONOUS GAS LEAKS INVOLVING FIRE OR FLAME IMPINGEMENT

The immediate concern is to protect exposed tanks by cooling. The primary objective is to shut off the flow of gas.

FLAMMABLE AND COMBUSTIBLE LIQUIDS (CLASS 3)

Emergencies involving flammable and combustible liquids could involve large-capacity highway

Figure 8.24 Additional cooling water is required if the pitch from the relief valve increases.

tank trucks, rail tank cars, industrial storage facilities or processes, or pipelines (Figure 8.25). Emergencies involving flammable liquid tanks tend to be more spectacular looking than those involving gases; however, these tanks do not pose the same threats as a pressurized vessel. Liquid tanks, as compared to gas tanks, are not prone to BLEVE.

Figure 8.25 Large flammable liquid fires can be spectacular. *Courtesy of the Addison (TX) Fire Department.*

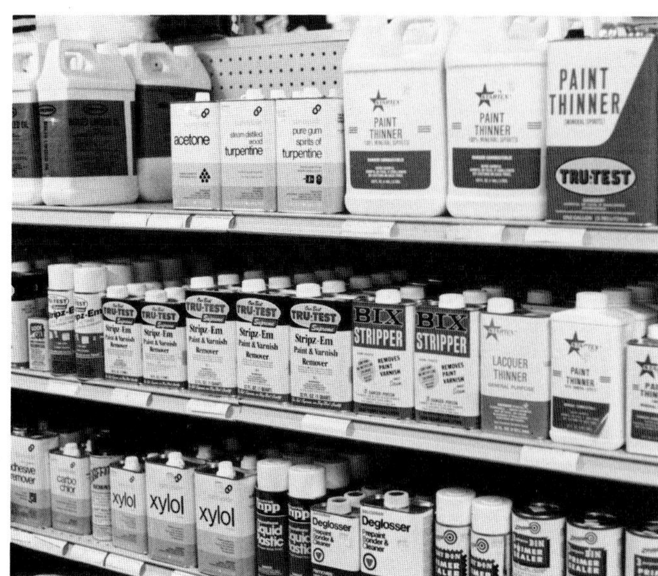

Figure 8.26 Any hardware store or paint store has a multitude of small flammable liquid containers.

Containers For Flammable And Combustible Liquids

Flammable and combustible liquids can be found in a multitude of containers such as metal cans, pails, drums, tanks, and pipelines. First responders must be familiar with each of these containers and the specific hazards they pose.

The most common containers used for flammable and combustible liquids are metal cans. These metal cans are very popular for storing paint thinners, solvents, camping fuel, and motor fuels (Figure 8.26). They are commonly found in large numbers in paint and hardware stores, residential garages, variety stores, service stations, and wholesale outlets. When metal cans are shipped, they are usually packaged in cardboard boxes (Figure 8.27).

Figure 8.27 Small flammable liquid containers are shipped in boxes.

The next size container is the pail (Figure 8.28). Pails are generally about 5 gallons (20 L) in size. The large bulk demand for solvents and thinners necessitates the use of these popular containers. Pails are normally delivered on wood pallets, and it is common for containers to be placed on a pallet and stacked three or four high.

Figure 8.28 Paint and other flammable liquids may come in 5-gallon (20 L) pails.

Flammable and combustible liquids are commonly shipped in drums (Figure 8.29). Most of the

Figure 8.29 Drums of flammable liquids are found in many facilities.

drums are constructed of metal, but some plastic drums are used. Bulk oils, thinners, and a wide variety of cleaning solvents, such as those used in automotive facilities, are commonly transported and stored in drums.

The DOT strictly forbids the use of glass containers for shipping flammable and combustible liquids. The practice of storing these substances in small glass containers is discouraged by fire prevention bureaus, OSHA, and insurance companies. However, first responders should realize that the private sector uses glass containers for storage of flammable and combustible liquids as well as for a variety of other chemicals. Additionally, these glass containers may not be properly labeled.

The DOT refers to tank trailers, tank trucks, and rail tank cars as bulk containers (Figure 8.30). Construction of highway containers may be steel, stainless steel, or aluminum, which is most extensively used. For rail, most construction is steel, although some aluminum, stainless steel, and nickel alloy cars are in service (Figure 8.31). The DOT and the CTC set the requirements for transportation tank design.

On-loading and off-loading of flammable and combustible liquid tanks is generally performed at the bottom of the tank. Top loading through manholes on highway tankers is being phased out by the industry. All connections, fittings, and valves are now being placed underneath the tank (Figure 8.32). Most new highway tankers are equipped with vapor recovery lines. Railcars are usually top loaded, and many are equipped with bottom unloading outlets (Figure 8.33).

Figure 8.30 A tank truck is a bulk container.

Figure 8.31 Most rail tank cars are steel.

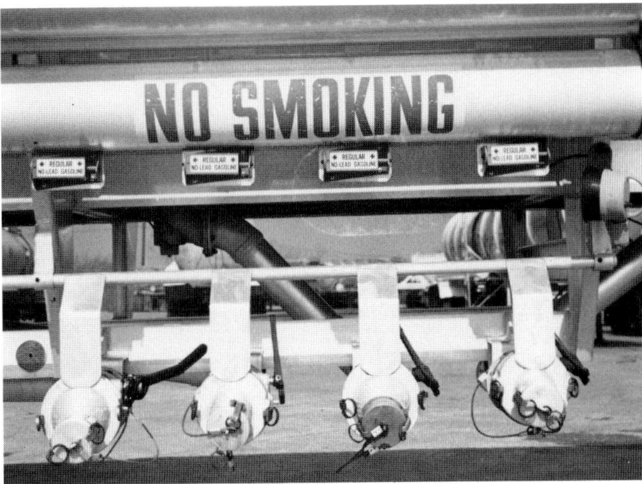

Figure 8.32 Fill connections for most tank trucks are beneath the tank.

Pipelines are used to transport large quantities of flammable and combustible liquids over long distances. Some pipelines are hundreds of miles long. They are primarily buried underground but may be exposed in some locations. They typically extend from one refinery or storage facility to another. Pumping stations are located in intermediate locations along the way (Figure 8.34). The DOT regulates all U.S. pipelines except those on the consumer's property.

Emergencies Involving Flammable Liquids

The following sections describe the emergencies involving flammable liquids and the immedi-

Figure 8.33 Discharge outlets for most rail tank cars are beneath the tank.

Figure 8.34 Liquid pipelines traverse many communities.

ate concern and primary objective that should be taken for each.

SPILLED FLAMMABLE/COMBUSTIBLE LIQUIDS NOT INVOLVING FIRE

The immediate concern is to prevent ignition of the fuel. The primary objective is to stop the flow of fuel. If the first responder prevents the fuel from igniting, the incident will be much easier to handle. The vapors from flammable liquids are often two to three times heavier than air. They flow and sink like a liquid. Most of the time they are invisible. The first responder must remove all ignition sources within a radius specified by the *ERG/IERG*. In those situations where there is a threat to life or exposures, the first responder should apply a blanketing layer of foam to the liquid pool (Figure 8.35). This helps to suppress the amount of vapors given off by the fuel. The area should be isolated and measures taken to evacuate all civilians beyond a specified perimeter.

When possible, the first responder should channel leaking liquids away from the incident scene and all exposures (Figure 8.36). If ignition does occur, the material can be allowed to burn under a controlled situation isolated away from on-scene activities. Street gutters, small drainage ditches, and dry creek beds may be used to channel the material to a collection point. First responders should dike storm drains or manholes along the collection route, and if any measurable quantity of the substance has already poured into the storm drain system, they must notify the appropriate public works agency (Figure 8.37). At the collection point, the material should be diked in a safe holding area. The first responder can apply foam to contain vapors and eliminate ignition at the collection point. Arrangements can be made later regarding removal and cleanup. The use of water must be restricted in order to minimize runoff problems.

SPILLED FLAMMABLE/COMBUSTIBLE LIQUIDS INVOLVING FIRE OR FLAME IMPINGEMENT

The immediate concern is to cool all exposures, including the tank itself. The primary objective is to stop the flow of fuel to enable extinguishment. First responders must realize that they alone may not have the capability to successfully complete this objective. However, first responders can initiate basic tactics that, in time, will lead to accomplishing this objective. The following tactics may be initiated:

- Lay initial hoselines (Figure 8.38).

Figure 8.35 Foam may be applied to spills to help prevent ignition.

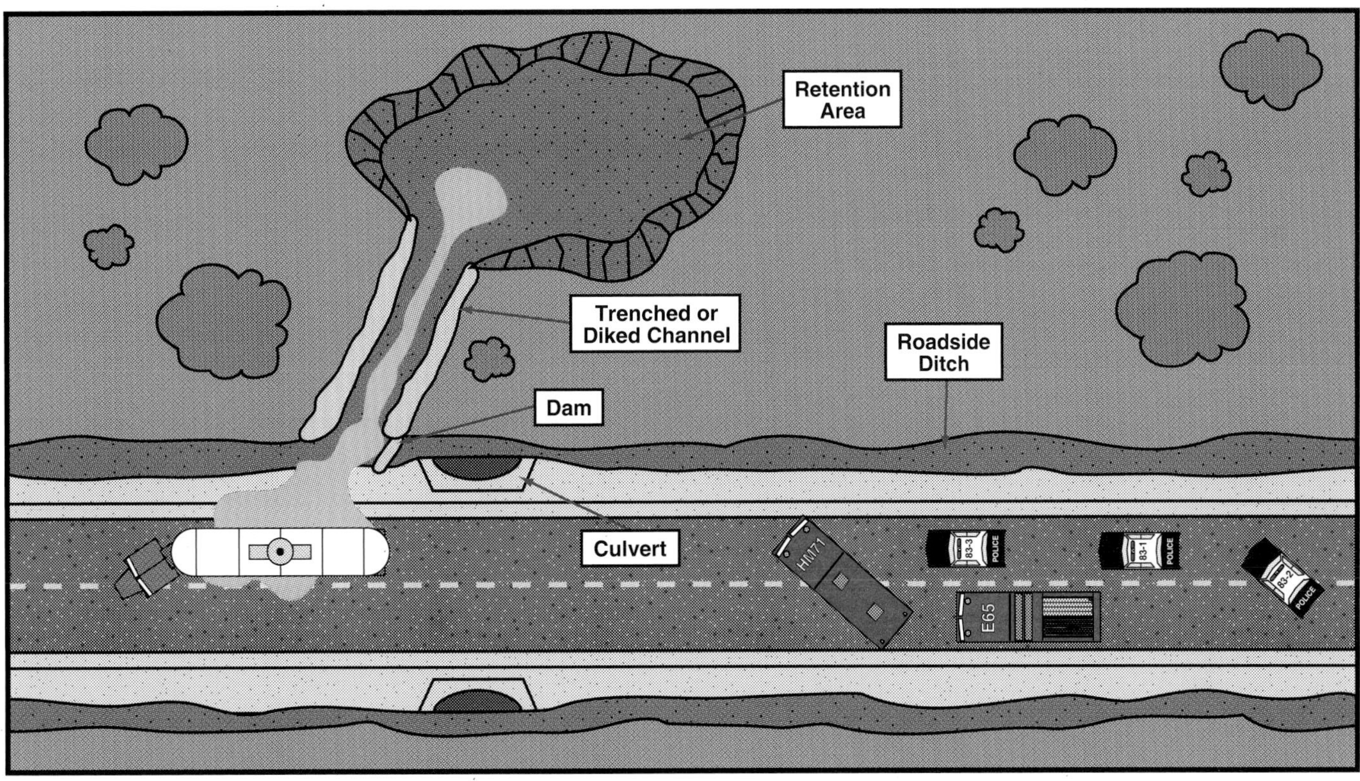

Figure 8.36 The first responder should channel leaking liquids away from the incident scene and from all exposures.

Figure 8.37 Prevent spilled products from entering storm sewers.

- Establish a continuous water supply of sufficient volume to control the incident.
- Protect exposures.
- Notify additional resources as necessary.
- Evacuate, isolate, and control the incident area.
- Establish an incident command system.
- Control flowing liquid material.

First responders can control small flammable liquid fires using common fire fighting equipment.

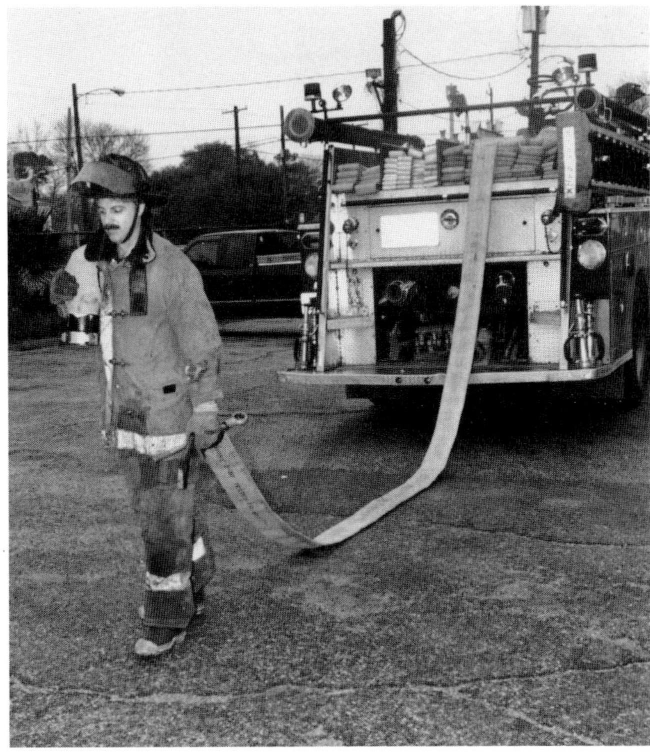

Figure 8.38 Engine companies need to lay supply lines in order to establish a water supply.

Larger fires and spills take more planning, effort, personnel, equipment, and time. These will require expertise beyond the level of a first responder.

FLAMMABLE SOLIDS, SPONTANEOUSLY COMBUSTIBLE MATERIALS, AND MATERIALS THAT ARE DANGEROUS WHEN WET (CLASS 4)

Accidents involving flammable solids, spontaneously combustible materials, and materials that are dangerous when wet are relatively rare. When an accident does occur, these materials can prove to be difficult to handle.

Containers For Class 4 Materials

A variety of containers are used for packaging Class 4 materials. Tubes, pails, steel and fiberboard drums, cardboard boxes, and bags are used for nonbulk packaging of the materials. Figures 8.39 a and b show some of the Class 4 containers. These containers are either secured tightly to prevent contact of the material with moisture or they are filled with an inert medium that excludes air from the material. White phosphorous and sodium are shipped in railroad tank cars.

Figure 8.39a A Class 4 container. *Courtesy of Warren Isman.*

Figure 8.39b A Class 4 container. *Courtesy of Warren Isman.*

Emergencies Involving Class 4 Materials

For the most part, extinguishment of Class 4 materials is not a primary objective for first responders. Actions of the first responders are generally limited to securing the scene, establishing zones and perimeters, evacuation, and calling for technical help. Because the immediate concerns of each Class 4 material may be different, the three divisions will be covered separately.

FLAMMABLE SOLIDS (DIVISION 4.1)

Metal powders, readily combustible solids that ignite by friction, and self-reactive materials that undergo a strong exothermal decomposition fall into the flammable solid class division of Type 4 materials. Also included are explosives that are wetted to suppress the explosive properties. The following sections describe the emergencies involving flammable solid materials and the immediate concern and primary objective that should be taken for each.

Spilled Flammable Solids Not Involving Fire

The immediate concern is to prevent ignition of the material. The primary objective is to isolate and confine the material until it can be removed. Spills of solid flammable materials are not too difficult to confine. Solid materials do not spread as do liquid and vapors. When several different materials are involved in a spill, the problem is more difficult than a single-material spill. It is more difficult because the different materials may react with each other.

Spilled Flammable Solids Involving Fire Or Flame Impingement

The immediate concern is to cool exposures. The primary objective is to control the fire by extinguishment or by controlled burning. If the spill is on fire and large quantities are involved, such as materials from a cargo van or boxcar, the fire may be impossible to extinguish. Tremendous volumes of water are required to extinguish large quantities of flammable solids. When the water supply is insufficient for extinguishment, the water should be used for exposure protection. First responders must then protect exposures until the fire burns out. If the fire can be extinguished, runoff water should be confined because it could contain residues that are harmful to the environment.

First responders are required to wear full-protective clothing and SCBA when they are in proximity to these fires. The products of combustion from combustible solids are highly toxic. Some materials, such as yellow and white phosphorous, can explode and scatter flaming fragments over a wide area.

Class D dry powders are special powders used to extinguish small metal fires. They can be applied by spreading the powder onto the fire by hand, by scoop, by shovel, or with an extinguisher (Figure 8.40). First responders are required to extinguish metal fires effectively with dry powder. The powder should not be used on reactive metals, on metals incompatible with the powder, or by *untrained* first responders.

Figure 8.40 Special agents must be used to extinguish Class D fires.

SPONTANEOUSLY COMBUSTIBLE MATERIALS (DIVISION 4.2)

Spontaneously combustible materials, also called pyrophoric materials, can ignite without an external ignition source after coming in contact with air. Spontaneously combustible materials can be either liquids or solids. The following sections describe the emergencies involving spontaneously combustible materials and the immediate concern and primary objective of each.

Spilled Spontaneously Combustible Materials Not Involving Fire

The immediate concern is to keep the material wet. The primary objective is to isolate and confine the material until it can be removed. When the material is exposed to air following a spill or because of the slightest breach of a container, ignition could occur immediately, or it could take up to five minutes. Employees who work with spontaneously combustible materials are trained to handle a release of the material. Laboratory employees are taught to immediately submerge the material in water. A serious fire will occur if an employee fails to respond properly to the release of spontaneously combustible materials. As in any haz mat incident, determining the material involved is important in developing the plan of action. The following questions should be asked:

- Is the material pyrophoric?
- Are there other materials nearby that are water reactive?
- What other chemicals are stored and how much?
- Can the area be isolated?
- Is everyone out of the area?

Spilled Spontaneously Combustible Materials Involving Fire Or Flame Impingement

The immediate concern is to protect exposures. The primary objective is to let the material burn until it is consumed. The situation becomes more serious when large quantities of pyrophoric materials are involved. There is no way that first responders can hope to bring the incident to a quick conclusion. First responders should not attempt to put out the fire.

MATERIALS THAT ARE DANGEROUS WHEN WET (DIVISION 4.3)

Some materials, such as magnesium phosphide, become spontaneously flammable or give off flammable or toxic gas when in contact with water. The following sections describe the emergencies involving materials that are dangerous when wet and the immediate concern and primary objective that should be taken for each.

Spilled Dangerous-When-Wet Materials Not Involving Fire

The immediate concern is to keep the material dry. The primary objective is to isolate and confine the material until it can be removed. Dangerous quantities of flammable gases are produced when some water-reactive materials get wet. Ignition sources must be controlled. First responders must wear full turnout gear and SCBA to protect themselves against toxic gases.

Spilled Dangerous-When-Wet Materials Involving Fire Or Flame Impingement

The immediate concern is to protect exposures. The primary objective is to let the material burn until it is consumed. If a dangerous-when-wet material is on fire, first responders should not attempt extinguishment with water. Instead, the area should be secured and the material allowed to burn. First responders generally do not have the resources needed to extinguish the material. The best attack is no attack.

OXIDIZERS AND ORGANIC PEROXIDES (CLASS 5)

Oxidizers (Division 5.1), such as perchloric acid, contain oxygen in their molecular structure and easily release the oxygen when heated. Most oxidizers are noncombustible but accelerate the burning of combustible materials. When mixed with organic materials, such as petroleum products, the mixture can ignite spontaneously. Organic peroxides (Division 5.2) are organic derivatives of the inorganic compound hydrogen peroxide. Some organic peroxides are shock sensitive, heat sensitive, and even light sensitive. Decomposition can be self-accelerating and result in an explosive pressure rupture of the container.

Types Of Containers

Packaging of oxidizers and organic peroxides runs a wide spectrum that is typical of hazardous materials. A common package is the plastic-lined, multi-ply paper bag (Figure 8.41). Metal tins and fiberboard, plastic, and metal drums are also used for packaging oxidizers. Portable, stationary, and applicator tanks contain oxidizers, which are made in slurry form for agricultural use. Liquid or slurry-form oxidizers are shipped in stainless-steel tank trucks. Dry-bulk tank trucks and railcars are used to move bulk loads of dry oxidizers. Many organic peroxides are limited in the quantity that can be shipped. The containers range in size from a few ounces (ml) up to 55-gallon (220 L) drums. Unlike similar containers used with other hazard class materials, the small organic peroxide containers are vented. Tank trucks may move some peroxides, but special permitting is required by DOT. Hydrogen peroxide solutions, however, are allowed to be shipped in tank cars made of aluminum.

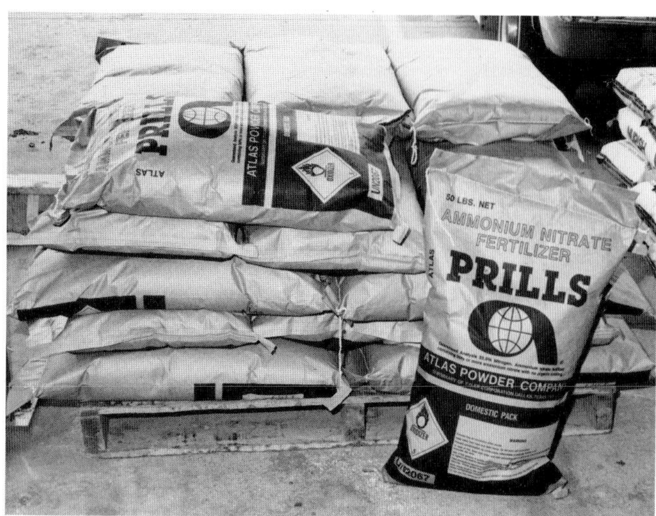

Figure 8.41 Oxidizers are shipped in multilayered paper sacks.

Emergencies Involving Oxidizers And Organic Peroxides

Oxidizing materials are unpredictable; they can react suddenly, violently, and without warning to friction or heat. If the first responder is unsure of what to do, he or she should withdraw and do nothing. If the first responder should take a wrong action, it can quickly accelerate the seriousness of an oxidizer incident. Tactics must be well planned. If there is any doubt as to the success of an operation, responders must stop and reassess the situation. The incident commander should never hesitate to order a total withdrawal when oxidizers are involved. The slightest hint of an incident worsening should initiate the withdrawal order. The following sections describe the emergencies involving oxidizers and organic peroxides and the immediate concern and primary objective that should be taken for each.

SPILLED OXIDIZERS NOT INVOLVING FIRE

The immediate concern is to prevent ignition by isolating combustibles from the material. The primary objective is to isolate and confine the material until it can be removed. First responders should make every effort in a safe manner to prevent contact of the oxidizer with combustible materials.The spill should be confined and nearby combustibles removed. This should not be done in the reverse order. Only the adjacent materials should be touched and removed (Figure 8.42). First responders must take care to avoid walking into the oxidizing material when isolating the spill. Contaminants on the sole of a boot coming in contact with the oxidizer or the friction of stepping on the material can cause ignition, sometimes explosively. Organic peroxides are especially prone to ignition.

SPILLED OXIDIZERS INVOLVING FIRE OR FLAME IMPINGEMENT

The immediate concern is to protect exposures. The primary objective is to control the fire either by extinguishment or by controlled burning. Extinguishment of some oxidizers is within the capabilities of a fire department, but the fire generally requires large volumes of water. Some organic peroxides are kept refrigerated because of their low self-accelerating decomposition temperatures (SADT). Elevated temperatures decompose the peroxides, and they become explosive. Involvement of these materials call for isolation and withdrawal. Attempting to get cooling water onto exposed containers of materials with a low SADT could be dangerous.

POISONOUS MATERIALS AND INFECTIOUS SUBSTANCES (CLASS 6)

Poisonous materials are those liquids and solids that are known to be toxic to humans or animals. These are Class 6, Division 6.1 materials. These materials include agricultural pesticides, cyanides, and even some exotic rocket fuels. An *infectious substance*, also called an etiological agent, is a microorganism (or its toxin) that causes human disease. These are Class 6, Division 6.2 materials. Examples of infectious substances are fluids or tissues infected with AIDS, rabies, and botulism.

Types Of Containers

Poisonous materials are packaged in all types of containers (Figures 8.43 a through c). They are shipped in bulk by intermodal portable tanks, tank trucks, railroad tank cars, barges, and marine tankers. Infectious substances are packaged in small vials that are measured in ounces (grams) and overpacked in strong containers for shipment (Figure 8.44).

Emergencies Involving Poisonous Materials

The greatest danger associated with poisonous substances is the health threat. First responders can be harmed by poisonous substances in any of the following ways:

Figure 8.42 Move exposed material away from oxidizers.

Figure 8.43a A typical poisonous material container.

Figure 8.43b A typical poisonous material container.

Figure 8.43c A typical poisonous material container.

- Physical contact with the material
- Inhalation of vapors
- Inhalation of material's products of combustion
- Contact with contaminated runoff water
- Contact with contaminated clothing

The following sections describe the incidents involving poisonous substances and the immediate concern and primary objective that should be taken for each.

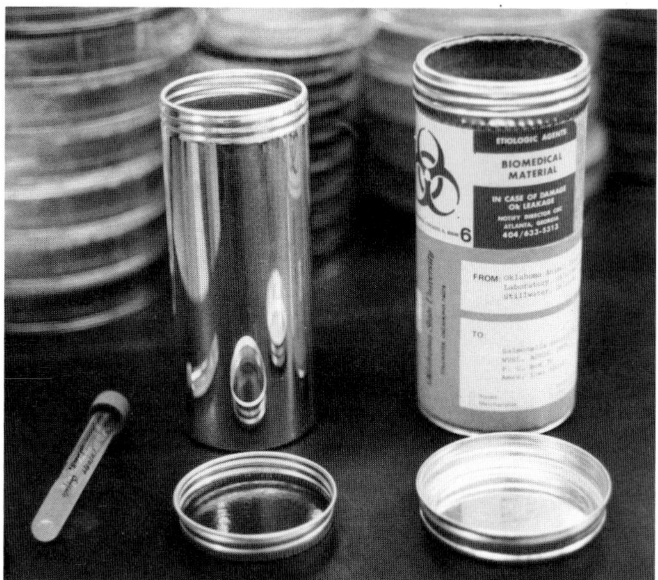

Figure 8.44 Infectious materials are overpacked.

SPILLED POISONOUS SUBSTANCES NOT INVOLVING FIRE

The immediate concern is to confine the spread of the material. The primary objective is to stop the flow and isolate the area. Spills can be confined by diking from a safe distance or diverting the spill to a safe collection point. Storm and sewer drains, waterways, and other sensitive areas must be protected (Figure 8.45). Solid materials can be confined by placing a salvage cover or plastic sheet over the spill. This will prevent wind from scattering the material. First responders must avoid contact with any poisonous substance, thus most con-

Figure 8.45 Sensitive areas should be protected.

finement and containment (stopping the flow) activities must be carried out by more highly trained personnel.

SPILLED POISONOUS SUBSTANCES INVOLVING FIRE

The immediate concern is to confine the spread of the material. The primary objective is to stop the flow, isolate the area, and let the material burn. In some cases, doing nothing is the best approach for fires involving certain poisonous materials (Figure 8.46). Contaminated runoff water from fires involving certain poisonous materials (e.g., pesticides) can be extremely damaging to the environment. Fire will destroy the toxic properties of some materials and reduce the cleanup effort. When it is feasible to fight a fire, first responders should confine the runoff for proper disposal. Protective clothing and SCBA must be worn by all responders around the scene because the products of combustion can be toxic. Poisonous substance fires must be fought from upwind.

Figure 8.46 In some cases, it is best to let the fire burn. *Courtesy of Mike Sanphy, Westbrook (ME) Fire Department.*

Emergencies Involving Infectious Substances

Most infectious substances will be found in and around hospitals, laboratories, and research centers. Most incidents will be small because of the limited quantities used and shipped. Above all, first responders must avoid contact with an infectious substance.

Regulated medical wastes are included with infectious substances under DOT regulations. First responders need to recognize the biomedical symbol and the distinctive plastic bag used to contain the waste (Figure 8.47). Hospitals, clinics, doctors' offices, and even some fire stations are locations where medical wastes are found. Medical wastes are transported by trucks from these locations to disposal sites. First responders responding to incidents involving medical wastes should treat the waste as an infectious substance. First responders should isolate the area and deny entry, avoid handling the waste, and call for help. Local health departments are often the regulating agency for medical waste.

The following sections describe the emergencies involving infectious substances and the immediate concern and primary objective that should be taken for each.

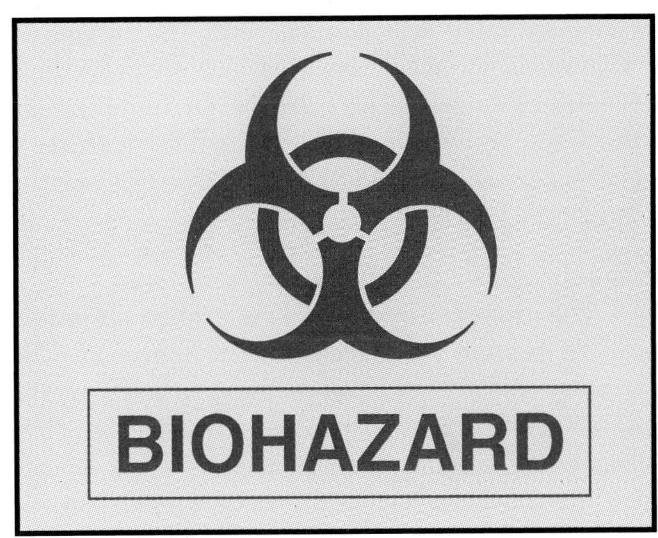

Figure 8.47 The biomedical symbol.

SPILLED INFECTIOUS SUBSTANCES NOT INVOLVING FIRE

The immediate concern is to protect life safety and confine the spread of the material. The primary objective is to isolate the area and deny entry. Damage to the outer container does not necessarily mean the inner vial has been damaged. However, the first responder should assume the worst. First responders should treat the incident as if the vial is damaged and the infectious substance has been released. The first responder should *never* handle the container to see if a vial is broken. An accepted way to treat infectious substance incidents is to cover the container with a liquid-bleach-saturated towel (Figure 8.48). Bleach will kill an infectious substance. First responders must wear full-protec-

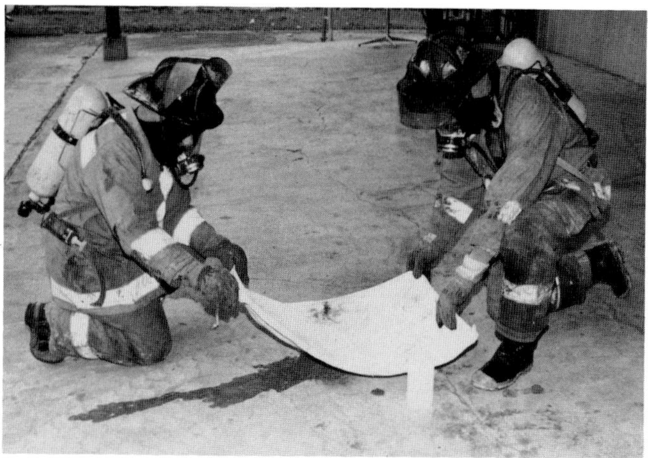

Figure 8.48 Cover infectious materials with bleach-soaked towels.

tive clothing, including SCBA, to protect them during infectious substance incidents. Bystanders who may have had contact with the infectious substances should be isolated and given medical assistance immediately. Advice on handling an infectious accident can be obtained from either a local hospital, CHEMTREC or CANUTEC, or the Center for Disease Control.

SPILLED INFECTIOUS SUBSTANCES INVOLVING FIRE

The immediate concern is to protect life safety and to confine the spread of the material. The primary objective is to isolate the area and to let the material burn. If the incident involves fire, first responders should protect exposures and let the fire burn out. The heat of the flame will destroy the infectious substance.

RADIOACTIVE MATERIALS (CLASS 7)

Radioactive materials are materials that spontaneously emit ionizing radiation. First responders should understand the three basic types of materials: alpha, beta, and gamma. Multiple types of radiation can be emitted from a single source.

Alpha radiation is the least penetrating because of its weight. The alpha particle travels but a few inches in the air, and clothing or human skin stops the travel. However, alpha particles can be inhaled or ingested.

Beta radiation is more penetrating than is alpha radiation. Beta particles will travel several yards in the air and can penetrate skin and clothing. Structural fire fighting clothing may not block beta particles. Aluminum foil will provide shielding against beta particles. Both alpha and beta particles will cause injury if they enter the body.

Gamma radiation is similar to X rays. Gamma rays are extremely penetrating and travel nearly the speed of light. Dense materials are required to shield against gamma radiation. For example, it takes 4½ inches (115 mm) of concrete to reduce gamma radiation by one-half.

Types Of Containers

Radioactive material packages are the strongest containers used to transport hazardous materials. For this reason, radioactive incidents are rare. The two categories of packaging for shipment are Type A and Type B packaging.

Type A packaging contains low-level commercial radioactive shipments. These containers include cardboard boxes, wooden crates, cylinders for compressed radiological gases such as xenon, and metal drums (Figure 8.49). Measuring devices, such as radiography instruments and soil density meters, contain radioactive materials and technically may be considered Type A packaging. Radiopharmaceuticals, those medicines that contain a radioactive material, are packaged in small quantities and generally limited to air transportation because of their short half-lives (degradation time).

Type B packaging is the strongest packaging and is used for more highly radioactive shipments. These containers include steel reinforced concrete casks, lead pipe, and heavy-gauge metal drums (Figure 8.50). They can survive serious accidents and fire without release of the radioactive material. Type B containers are designed to carry fissionable materials, high-grade raw materials, nuclear fuels (both new and spent), and highly radioactive metals.

Emergencies Involving Radioactive Materials

Time, distance, and shielding are the three ways by which first responders can protect themselves against exposure to radiation (Figure 8.51). When the first responder is exposed to radiation for a short period of time, he or she will receive less exposure to the radiation. When the first responder is farther from the source, the level of exposure will be less. Appropriate shielding also reduces the amount of exposure. Relatively high levels of radioactivity do not necessarily preclude a rescue from being under-

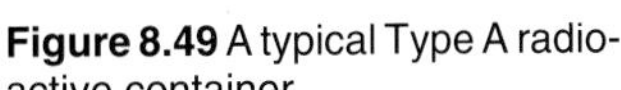

Figure 8.49 A typical Type A radioactive container.

Figure 8.50 A typical Type B radioactive container.

taken. First responders should be able to get in, get the victim, and get out without significant harm. Time is the factor.

Dose Rate x Exposure Time = Total Dose

The following sections describe the emergencies involving radioactive materials and the immediate concern and primary objective that should be taken for each.

SPILLED RADIOACTIVE MATERIALS NOT INVOLVING FIRE

The immediate concern is to confine the spread of the material. The primary objective is to isolate the area and deny entry.

SPILLED RADIOACTIVE MATERIALS INVOLVING FIRE

The immediate concern is to confine the spread of the material. The primary objective for a fire involving radioactive materials is to extinguish the fire if it can be done without risk to the responder. Otherwise, the fire should be left to burn out. Letting a fire burn is the definite approach when fire involves nuclear fuel, waste fuel, or military weapons. If a fire can be put out, first responders should do as little overhaul as possible after extinguishment. They should withdraw from the scene and await the arrival of radiation technicians.

Technical help is definitely needed at incidents involving radioactive materials. Nothing should be removed from the site until checked by technicians for contamination.

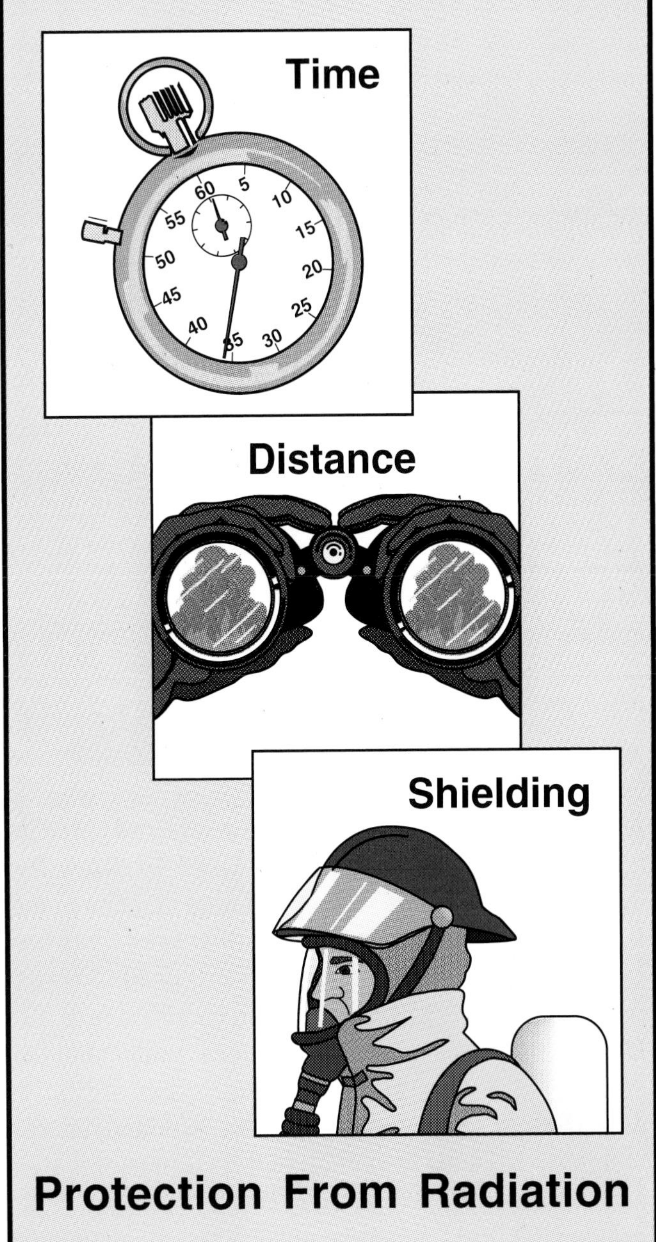

Figure 8.51 Time, distance, and shielding are the best defenses against radiation.

CORROSIVES (CLASS 8)

Corrosives are materials that corrode, degrade, or destroy human skin, aluminum, or steel. Corrosives are either acids or bases. The terms *caustic* and *alkaline* are also used to refer to base materials. The pH of a material is used to determine whether it is an acid or a base. Acids have a pH of 1 through 6, and bases have a pH of 8 through 14. A pH of 7 is neutral.

Types Of Containers

Corrosives come in a wide variety of containers. The containers range in size from glass/plastic bottles and carboys to plastic drums (Figures 8.52 a through c). Fiberboard drums and multilayered

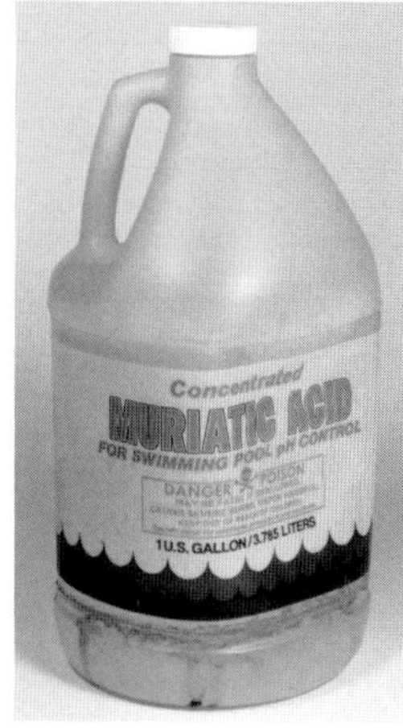

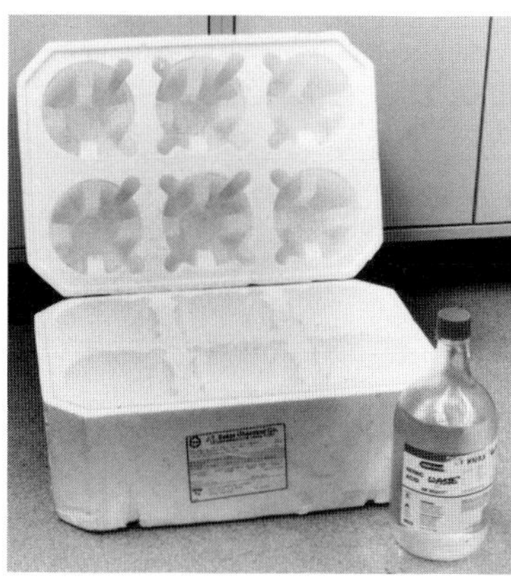

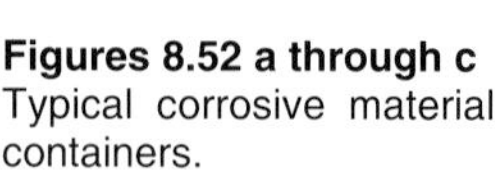

Figures 8.52 a through c Typical corrosive material containers.

paper bags are used for acid materials and caustics in dry form. Wax bottles are used to store hydrofluoric acid because the acid also attacks glass. Intermodal portable tanks, tank trucks, railroad tank cars, barges, and pipelines are used to transport bulk shipments of corrosives (Figure 8.53). Because of the density of corrosives, tanks that are smaller than those used for other types of liquids are used to transport corrosives. Corrosives can weigh up to twice that of an equal volume of water. Maximum capacity for tank trucks is about 6,000 gallons (24 000 L) and about 24,000 gallons (96 000 L) for railroad tank cars.

Figure 8.53 A corrosive rail tank car. The orange stripe around the car denotes that this car carries hydrofluoric acid. Sulfuric acid cars have a black stripe.

Emergencies Involving Corrosive Materials

Structural turnout clothing is quite limited in protecting first responders from corrosives. Vapors can permeate the turnout clothing and irritate the skin, especially the damp areas of the body. Irritation can be severe and can cause skin damage. Incidents involving corrosives can be serious.

The following sections describe the emergencies involving corrosive materials and the immediate concern and primary objective that should be taken for each.

SPILLED CORROSIVE MATERIALS NOT INVOLVING FIRE

The immediate concern is to confine the spread and not to dilute the material. The primary objective is to shut off flow, isolate the area, and deny entry. First responders should confine the spread of the material if it can be done without coming into contact with the material or its vapors. Corrosives should be prevented from reaching organic materials or other corrosives. Water is not to be used because it will only worsen the problem. It becomes corrosive itself after contact. Dense vapor clouds are common with some corrosive spills, and controlling vapors is part of confinement. Foam may be used in reducing vapor production. Fog streams can be used to disperse or redirect vapors, but the water must be prevented from contacting either the container or the spill.

Scattering of solid corrosive materials by wind currents can be prevented by covering the spill with salvage covers or plastic sheets. Corrosives should be kept from entering storm and sewer systems, creeks, canals, bayous, and rivers. Authorities should be notified if corrosives have entered a waterway. Corrosives are environmentally damaging. Emergency decontamination is needed for victims who have come in contact with a corrosive. Clothes must be removed immediately and the victim flushed with water thoroughly. Medical personnel need to know the name of the corrosive so that proper treatment can be undertaken.

SPILLED CORROSIVE MATERIALS INVOLVING FIRE

The immediate concern is to confine the spread of the material and protect exposures. The primary objective is to shut off the flow, isolate the area, and deny entry.

Some corrosives are flammable and first responders will need to consider ignition sources. Others are strong oxidizing agents and can ignite organic materials. The primary objective when a spill is burning is to protect exposures — not to extinguish the fire. Water can worsen the situation. Smoke from burning corrosives can be corrosive and should be avoided. Even with all exposed areas of the body covered, the first responder can still be harmed. Smoke from burning corrosives (and especially the vapors) can permeate fire clothing.

HAZ MAT

Chapter 9

Photo Courtesy of Scott D. Christiansen, Minot, N.D.

Decontamination Techniques

LEARNING OBJECTIVES

This chapter provides information that will assist the reader in meeting the objectives from NFPA 472, *Standard for Professional Competence of Responders to Hazardous Materials Incidents* and 29 CFR 1910.120 that are listed below. The objective numbers are also noted directly in the text in the sections where they are addressed. Objectives in the list below that are denoted with an asterisk (*) are global in nature and are covered by reading the chapter in its entirety.

OPERATIONAL LEVEL

- Know how to implement basic decontamination procedures. [29 CFR 1910.120(q)(6)(ii)(E)]*

NFPA 472: 3-2.3.1.2 Identify the differences among the following terms:

(a) Exposure and hazard;

(b) Exposure and contamination; and

(c) Contamination and secondary contamination.

NFPA 472: 3-3.4 The first responder at the operational level shall identify emergency decontamination procedures.

NFPA 472: 3-3.4.1 Identify ways that personnel, personal protective equipment, apparatus, and tools and equipment become contaminated.

NFPA 472: 3-3.4.2 Describe how the potential for secondary contamination determines the need for emergency decontamination procedures.

NFPA 472: 3-3.4.3 Identify the purpose of emergency decontamination procedures at hazardous materials incidents.

NFPA 472: 3-3.4.4 Identify the advantages and limitations of emergency decontamination procedures.

NFPA 472: 3-4.1.3 Identify the considerations associated with locating emergency decontamination areas.

NFPA 472: 3-4.1.4 Demonstrate the ability to perform emergency decontamination.

Chapter 9

Decontamination Techniques

Decontamination, commonly referred to as decon, is performed when a victim or responder leaves the hot zone. Proper decontamination prevents the spread of contaminants, known as secondary contamination, beyond the hot zone. Everyone in the hot zone is subject to contact with the hazardous material and can become contaminated. Because of this potential, everyone who goes into the hot zone should pass through decon upon leaving the hot zone. First responders should understand how people, equipment, and the environment are contaminated. First responders at the operational level must be able to assist in decontamination techniques, which include selecting a decon site, setting up the corridor, and performing both basic and emergency decon.

CONTAMINATION

[NFPA 472: 3-3.4.1]

Contamination is the transfer of a hazardous material to persons, equipment, and the environment in greater than acceptable quantities. Contamination occurs in the hot zone. First responders become contaminated by walking in a spill, by walking through a vapor cloud, or by touching the material. Tools used by haz mat technicians to tighten a valve or to plug a leak will be contaminated (Figure 9.1). Even vehicles can be contaminated if driven through a contaminated area. First responders and the environment can be contaminated by the dusts, particles, gases, fumes, vapors, mists, and runoff of hazardous materials. Smoke and other products of combustion are sources of contamination. To avoid being contaminated, first responders should stay away from the smoke of any burning hazardous material. The environment becomes contaminated when materials soak into the ground or are allowed to flow into storm drains and into waterways.

[NFPA 472: 3-2.3.1.2 (b)]

Exposure is the process by which people, equipment, and the environment are subjected to or

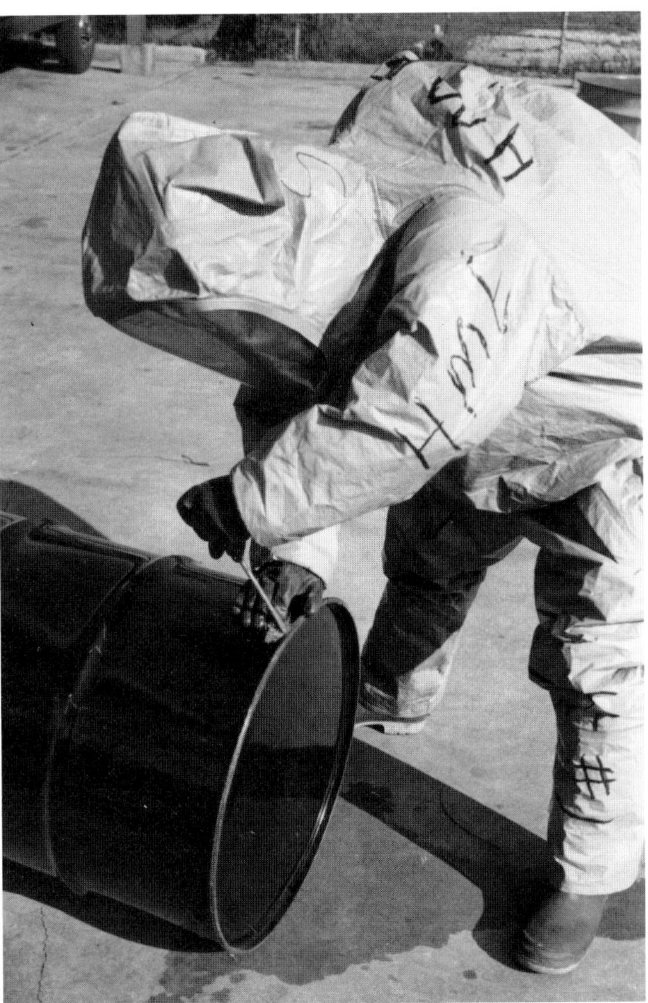

Figure 9.1 The tools used to plug a leak will be contaminated.

come into contact with a hazardous material. Being exposed to a hazardous material does not necessarily mean the exposed person or object is contaminated — it depends on the magnitude of the exposure. For example, employees in the workplace are allowed to be exposed to 10 parts per million (ppm) of hydrogen sulfide during the course of a normal work day. This is the threshold limit value, time-weighted average (TLV-TWA) set by the federal Occupational Safety and Health Administration (OSHA) for hydrogen sulfide. Threshold limit values relate to the concentration of a material in the air; it is not the concentration of the material in stored form. Threshold limit values are the same whether a solution contains 10 percent or 98 percent of a material. The values apply to the material. It takes a larger spill of a weak solution to produce a harmful exposure; a more concentrated solution of the material takes a smaller spill to produce a harmful exposure.

[NFPA 472: 3-2.3.1.2 (a)]

The magnitude of an exposure is dependent on the duration of the exposure and the concentration of the hazardous material. As cited in the previous example, an employee is not harmed with an exposure of 10 ppm or less of hydrogen sulfide over a period of eight hours. At 15 ppm (the TLV-STEL of hydrogen sulfide), there is no harmful effect for a short exposure of 30 minutes. The threshold limit value-ceiling (TLV-C) for hydrogen sulfide is 20 ppm. It is considered a harmful exposure in the workplace only when concentrations exceed the TLV-C for that material. Threshold limit values of some materials are much higher than that of hydrogen sulfide. The higher the TLV, the less hazardous the material. A low TLV means that the material is harmful even in small quantities.

Secondary Contamination

[NFPA 472: 3-2.3.1.2 (c)]; [NFPA 472: 3-3.4.2]

Secondary contamination is the contamination of people, the environment, or equipment outside the hot zone (Figure 9.2). The contaminant is carried from the zone by workers' clothing or tools and in air currents and runoff. A victim of a haz mat incident who is rushed to an ambulance without being decontaminated can contaminate the ambulance, EMS personnel, the emergency room, and the doctor and nurses treating the victim (Figure 9.3). Workers leaving the hot zone, if not decontaminated, can contaminate whomever they touch. Their tools will also be a source of secondary contamination when laid on the ground in the cold zone or on the running board of fire apparatus. Both emergency decontamination and standard decontamination prevent secondary contamination.

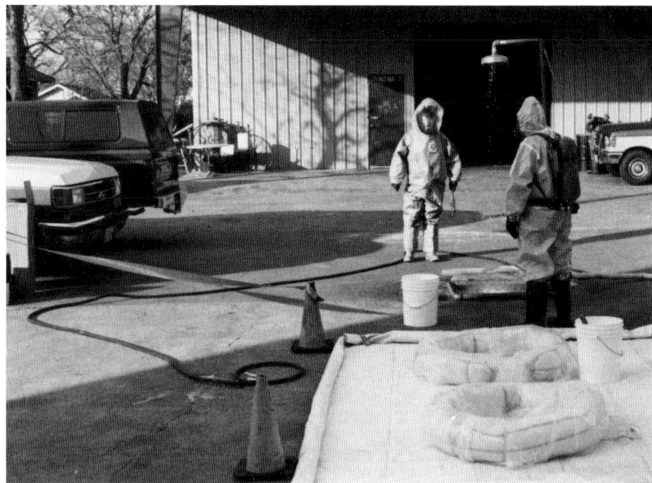

Figure 9.2 Secondary contamination occurs when either the responder or the victim leaves the hot zone.

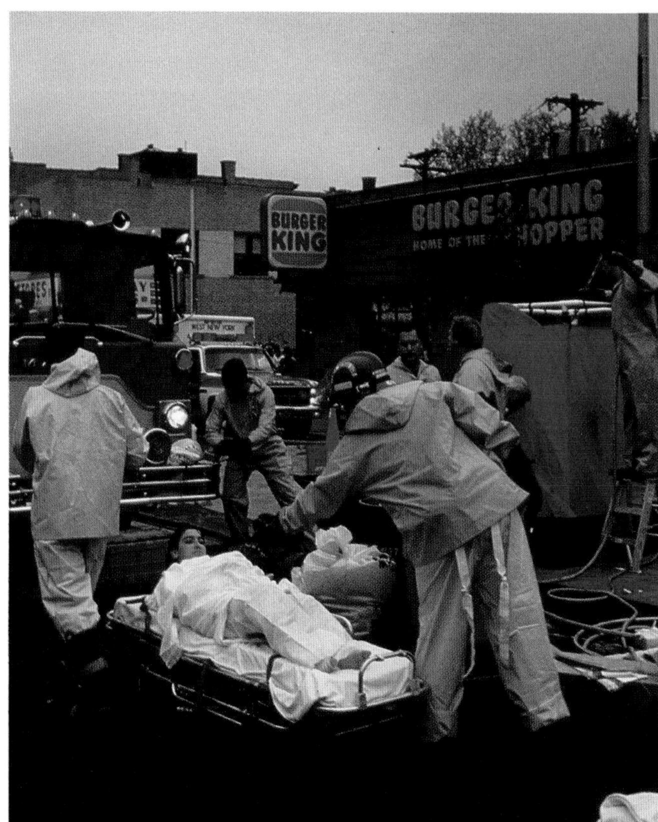

Figure 9.3 Contaminated victims who are rushed from the scene can spread the contamination to EMS personnel. *Courtesy of Ron Jeffers.*

DECONTAMINATION

Decontamination is the removal of contaminants from workers' personal protective equipment and tools. Decontamination occurs after personnel leave the hot zone and enter the warm zone. Decon minimizes the chance of secondary contamination and is an important function at all haz mat incidents. Decontamination occurs in the warm zone decontamination corridor that runs from the hot zone to the cold zone (Figure 9.4). First responders should be able to select an appropriate site and be able to perform both basic decontamination and emergency decontamination procedures.

BASIC DECONTAMINATION STRATEGIES

Before allowing personnel to enter the hot zone, the incident commander must determine the form of decon that will be used and make the appropriate preparations. The following are the four basic methods of decontamination that the first responder may use:

- Dilution
- Absorption
- Chemical Degradation
- Isolation and Disposal

Any procedure for decontaminating people, the environment, or property must encompass one of these four methods. The following sections briefly describe each of these methods.

DILUTION

Dilution is the process of using water to flush the contaminant from the contaminated victim or object. Dilution is advantageous because of the accessibility, speed, and economy of using water. However, there are some serious disadvantages as well. Depending on the involved material, water may cause a reaction and create even more serious problems. Another problem with water is knowing what to do with it after it has been used. It must be confined and then disposed of properly.

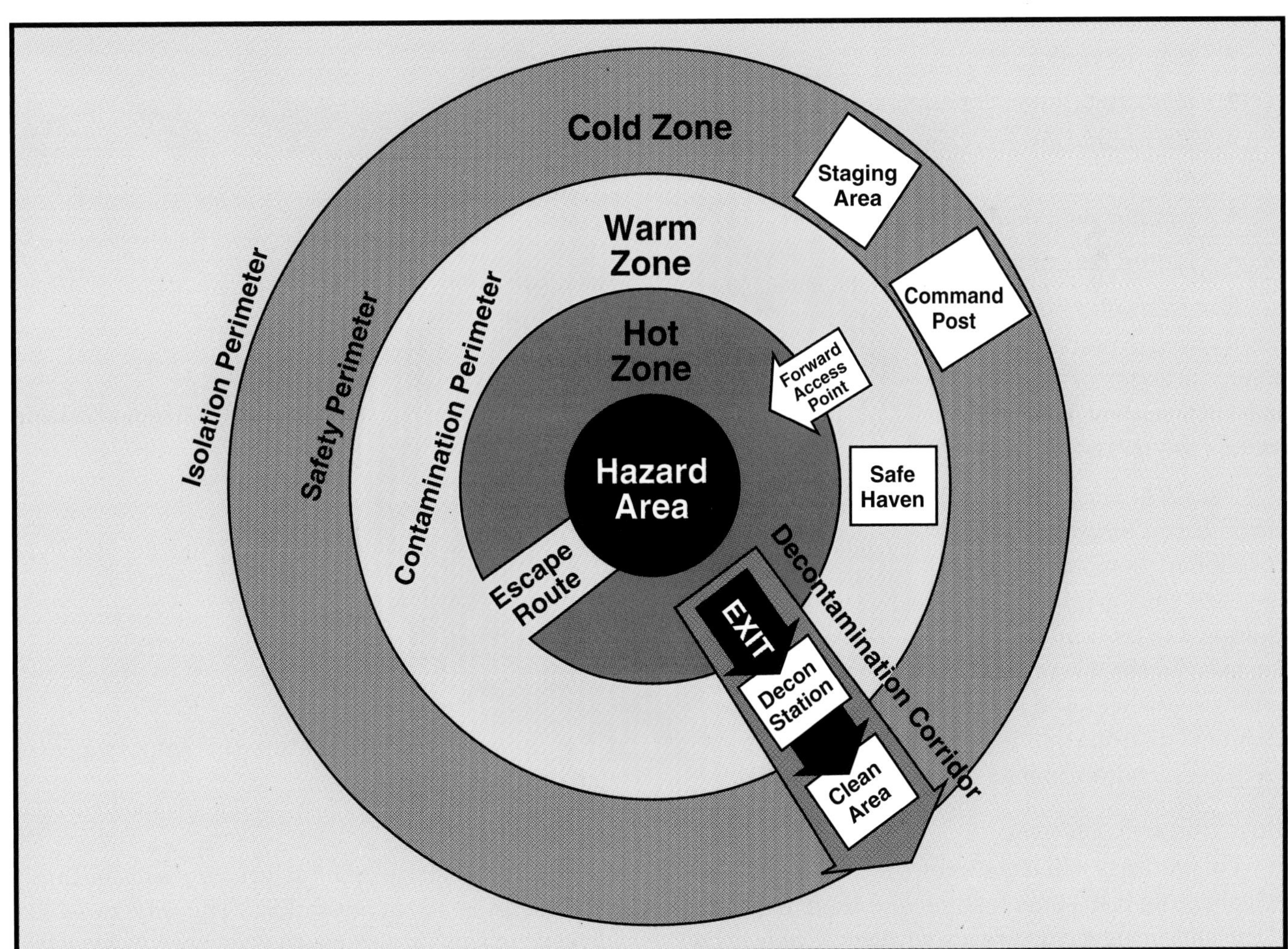

Figure 9.4 The decon corridor runs from the hot zone, through the warm zone, and into the cold zone.

ABSORPTION

Absorption is the process of picking up a liquid contaminant with an absorbent. Absorbents are inert materials; that is, they have no active properties. Some examples of absorbents are soil, diatomaceous earth, vermiculite, sand, or other commercially available materials.

Most absorbents are inexpensive and readily available. They work extremely well on flat surfaces. Some of their disadvantages are that they do not alter the hazardous material, they have limited use on protective clothing and vertical surfaces, and their disposal may be a problem.

CHEMICAL DEGRADATION

Chemical degradation is the process of using another material to change the chemical structure of the hazardous material. There are several materials commonly used to change the chemical structure of a hazardous material:

- Household bleach (sodium hypochlorite)
- Isopropyl alcohol
- Hydrated lime (calcium oxide)
- Household drain cleaner (sodium hydroxide)
- Baking soda (sodium bicarbonate)
- Liquid detergents

For example, household liquid bleach is commonly used to neutralize spills of etiological agents. The interaction of the bleach with the agent almost instantaneously kills the dangerous germs and makes the material safer to handle.

Using chemical degradation can reduce cleanup costs; it can also reduce the risk posed to the first responder. Some of the disadvantages of the process are that it takes time to determine the right chemical to use and to set up the process. It can be harmful to the first responder in that the process could create heat and toxic vapors. For this reason, chemical degradation is never used on people who have been contaminated.

ISOLATION AND DISPOSAL

This process will isolate the contaminated item by collecting it in some fashion and then disposing of the materials in accordance with applicable regulations and laws. All equipment that cannot be decontaminated properly must be disposed of correctly (Figure 9.5). All spent solutions and wash water must be collected and disposed of properly. Disposing of equipment may be easier than decontaminating it; however, this can be very costly in circumstances where large quantities of equipment have been exposed to the material.

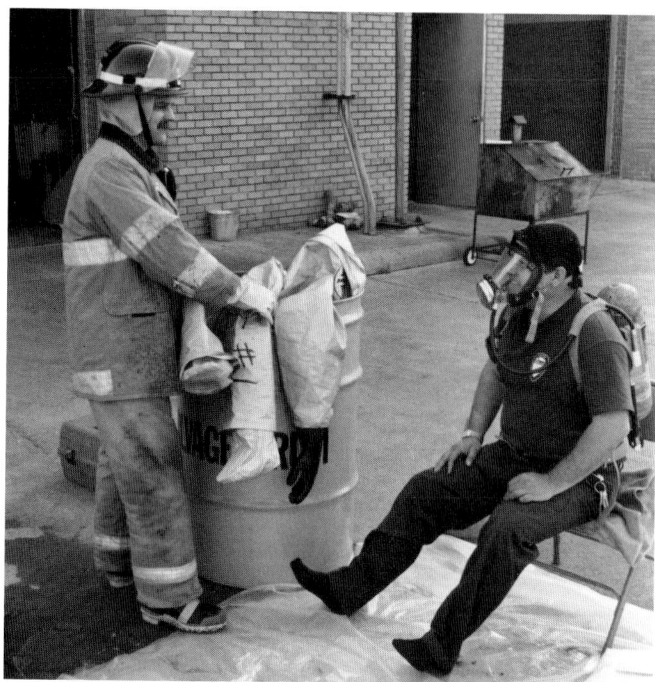

Figure 9.5 All equipment that cannot be decontaminated properly must be disposed of correctly.

Site Selection

[NFPA 472: 3-4.1.3]

Several factors are considered when choosing the decontamination site:

- Accessibility
- Surface material
- Lighting
- Drains and waterways
- Water
- Weather

The following sections detail the importance of each of these factors.

ACCESSIBILITY

The first priority in the selection of a decontamination site is its accessibility. The site must be adjacent to the hot zone so that persons exiting the hot zone can step directly into the decontamination corridor. The adjacent site eliminates the chance of contaminating clean areas. It also puts the decontamination site as close as possible to the actual incident. Time is a major consideration in the selection of a site. The less time it takes the person to get to and from the problem, the longer the person can work. There are four crucial time periods:

- Travel time in the hot zone
- Time allotted to work in the hot zone
- Travel time back to the decontamination site
- Decontamination time

The decontamination site should ideally slope toward the hot zone (Figure 9.6). Anything that may accidentally get released in the decontamination corridor would drain toward and into the contaminated hot zone. With an opposite slope, contaminants could flow into a clean area and spread contamination. Finding the perfect topography is not always possible, and first responders may have to place some type of barrier to ensure confinement of an unintentional release. Diking will prevent accidental contamination away from the site.

SURFACE MATERIAL

It is best if the site has a hard, nonporous surface. Hard surfaces prevent ground contamination. If a hard-surfaced driveway, parking lot, or street is not accessible, some type of impervious covering may be used to cover the ground. Salvage covers or plastic sheeting will prevent contaminated water from soaking into the earth. A cover should be used to form the decontamination corridor whether or not the surface is porous (Figure 9.7).

Figure 9.7 Lay down a tarp for the decontamination corridor.

LIGHTING

The decontamination corridor must have adequate lighting to help reduce the potential for injury to personnel in the area. Selecting a decontamination site illuminated by street lights, flood lights, or other type of permanent lighting reduces the need for portable lighting. If permanent lighting is unavailable or inadequate, portable lighting is required (Figure 9.8).

DRAINS AND WATERWAYS

Locating a decontamination site near storm and sewer drains, creeks, ponds, ditches, and other waterways should be avoided. If this is not possible,

Figure 9.6 The decontamination site should slope toward the hot zone.

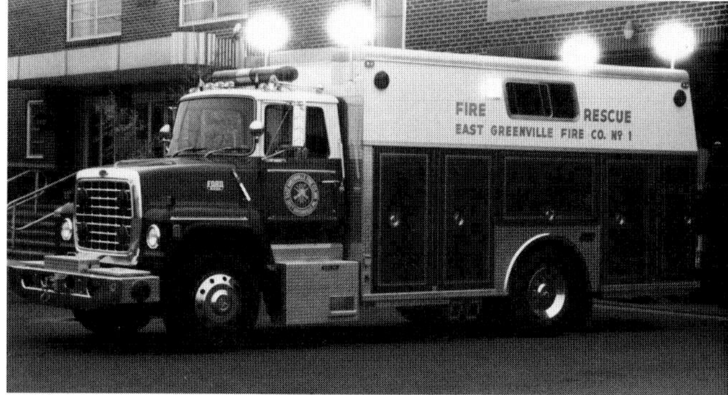

Figure 9.8 Floodlights on the apparatus are used to light the area.

a dike can be constructed to protect the sewer opening, or a dike may be constructed between the site and a nearby waterway. All environmentally sensitive areas should be protected.

WATER

Water must be available at the decontamination site. This is probably not a problem for fire departments. Fire apparatus carry water on board, or the apparatus can easily be hooked to fire hydrants (Figure 9.9). Water is more of a concern for private response groups and cleanup companies that are working without the assistance of the fire department. They may not have the ability to provide adequate amounts of water to the scene.

Water and detergent are the basic materials used for decontamination. They dilute and remove most hazardous materials. Some jurisdictions use other solutions that degrade the contaminant. For example, weak hydrochloric acid solutions are sometimes used to neutralize caustic materials. Conversely, sodium carbonate solutions may be used to neutralize acids. A number of different solutions may be used. The use of such solutions, however, would normally take place in a formal decontamination setting and require a higher level of training than that of the first responder.

Figure 9.9 Pumpers carry water on board.

WEATHER

The decontamination site should be upwind of the hot zone. Upwind sites help to prevent the spread of airborne contaminants into clean areas. Wind currents will not blow mists, vapors, powders, and dusts toward first responders if the decontamination site is upwind. Ideally, during cold weather, the site should be protected from blowing winds, especially near the end of the corridor. The user should be shielded from the cold winds when removing protective clothing. Some jurisdictions overcome this problem by setting up a portable decontamination shelter (Figure 9.10). Other jurisdictions have decontamination trailers that are set up at the scene (Figure 9.11). Some jurisdictions perform decontamination in a remote building, such as a fire station, away from the site. However, this extends travel time and may not be practical.

Figure 9.10 A commercially available decontamination shelter. *Courtesy of Zumro, Inc.*

Figure 9.11 Some agencies have decontamination trailers. *Courtesy of Ron Jeffers.*

Decontamination Corridor

The decontamination corridor should be established before any work is performed in the hot zone. First responders are often involved with setting up and working in the decontamination corridor. The types of decontamination corridors vary as to the numbers of sections or steps used in the decontamination process. Corridors can have a few steps, or they can be more complex. For a discussion of a more formal decontamination, see Fire Protection Publications' **Hazardous Materials Response Team Leak and Spill Guide** and **Hazardous Materials: Managing the Incident** manuals. First responders must understand and be trained in setting up the type of decontamination established by the jurisdiction. Basic decontamination, which is usually near the lower end of the steps, will be covered in this section.

The decontamination corridor may be staked out with barrier tape, safety cones, or other items that will visually identify it (Figure 9.12). Coverings, such as a salvage cover or plastic sheeting, are used to form the corridor (Figure 9.13). The protective covering ensures against environmental harm if contaminated rinse water splashes from a containment basin. Containment basins can be constructed of salvage covers and fire hose or ladders (Figure 9.14). Some departments use wading pools or portable drafting tanks as containment basins (Figure 9.15). Also needed at the site are a recovery drum or other type container and plastic bags for stowing the contaminated tools and personal pro-

Figure 9.12 Use cones to mark the corridor.

Figure 9.13 A finished decontamination corridor.

Figure 9.14 In this case, the catch basins were made with absorbent pigs and plastic sheets.

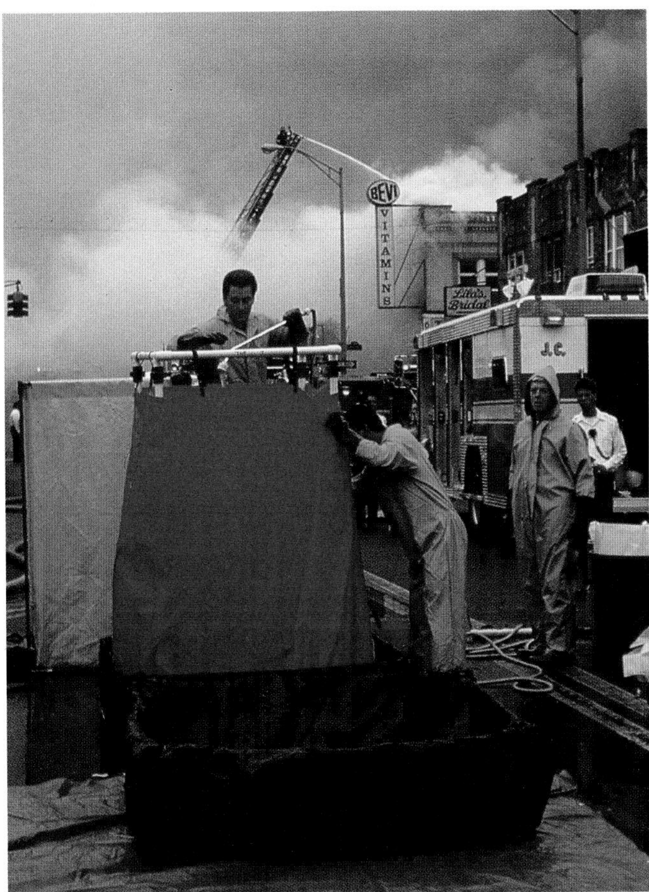

Figure 9.15 Wading pools may be used for catch basins. *Courtesy of Ron Jeffers.*

tective equipment. A low-volume, low-pressure hoseline, such as a garden hose or booster line, is ideal for decontamination (Figure 9.16). Pump pressure at idling speed is adequate and minimizes splashing.

Figure 9.16 A booster or garden hose may be used to rinse the victim.

How first responders are protected when working in the decontamination area will depend on the hazards of the material. Standard firefighter personal protective clothing and SCBA may be adequate. In some cases, chemical protective clothing may be necessary. In either case, chemical gloves are necessary; fire fighting gloves should not be used in decontamination. Because there is a possibility that first responders in decontamination operations will become contaminated themselves, they should pass through decontamination before leaving the corridor.

Decontamination Procedures

[29 CFR 1910.120 (q)(6)(ii)(E)]

First responders must know what to do when assigned to the decontamination corridor. The following is an example of a simple, basic decontamination procedure:

Step 1: The first responder removes the majority of contamination from the victim and from any tools the victim is carrying (Figure 9.17). This step is commonly referred to as gross decontamination. This is usually done by hosing down the victim. A decontamination worker is needed to hose the victim unless a portable shower is available that can be operated by the victim. Some type of catch basin should be provided to confine and hold the water. This step can be skipped if the victim is not badly contaminated.

Figure 9.17 First, rinse the bulk of the material off the victim.

Step 2: The first responder discards the tools and equipment near the edge of the corridor (Figure 9.18). If gross decontamination (Step 1) is not needed, this becomes the first step instead. Technical expertise may be needed later for advice on the decontamination or disposal of the contaminated tools and equipment.

Step 3: The person being decontaminated steps into the rinse area. A decontamination worker scrubs down the contaminated person with detergent and water (Figure 9.19). The worker should give special attention to areas that are easily missed

Figure 9.18 Discard tools into collection containers.

Figure 9.19 Scrub remaining materials from the victim.

such as the folds in the chemical suit, under the arms, and in the crotch. This water should be kept and analyzed before release, although rinse water at this step is seldom more than slightly contaminated.

Step 4: After wash down, the person steps from the catch basin into the final area where clothing and SCBA are removed (Figure 9.20). A second decontamination worker assists at this point by helping to remove the boots, the personal protective clothing, and last, the SCBA. Undergloves and SCBA are always the final items removed. The contaminated clothing is left in the decontamination corridor for decontamination or disposal.

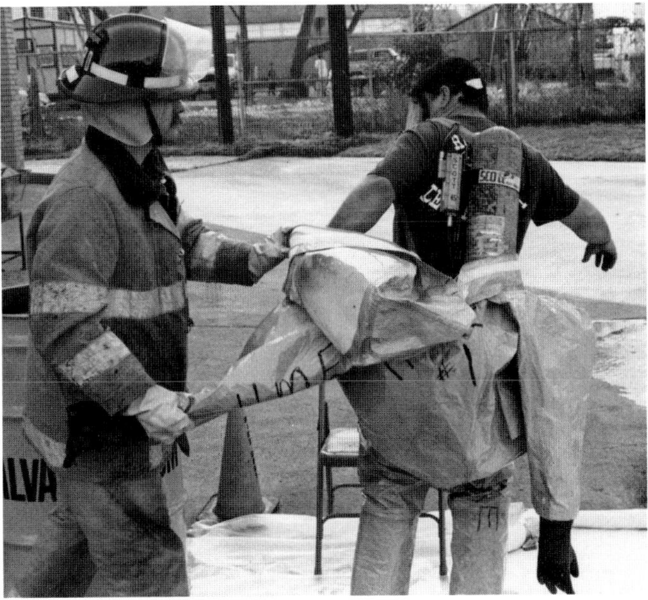

Figure 9.20 Remove the protective clothing and SCBA carefully.

Emergency Decontamination

[NFPA 472: 3-3.4]; [NFPA 472: 3-3.4.3]; [NFPA 472: 3-4.1.3]; [NFPA 472: 3-4.1.4]

Emergency decontamination is the physical process of immediately ridding dangerous contaminants from individuals. Emergency decontamination is needed for the following conditions:

- Special protective clothing fails.
- First responders accidentally become contaminated.
- Victims need immediate medical attention.

The goal is to remove the threatening contaminant as quickly as possible. There is no regard for the environment or property.

Emergency decontamination and gross decontamination are sometimes used synonymously. Procedures are generally the same; the difference is where the decontamination occurs. Gross decontamination takes place at the decontamination corridor and is the first step in a series of procedures; emergency decontamination takes place anywhere (Figure 9.21). Emergency decontamination usually occurs early in an incident before basic decontamination operations have been established. Even when a decontamination corridor is in place, emergency decontamination may be necessary for threatening situations.

Emergency decontamination may be necessary for both the victim and the rescuers. If either is contaminated, the individual is stripped of his or her clothing and given a quick wash down. Victims may need immediate medical treatment and cannot wait for a formal decontamination corridor to be established.

Figure 9.21 A portable shower may be provided for responders to perform emergency decontamination on themselves if necessary.

[NFPA 472: 3-3.4.4]

A limitation of emergency decontamination is that it is a quick fix. Removal of all contaminants may not occur, and a more thorough decontamination will have to follow. Emergency decontamination can definitely harm the environment. However, the advantage of eradicating a life-threatening situation far outweighs any negative effects that may result from emergency decontamination.

Emergency decontamination may be necessary even after basic decontamination is established. For example, a haz mat technician accidentally slips into a ditch of hydrochloric acid that came from an overturned tank truck. The acid fills the technician's boot and penetrates the boot seam of the protective suit. When the technician's foot begins to burn, it is quicker to strip him or her down and have a nearby hoseline brought in than for the technician to hobble to the decontamination corridor. Emergency decontamination will minimize this type of injury.

There are times when what appears to be a "normal" incident turns out to involve hazardous materials. First responders may become contaminated before they realize what they have gotten into. When this situation occurs, first responders should withdraw immediately. They should get emergency decontamination and remove their turnout clothing, even if there is no apparent contamination. The first responders should stay isolated until someone with expertise can ensure them that they have been adequately decontaminated.

Appendix A

Agencies Available For Help and Information

Seldom is one unit of an emergency response organization able to handle a haz mat incident alone. Often a dozen or more agencies are needed to help. Teamwork among the agencies involved is a must. Mutual aid agreements outlining responsibilities must be set up before a haz mat incident occurs to ensure that mass confusion does not interfere with bringing the incident under control. Every agency must understand its value so that the agency can be used to its full extent. Each agency must also know its limitations to avoid confusion in offering more than it is able to provide. Each agency must fully understand the chain of command at a haz mat incident to avoid jurisdictional and power disputes during the decision-making process. A planning committee should determine the chain of command when the mutual aid agreements are established.

Agencies involved in handling haz mat incidents are available at the national, regional, and local levels. At all three levels, several agencies are extremely valuable in providing immediate information to on-scene emergency personnel. Others can provide guidance, leadership, or the equipment needed to control the incident. Many are government agencies funded by local, state, or federal taxes, while others are privately funded.

NATIONAL AGENCIES

Bureau Of Explosives

Emergency Phone Number — Contact CHEMTREC — (800) 424-9300

The Bureau of Explosives is a division of the Association of American Railroads (AAR). The Bureau will send an expert consultant to railroad incidents involving explosives to advise the incident commander on how to handle the situation. The AAR has also published an on-site guide, *Emergency Handling of Hazardous Materials in Surface Transportation*. This publication is also available on software for computer applications. The telephone number for information on publications is (202) 639-2222.

Canadian Transport Emergency Centre (CANUTEC)

Emergency Phone Number — Call Collect — (613) 996-6666

CANUTEC, the Canadian Transport Emergency Centre, is Canada's equivalent to CHEMTREC. Their capabilities are similar to those described for CHEMTREC. Bilingual (English/French) information is available. Emergency officers have access to a file of more than 25,000 products.

Center For Disease Control

Emergency Phone Number — (404) 633-5313

The Center for Disease Control in Atlanta, Georgia, operated by the U.S. Public Health Service, has information on etiologic agents.

Chemical Transportation Emergency Center (CHEMTREC)

Emergency Phone Number — 800-424-9300

CHEMTREC, the Chemical Transportation Emergency Center, provides information and technical assistance to those involved in or responding to chemical or haz mat emergencies. The Center's primary mission is to help with transportation incidents, but it also helps with chemical and haz mat emergencies in other incidents. The Center was established in 1971 as a public service of the Chemical Manufacturers Association (CMA) in Washington, D.C.

CHEMTREC operates in two stages. First, when the Center learns the name of the chemical involved, it gives immediate advice on the nature of the product and the necessary steps required to handle the early stages of the problem. Second, CHEMTREC promptly contacts the shipper of the material involved for more detailed information and appropriate follow-up, including on-scene assistance from the shipper/manufacturer if necessary. CHEMTREC operates 24 hours a day, 7 days a week, to take calls on the emergency telephone number.

An emergency reported to CHEMTREC is taken by communicators on duty who record the details by entering the information into a computer and by recording the conversation. They question the caller to get as much information on the problem as possible. Then, working with a library of more than 1,000,000 product specific material safety data sheets, they can provide the best available information on the chemical(s) reported to be involved, giving specifics on hazards and what to do or what not to do in the case of spills, leaks, fire, or exposure.

Communicators immediately notify the shipper by phone or by an electronic data transmission system relaying what is known about the emergency. The responsibility for further action, including dispatching personnel to the scene, passes to the shipper.

Identifying both the product and the shipper is important to reduce the time needed to provide information and help. Shipping papers are carried by truck drivers and can be found in the engine or caboose of trains. Railcar and truck/trailer numbers and carrier names are useful in tracing unknown cargoes.

The second stage of assistance is more difficult when the shipper is unknown, but communicators have other resources. For example, CHEMTREC can call on the U.S. Department of Energy (DOE) for help with incidents involving radioactive materials. There are also trade associations with mutual aid programs for many products such as chlorine, hydrocyanic acid, vinyl chloride monomer, hydrogen fluoride, and phosphorus. In this type of program, one producer will service the field emergencies involving another producer's products.

In 1981, CHEMTREC installed a telecommunications bridge to improve its emergency response capabilities. The on-scene commander facing a multishipper incident can set up a conference call with shippers and others involved, freeing the incident commander from making one call at a time. This brings about a rapid delivery of information to the on-scene commander.

Because CHEMTREC receives many calls that are unrelated to emergencies, the CMA initiated a nonemergency phone line for information about chemicals and chemical products. The center gives callers assistance in contacting the manufacturer of the product. Over 2,000 chemical companies are participating in the program. The line is answered between 9:00 a.m. and 6:00 p.m. (Eastern standard time), Monday through Friday. Those seeking information should call (800) CMA-8200.

Chlorine Emergency Plan

Emergency Phone Number — Call CHEMTREC — (800) 424-9300

The Chlorine Emergency Plan was developed by the Chlorine Institute, 2001 L Street NW, Washington, D.C. 20036. Under the plan, the chlorine producer nearest the chlorine incident responds to give information and help. CHEMTREC contacts the emergency plan only if chlorine is involved.

Federal Emergency Management Agency (FEMA)

The Federal Emergency Management Agency (FEMA) helps communities prepare for and mitigate large-scale disasters. FEMA also provides training to responders through courses conducted at the National Emergency Training Center at Emmitsburg, Maryland and through field courses. FEMA has no emergency number but may be reached during business hours at (301) 926-5110.

National Animal Poison Control Center Of The University Of Illinois

Emergency Phone Numbers —

- (900) 680-0000 — $20 for the first five minutes and then $2.95 for each additional minute. Billing is through the phone company.
- (800) 548-2423 — $30 per call. The person must have a credit card number available when the call is made.

This organization provides 24-hour consultation in diagnosis and treatment of either suspected or actual animal poisonings or chemical contamination. It staffs an emergency response team to rapidly investigate such incidents in North America and performs laboratory analyses on feed, animal specimens, and environmental materials for toxicants and chemical contaminants.

Ortho/Chevron

Emergency Phone Number — (415) 233-3737

The Ortho/Chevron Company operates a 24-hour emergency number for information and help when either an Ortho chemical or Chevron petroleum product is involved in an incident.

Union Carbide Emergency Response System

Emergency Phone Number — (304) 744-3487

Union Carbide helps with any emergency involving Union Carbide shipments and may also be reached through CHEMTREC.

Texas Tech University Pesticide Hot Line

Emergency Phone Number — (800) 858-7378

The National Pesticide Telecommunications Network provides information on pesticide-related health/toxicity/minor cleanup questions to physicians, veterinarians, fire departments, government agency personnel, and the general public. This line is answered from 8:00 a.m. to 6:00 p.m. (Central standard time).

Transportation Emergency Assistance Program

The Canadian Chemical Producers Association operates the Transportation Emergency Assistance Program. This program provides information by phone or provides help in the field from regional teams. The Transportation Emergency Assistance Program can be contacted at the following regional telephone numbers:

Atlantic Provinces and Eastern Quebec (819) 537-1123

Southwestern Quebec (514) 373-8330

Eastern Ontario (613) 348-3616

Central Ontario (416) 356-8310

Southwestern Ontario (519) 339-3711

Northern/Western Ontario (705) 682-2881

Manitoba, Saskatchewan, Alberta (403) 477-8339

British Columbia (604) 929-3441

U.S. Armed Forces

The U.S. Army, Navy, Air Force, and Marines may help handle hazardous materials incidents. The military may provide the following assistance:

- Personnel to help with evacuations
- Air and water rescue
- Medical transport
- Expert help for incidents involving radioactive and nuclear materials, explosives, and many poisonous chemicals

One element of the Army is its explosives ordnance disposal unit (EODU) that responds at any hour to any incident involving military explosives, including bombs, TNT, and ammunition. EODUs are prohibited from handling situations involving commercial or nonmilitary ordnance unless there is a severe life hazard to the surrounding public. The Army has divided the country into regions and has assigned an EOD response team to each region. The units are based at regional Army posts.

A national phone number is not available for use in contacting any branch of the military. Responders should have the phone number of the closest active military installation available to them. In many cases, this installation will be able to provide the necessary assistance. If not, the military personnel will then contact the appropriate resources.

U.S. Coast Guard

Emergency Phone Number — (800) 424-8802

The U.S. Coast Guard plays a major role in haz mat incidents that occur on navigable waterways. The Coast Guard responds to spills on waterways. It operates the National Response Center, which also handles communications for the Department of Transportation (DOT) and the Environmental Protection Agency (EPA). The Coast Guard has developed the Chemical Hazards Response Information System that is a sophisticated information service that covers about 900 chemicals. It consists of the following four manuals:

A Condensed Guide to Chemical Hazards

Hazardous Chemical Data Manual

Hazard Assessment Handbook

Response Methods Handbook

In addition to the manuals, it also includes a group of regional data bases for planning and a computerized model for assessing the potential effects of an incident.

When hazardous materials are spilled or released in a navigable waterway, the Coast Guard responds with teams and equipment to help recover the materials and reduce pollution. The Coast Guard coordinates control activities. They also have a portable laboratory to help during the control of waterway spills.

The National Response Center is the Coast Guard's headquarters for emergency incidents. It receives the reports required by federal law whenever a hazardous material release threatens American waters. The legal reporting requirements include the following:

- The person in charge of the vessel, facility, or vehicle must report discharges of oil or other hazardous substances.
- The carrier must report transportation incidents.
- The coordinator of a facility that treats, stores, or disposes of hazardous wastes must report any releases.
- Hazardous waste discharges at abandoned dump sites should be reported by anyone who sees them.

The National Response Center works with CHEMTREC to issue information for use at haz mat incidents. Both operate in much the same way. NRC, however, provides its information to federal coordinators and other government agencies; whereas, CHEMTREC passes on its data to carriers and response services. NRC will also contact the federal on-scene coordinator assigned to the area in which the incident occurred.

U.S. Department Of Energy (DOE)

The Department of Energy provides both information and response teams to help with radiological emergencies. DOE has divided the country into eight regions, each with its own headquarters phone number. They do not accept calls from first responders, but can be contacted by CHEMTREC if needed.

U.S. Department Of Transportation (DOT)

Emergency Phone Number — (800) 424-8802

The Department of Transportation develops the regulations governing hazardous materials during transport. DOT regulates shipping containers and labels and placards for hazardous materials and sets requirements for shipping documents. For information on federal regulations call the DOT Hot Line at (202) 366-4488. DOT also publishes the *Hazardous Materials Emergency Response Guidebook* (see Chapter 3). The following haz mat incidents during transport must be reported to DOT as soon as possible:

- An incident that involves death or injury
- An incident with damage in excess of $50,000
- An incident involving radioactive materials or etiologic agents
- An incident that continues to endanger life at the scene

U.S. Environmental Protection Agency (EPA)

Emergency Phone Number — (800) 424-8802

The Environmental Protection Agency sponsors an emergency response center that provides immediate information on hazardous materials. The center also alerts regional teams that respond to help in emergencies. Assistance includes advice and counsel, training, safety recommendations, and certain operational activities.

The agency's emergency response is financed by the "Superfund," a billion-dollar fund Congress established under the Comprehensive Environmental Response, Compensation and Liability Act (CERCLA) of 1980. The fund is financed by industry and the federal government. The law also lets the government seek to recover response costs for those responsible for the emergency. For information on problems encountered by small-quantity generators of hazardous waste, call the EPA Small Business Hot Line at 800-368-5888.

REGIONAL AGENCIES

Regional agencies are limited to a particular state or region of the country.

Disposal Services

Some private companies package, transport, and dispose of hazardous waste. Company personnel are highly trained and take special precautions in handling hazardous materials. Emergency responders should be familiar with companies in their area that are available to perform these services.

Emergency Preparedness Or Emergency Service Disaster Agency

Many counties and/or states have agencies that organize preparations for civil disasters. Titles vary but the functions are basically the same. In some states the emergency preparedness office has the authority to handle haz mat incidents. In years past, these organizations were called Civil Defense.

Emergency preparedness personnel can help with evacuation, rescue, traffic control, first aid, fire fighting, and radiological monitoring. Personnel can work in refuge centers and find necessities such as water, food, and blankets for victims. They can locate emergency equipment such as generators, lighting, and public address systems. Emergency preparedness is also responsible for the area's emergency preparedness plan. Contact the county emergency preparedness office to find out the local organization's functions and determine its ability to help with a haz mat incident.

Environmental Protection Agencies

The state environmental protection agency can provide technical and legal advice on pollution hazards. State agency directives may change the control actions the incident commander implements.

Fish And Game Departments, Water Resource Departments, Or Departments Of Natural Resources

The fish and game department, department of natural resources, or water resource departments can provide technical assistance for controlling pollution and cleaning up contamination in streams, lakes, waterways, and other fish and game reserves.

Forestry Departments

The state forestry department must be notified of any incident that is located in or threatens any state or national forest. The department can help with information on pollution hazards and with personnel for fire fighting.

Highway Departments

The state highway department can provide heavy equipment, such as graders, bulldozers, tankers, and dump trucks, for transportation incidents on state and interstate highways. The highway department can also supply large amounts of sand and dirt for use as extinguishing or control agents.

Highway Patrols (State Police)

Each state has an organization responsible for enforcing the law on state and interstate highways. The name varies from state to state, but each organization operates basically the same way. The highway patrol (state police) can be used for evacuation, traffic control, and other duties designated by law. In many areas, the highway patrol (state police) is under written agreement to offer help to county and local police or sheriffs' departments during a haz mat emergency. In some states, the state law enforcement agency has the responsibility to provide haz mat experts to coordinate incident control.

Industry And Carriers

The regional industry that makes or processes a hazardous material will have the best advice on the material. Many companies and carriers will send representatives, who are experts on the products they carry, to the scene of an emergency. Carriers can also find replacement transport vehicles to remove or transfer the materials from an incident.

National Guard

The National Guard can be called upon for the same kind of help as the military offers. The Guard can provide personnel to help with evacuations, air and water rescue, medical transport, decontamination showers, heavy equipment, traffic and people control, and refuge centers. In most states, the Guard can only be activated by the Governor.

Nuclear Regulatory Agencies

Some states, in agreement with the Nuclear Regulatory Commission (NRC), are allowed to license and regulate the producers and users of radioactive materials. Agency names can vary, but all of the agencies perform the same function. In states not in agreement with the NRC, the NRC handles licensing and regulation.

Railroads

Railroads are involved in any haz mat incident that occurs on their railway. The railroads have information on the identity of the substance and the manufacturer or supplier. Some railroads employ haz mat personnel who offer on-site help. Railroads can also supply heavy-duty equipment like rail-mounted hoists, water tanks, and commodity transfer equipment.

State Or County Health Department

Many state and county health departments regulate special landfills for dumping hazardous materials. These departments can provide information about the pollution dangers of many hazardous materials and about their disposal. Failure to properly dispose of hazardous materials may cause further incidents that can have catastrophic effects on the public and the environment.

Western Oil And Gas Association (WOGA)

The Western Oil and Gas Association (WOGA) is a conglomeration of oil and gas companies that provide training for industrial workers. WOGA can provide technical information and help in handling haz mat incidents involving oil and gas products.

LOCAL AGENCIES

Local agencies are in or around the local jurisdiction where haz mat incidents may occur.

Contractors

Contractors can provide heavy equipment, such as bulldozers, graders, cranes, dump trucks, and other equipment or materials, to assist in control of the incident.

Emergency Medical Services (EMS)

Local emergency medical services useful in a haz mat incident include municipal or private ambulance companies, hospitals, and doctors. If an incident causes or threatens to cause a high number of casualties, these medical services must be coordinated to implement medical aid immediately and efficiently.

Fire Departments

Local fire departments under mutual aid agreements can provide additional personnel, equipment (pumpers, tankers, air compressors, generators, and other special items), extinguishing agents, and relief personnel. Outside fire departments can also protect the local jurisdiction while personnel are committed to the haz mat incident.

Flood District Office

Many areas subject to flooding have a local flood district office that can provide advice and flood warnings during haz mat incidents. Flood district offices can help determine the route of runoff from an incident and what locations may be involved in flooding. If a facility manufacturing, processing, or storing hazardous materials is in the potential flood area, the responders must determine if the materials are water reactive or if they can create severe pollution or health hazards when released during a flood or when subjected to large applications of water.

Food Services

Local restaurants or food suppliers can feed emergency personnel and evacuees. Food should not be allowed near the hazardous material. All food should be kept in a special uncontaminated clean area.

Industry

Local industries may be able to provide equipment, extinguishing or control agents, and expertise in handling hazardous materials. Industry's ability to help depends on the type of industry in the area, the type of hazardous material involved, and the resources they have available.

Police And Sheriffs' Departments

The local police or sheriff's department can assign personnel for evacuation, traffic control, and other duties designated by law or ordinance. Law enforcement officers can staff barrier points to authorize or forbid entry into the incident support area. They can also help locate property or equipment owners who may be needed in handling an incident. In some states, the sheriff may have the responsibility of providing haz mat experts to coordinate the incident control.

Public Utilities

Public utilities include public and private energy and communications utilities that provide water, natural gas, telephone, and electrical service. The companies can provide utilities, or in the case of a major hazard, they can shut off utilities in the incident area.

Red Cross

The Red Cross can give first aid, operate refuge centers, provide for physiological needs, and reunite families. In some localities, the Red Cross also provides emergency canteen services for responders and victims.

Salvation Army

The Salvation Army can operate refuge centers and help evacuees. In some localities, the Salvation Army also provides canteen services for responders and victims.

Sewage Treatment

The sewage treatment plant can decide if runoff from a certain hazardous material must be disposed of by means other than the public sewage system. The sewer department must be notified if runoff from any hazardous material has accidentally entered the sewer system. Once there, the runoff may cause even greater hazards, including explosion hazards. The sewer department and the sewage treatment plant must also be notified of the exact material entering the system. Many hazardous materials cannot be neutralized by the chemicals used in the standard sewage treatment facility. Depending on the hazardous material, neutralizing chemicals may be added to reduce the hazards. If this is not possible, and with many hazardous materials it is not, other means will have to be used to control the contamination. When any one of many hazardous materials enters a sewer system, the incident is a long way from being controlled. In fact, the incident can be worsened.

Towing Service

Local towing companies can move threatened vehicles or upright an overturned haz mat transport vehicle. Extreme care must be used to avoid exposing the tow truck driver to the hazardous substance. All personnel should stay clear of any tow cable under tension. A loaded MC 306 should never be placed in the upright position without off-loading.

Transportation

Local transportation companies or carriers may be able to provide equipment such as haz mat containers, vans, refrigerated units, or replacement vehicles for commodity transfer. Mass transit companies may be able to transport evacuees.

Appendix B
Notification And Reporting Requirements

Section 304 requires reporting of releases of the extremely hazardous substances listed in Section 302 of SARA Title III, as well as any substances that require notification under Section 103(a) of the Comprehensive Environmental Response, Compensation, and Liability Act of 1980 (CERCLA). Releases resulting in an exposure to a person while at a fixed facility may be exempted from being reported. It is reportable, however, if the amount of material exceeds reportable quantity established by CERCLA.

The following agencies should be notified by phone or radio in the listed order:

- Local fire department and/or police/sheriff
- Local emergency planning committee
- State emergency response committee
- National Response Center

The following information, to the extent known at the time, should be provided during the notifications:

- Chemical name
- Whether the substance is on either the SARA or the CERCLA lists
- Estimate of quantity released
- Time and duration of release
- Whether release was made to air, land, or water
- Acute or chronic health risks associated with the release and what medical attention is necessary for exposed individuals
- What precautions should be taken including evacuation
- Name and phone numbers of contacts for further information

A written follow-up should be provided to the National Response Center (NRC), the State Emergency Response Commission (SERC), and the Local Emergency Planning Commission (LEPC) as soon as possible after the incident. The report should include information on any actions taken to control and contain the release and any information regarding medical treatment for exposed individuals.

The following section lists reporting information for individual states and territories at the time this manual was published. These requirements may change at any time, and responders should keep abreast of this information for their jurisdiction. Jurisdictions not listed in this section made no information available for publication.

ALABAMA

Who to notify: Alabama Department of Environmental Management and the Department of Public Safety (Public Safety Command)

Address: Alabama Department of Environmental Management
Field Operations Division
1751 Cong. W. L. Dickinson Drive
Montgomery, AL 36130

Phone: Emergency Management — (800) 843-0699
Department of Environmental Management — (205) 260-2700

Reporting requirements: As per SARA Title III, Section 304, with written report provided NOT LATER THAN 5 working days.

ALASKA

Who to notify: Department of Public Safety who contacts the Department of Environmental Conservation (DEC), or the DEC can be contacted directly at the number listed below. The Coast Guard must be notified if any release reaches water.

Address: Department of Environmental Conservation
Central Office
410 Willoughby Street
Juneau, AK 99801

Phone: (907) 465-5000

Reporting requirements: Article 3, 18 AAC 75.300 Discharge Notification Requirement states that a person must provide notification as soon as he or she has knowledge of any of the following:

- A release of any hazardous material other than oil
- A release of oil to water
- A release of oil exceeding 55 gallons to land outside a secondary containment area or structure

It also states that notification is required within 48 hours of a release of oil to land under any of the following conditions:

- The amount is between 10 and 55 gallons
- It exceeds 55 gallons but is contained in a secondary containment area

Finally, written notification is required, on a monthly basis, of any discharge of oil to land between 1 and 10 gallons. All quantities given above include cumulative amounts. Requirements for written reports can be obtained by contacting the Department of Environmental Conservation.

AMERICAN SAMOA

Who to notify: Territorial Emergency Management Coordination Office

Address: American Samoa Territorial Emergency
Management Coordination Office
Office of the Governor
Department of Public Safety
Pago Pago, American Samoa 96799

Phone: 011-684-633-4116 (Ask for Commissioner or Deputy Commissioner)

Reporting requirements: None provided.

ARIZONA

Who to notify: For facility incidents, contact Arizona Department of Environmental Quality's (DEQ) Emergency Response Unit; for transportation incidents on state/interstate highways, contact the Department of Public Safety's (DPS) Duty Officer; for other transportation incidents, call the operator or 911.

Address: Arizona Division of Emergency Services (DES)
Hazardous Materials Section
5636 East McDowell Rd.
Phoenix, AZ 85008

Phone: DEQ — (602) 257-2330
DPS — (602) 223-2212 (24-hour number)

Reporting requirements: As per SARA Title III with the following additions:

- During the verbal report, provide the specific location of the release.
- During the written report, provide an explanation of the measures that have been or will be taken at the facility to avoid a recurrence of similar releases.
- Written reports should be made by the State On-Scene Coordinator (SOSC) to the Arizona Division of Emergency Services and by the owner/operator to the SERC and LEPC. Reports to DES must be made within 15 days and reports to the SERC and LEPC must be made within 30 days with any follow-up information reported within 7 days of becoming known.

ARKANSAS

Who to notify: Office of Emergency Services (OES)

Address: SERC
c/o: Department of Pollution Control and Ecology (DPCE)
8001 National Drive, P.O. Box 8913
Little Rock, AR 72219-8913

Phone: OES — (800) 322-4012
DPCE — (501) 562-7444

Reporting requirements: Generally, Arkansas is set up to follow SARA Title III requirements. The Office of Emergency Services must be notified about ANY incident involving hazardous materials or substances. OES will notify SERC and DPCE. Written reports go to the DPCE. Any incidents involving hazardous waste must be reported to the Hazardous Waste Delegation through the Department of Pollution Control and Ecology.

For incidents involving transport of materials, the following agencies may become involved:

- State Police
- Department of Pollution Control and Ecology
- Office of Environmental Management

For incidents involving storage and/or disposal of materials, the following agencies may become involved:

- Department of Pollution Control and Ecology's Hazardous Waste Division
- Office of Environmental Management

CALIFORNIA

Who to notify: Office of Emergency Services (OES)
Hazardous Materials Division

Address: Office of Emergency Services
Hazardous Materials Division
2800 Meadowview Road
Sacramento, CA 95832

Phone: Incident call in — (916) 262-1621
Haz mat unit — (916) 262-1750
In state — (800) 852-7550

Reporting requirements: Immediate verbal report of release or imminent release is required for any material whose quantity, concentration, or characteristics pose a risk or potential risk to human health and safety or to the environment. This report shall include the following:

- Location of the incident or threatened incident
- Name of person reporting
- Hazardous material involved
- Estimate of the quantity of the material involved
- Potential hazards presented by the material

If the release or threatened release poses no threat to public health, safety, or the environment, the immediate reporting requirement does not apply.

When Section 304 of SARA Title III requires a written follow-up report, it must be on a form published in Title 19 of the California Code of Regulations, Section 2705(c). Administering agencies should require monthly written reports to OES using the California Hazardous Material Incident Reporting System (CHMIRS).

Any incidents occurring during transportation on a highway requires notification of the California Highway Patrol.

COLORADO

Who to notify: State Health Department for any of the following:

- Releases meeting SARA notification requirements
- A release of any quantity to water or to any drainage system leading to water (including storm sewer runoff, dry gullies, groundwater, standing water, streams, etc.)
- A release of petroleum, exceeding 25 gallons, from an underground storage tank

  Notify the Oil and Gas Conservation Commission for any incidents involving an oil or gas well.

Address: State Department of Health
Director of Emergency Management Unit
4210 East 11th Avenue
Denver, CO 80220

Phone: State Department of Health — (303) 692-2000
State Department of Health Emergency Number — (303) 756-4455
Department of Natural Resources — (303) 894-2100

Reporting requirements: As per SARA Title III, Section 304 with the additional requirements listed above.

CONNECTICUT

Who to notify: SERC/Department of Environmental Protection (DEP) and community response coordinator

Address: Connecticut Department of Environmental Protection
Bureau of Waste Management
Oil and Chemical Spill Response Division
165 Capitol Avenue
Hartford, CT 06106

Phone: SERC/DEP — (203) 566-3338

Reporting requirements: As per federal guidelines (SARA Title III). Additionally, any party responsible for transport vehicles, transport terminals, or fixed facility who causes or becomes aware of ***any*** discharge of oil or petroleum or chemical liquids; of solid, liquid, or gaseous products; or of hazardous wastes, must immediately report the facts to the SERC/DEP. Follow-up written reports must be submitted within 24 hours to DEP Hazardous Materials Management Unit. Written reports must include the following information:

- When the incident occurred
- Where the incident occurred
- How the incident occurred
- Whose control the chemical or petroleum product was in at the time of the incident
- Who the owner of the property is where spill occurred
- When the incident was verbally reported to DEP
- Who the incident was reported by and who they represent
- What chemicals or petroleum products were released
- Whether they were on an Extremely Hazardous Substance list or a CERCLA list; if so, include an MSDS for each chemical involved
- What the quantity of material released was
- Whether any of the release traveled beyond the property line (materials entering groundwater are considered to have gone beyond the property line)
- Actions taken to respond to and contain the release
- Actions being taken to prevent reoccurrence of an incident of this type
- Whether injuries occurred as a result of the incident (include the names, addresses, and phone numbers of the injured and describe their injuries)
- Appropriate advice regarding medical attention necessary for exposed individuals
- Known or anticipated health risks, acute or chronic, associated with the release of this chemical or medical advice that should be communicated
- Whether the incident has been cleaned up by the time this report was submitted (If not, what are anticipated remedial actions and their duration?)

Address written reports as shown above.

DELAWARE

Who to notify: Environmental Response Branch, Division of Air and Waste Management, Department of Natural Resources and Environmental Control (DNREC)

Address: Director
Division of Air and Waste Management
Department of Natural Resources and Environmental
Control (DNREC)
P.O. Box 1401, 89 Kings Highway
Dover, DE 19903

Phone: In state 24-hour hot line — (800) 662-8802
Out of state 24-hour hot line — (302) 739-5072

Reporting requirements: The following is excerpted from the Delaware DNREC regulations Sections 1-3:

"Any person who causes or contributes to the discharge of an air contaminant into the air or a pollutant, including petroleum substances, into surface water, groundwater or land, or disposal of solid waste in excess of any DRQ (Delaware Reportable Quantity) specified under this regulation, shall report such discharge to the Department immediately upon discovery of said discharge and after activating the appropriate emergency site plan unless circumstances exist which make such a notification impossible. A delay in notification shall not be considered to be a violation of this regulation when the act of reporting may delay the mitigation of the discharge and/or the protection of public health and the environment. Discharge or disposal in compliance with a validly issued state permit(s) or in compliance with other state or federal regulations is exempt from this reporting requirement."

Discharges wholly contained within a building are exempt from these requirements. If the discharge results in injuries outside the workplace that require medical treatment or result in death to anyone affected by the discharge, notification must be made. This provision does not apply if it takes longer than 7 days to learn about the injury, but it does apply whenever the knowledge of a death attributable to a discharge is obtained.

During notification, the information given should be the same as that required by SARA plus the following:

- Include the facility name and/or location of the discharge.
- Include the type of incident, e.g. discharge, fire, explosion.
- Include the Chemical Abstract System (CAS) number for the chemical or constituent chemicals of a mixture.
- Indicate whether the chemical or chemicals are listed in Table I of the Delaware Regulation for the Management of Extremely Hazardous Substances.
- Give the name of the reporting person and a call-back number.
- Indicate whether or not this is a complete report. Incomplete reports must be completed when the information is available but in no case more than 24 hours later than the initial contact.

If not already required by SARA Title III, Section 304, and at the DNREC's discretion, the DNREC may require a written report within 30 days or less. It should include and update the above information, include the additional information required under SARA, and the following information:

- Name, address and phone number of the owner or operator
- Anticipated environmental impact
- An evaluation of all pertinent prevention and response plans and policies in light of the discharge and the owner's or operator's response thereto
- A detailed identification of the pathway through which the discharge to the environment occurred with drawings, if necessary, to clearly explain this path
- Measures proposed to prevent such a discharge from occurring in the future and to remedy the shortcomings in the prevention, detection, response containment, and cleanup or removal plan components

DISTRICT OF COLUMBIA

Who to notify: Mayor's Command Center

Address: District of Columbia Office of Emergency Preparedness
2000 14th Street, N.W.
8th Floor
Washington, D.C. 20009

Phone: Information Center — (202) 727-6161

Reporting requirements: As per SARA Title III, Section 304. The Mayor's Command Center will coordinate all necessary response.

FLORIDA

Who to notify: Division of Emergency Management/State Warning Point/Department of Environmental Regulations Hazardous Waste Division

Address: Florida State Emergency Response Commission
Division of Emergency Management Hazardous Materials Planning
2740 Centerview Drive
Tallahassee, FL 32399

Phone: Information Hot Line Center — (904) 488-1320

Reporting requirements: As per SARA Title III with the following modifications. Notification of the release of a reportable quantity of solid particles is not required if the mean diameter of the particles is larger than 100 micrometers (0.004 inches) for the following materials:

- antimony
- arsenic
- beryllium
- cadmium
- chromium
- copper
- lead
- nickel
- selenium
- silver
- thallium
- zinc

GEORGIA

Who to notify: Environmental Protection Division (EPD) through the Department of Natural Resources (DNR) Emergency Operations Center

Address: EPD/DNR
7 Martin Luther King Jr. Dr.
Suite 139
Atlanta, GA 30334

Phone: Dispatch Center — In state (800) 241-4113 or (404) 656-4863 (both numbers are 24-hour numbers)

Reporting requirements: Any "Oil" or "Hazardous Substance" (as defined by Chapter 14 of Title 12 of the Official Code of Georgia Annotated, including amendments) that is released, of an unknown quantity or in excess of the reportable quantity (RQ), must be reported immediately, as per SARA Title III, Section 304.

GUAM

Who to notify: Civil Defense/Guam Emergency Services Office

Address: State Emergency Response Commission
c/o Civil Defense/Guam Emergency Services Office
P.O. Box 2877
Agana, Guam 96910

No other information was provided.

HAWAII

Who to notify: Hawaii State Emergency Response Commission (SERC)

Address: Office of Hazard Evaluation and Emergency Response
Hawaii State Department of Health
Five Water Front Plaza, Suite 250
500 Ala Moana Blvd.
Honolulu, HI 96813

Phone: State Civil Defense Dispatch Center — (808) 734-2161 (24 hours)

Reporting requirements: As per SARA Title III, Section 304. If release occurs during transport, contact an operator or call 911.

IDAHO

Who to notify: State Emergency Medical Services Communication Center

Address: Idaho Emergency Response Commission
1410 North Hilton
Boise, ID 83706

Phone: In state — (800) 632-8000
Dispatch Center — (208) 327-7422

Reporting requirements: As per SARA Title III, Section 304. Send written reports to the address above.

ILLINOIS

Who to notify: Illinois Emergency Services and Disaster Agency (IESDA)

Address: SERC
Supervisor, Hazardous Materials Compliance and Enforcement
Illinois Emergency Management Agency
110 East Adams Street
Springfield, IL 62706

Phone: SERC — (800) 782-7860
Dispatch Center — (217) 782-7860

Reporting requirements: Reporting requirements are similar to SARA Title III, Section 304 requirements. The different circumstances, under which reporting an incident is necessary, can be found in 29 Illinois Administrative Code Chapter I, Section 430.30, Subchapter (d). Information required for the verbal report is the same as SARA with the following additions:

- Specific location of the release
- Name and phone number of the reporting person

The written requirements are the same as SARA Title III, Section 304.

INDIANA

Who to notify: Department of Environmental Management

Address: Indiana Emergency Response Commission
Department of Environmental Management
Office of Environmental Response
5500 West Bradbury Avenue
Indianapolis, IN 46241

Phone: Dispatch Center — (317) 233-7745

Reporting requirements: Immediate notification is required under the guidelines provided by SARA Title III, Section 304.

IOWA

Who to notify: Department of Natural Resources (DNR); Disaster Services Division; Department of Public Defense (DPD); the local police department or county sheriff

Address: DNR Emergency Response

900 East Grand Avenue

Des Moines, IA 50319-0034

Phone: DNR Reporting Center — (515) 281-8694

Reporting requirements: Actual, imminent, or probable release of a hazardous substance, which presents a danger to public health or safety, must be reported as soon as possible and no later than 6 hours after discovery or onset of the hazardous condition. The report should include as much of the information, as time allows to gather, that is required in the written follow-up. The following information should be included in the written follow-up:

- Exact location of incident
- Time and date of onset or discovery of hazardous condition
- Name, manufacturer's name, and volume of hazardous material or hazardous contaminant within a material
- Medium into which the release occurred
- Name, address, and phone number of party responsible for hazardous condition
- Time and date of verbal report to the DNR
- The weather conditions at the time of the incident
- Name, address, and phone number of party reporting the incident
- Name and phone number of the person closest to the scene who can be contacted for further information and action
- Any other information, including circumstances leading to the incident, visible effects of incident, and containment measures that may assist in proper evaluation by the DNR

At the time of the telephone report, an incident number is issued. This number must be included in the written report. The written report must be made within 30 days to the address listed above.

KANSAS

Who to notify: Division of Emergency Preparedness (DEP); Department of Health and Environment

Address: Kansas State Emergency Response Commission
c/o Right-to-Know Program
Bureau of Environmental Health Services
Kansas Department of Health and Environment
109 SW 9th, Suite 501
Topeka, KS 66612

Phone: Notifier number — (913) 296-3176 (24 hours)
Main office — (913) 266-1000

Reporting requirements: As per SARA Title III, Section 304. All spills of a potentially polluting material (including sewage, hazardous chemicals, petroleum, etc.) must be reported to Kansas Department of Health and Environment/Bureau of Environmental Remediation at (913) 296-1662. Send written follow-up for SARA Title III requirements to the address above.

Also, the Kansas Corporation Commission (KCC) receives spill information for the Oil and Gas Industry (crude oil, brine spills, etc.) at:

KCC
1500 S. W. Arrowhead
Topeka, KS 66604-4027
Telephone: (913) 271-3100

KENTUCKY

Who to notify: Kentucky Disaster and Emergency Services

Address: ATTN: KYERC
Boone Center
Frankfort, KY 40601-6168

Phone: Kentucky Disaster and Emergency Services office —(502) 564-8660
24-hour Communication Center — (502) 564-7815

Reporting requirements: No information was provided.

LOUISIANA

Who to notify: Emergency Response Commission via Office of State Police, Transportation and Environmental Safety Section.

Address: SERC
c/o Department of Public Safety and Corrections
Office of State Police
Transportation and Environmental Safety Section
Box 66614
Baton Rouge, LA 70896

Phone: Reporting Center — (504) 925-6595 (collect calls accepted 24 hours a day)

Reporting requirements: Materials that meet the following criteria must be reported immediately if they are released beyond the site of the facility and they exceed the reportable quantity:

- Any material and its reportable quantity that appears on the most current list of:
  - Extremely Hazardous Materials
  - CERCLA Hazardous Substances as established by the EPA
  - Hazardous Substances and Reportable Quantities as established by the Department of Transportation
- Any material, which requires maintenance of an MSDS under the OSHA Standard found in 29 CFR 1910.1200 and does not appear in one of the above lists, must be reported if the material released exceeds the reportable quantity of 5,000 pounds established by the Louisiana Department of Public Safety and Corrections. Exceptions are:
  - Compressed or refrigerated flammable gases will have a reportable quantity of 100 pounds.
  - All flammable liquids will have a reportable quantity of 100 pounds.
  - All other liquids requiring maintenance of an MSDS will have a reportable quantity of 500 pounds

Information must be given as per SARA Title III, Section 304, and should also include the following:

- Name of company
- Where incident occurred
- Substance's degree of hazardousness and whether it was a solid, liquid, or gas
- Substance's hazard class and other identifier
- Details of incident
- Whether release resulted in a fire, injury, or fatality
- Whether responsible state and local agencies have been notified

Follow-up written reports are required within 5 days of a release. They should follow the guidelines established in SARA Title III, Section 304, and contain the above information. Reports must be sent to the organizations listed under "Who to notify" above.

The Office of State Police, Transportation and Environmental Safety Section will coordinate emergency response activities.

MAINE

Who to notify: Maine State Police for SERC and DEP notification; County Sheriff's Office; Local Fire Department

Address: Maine Emergency Management Agency
Chairman of SERC
State House Station 72
Augusta, ME 04333

Phone: In state only — (800) 452-8735
Reporting office — (207) 287-4080

Reporting requirements: Any release beyond the facility boundary and/or reportable quantity of a CERCLA Hazardous or Extremely Hazardous substance. The initial report should contain all the information required by SARA Title III, Section 304, and the following:

- Date and time of incident
- Company name and location
- Type of incident (fixed or transportation; if transportation, give truck/rail car number)
- DOT ID, hazard class, and the material's CAS number
- Physical state of the material when stored and after release and the type of container involved
- Whether the release is ongoing or over and whether it was contained
- Wind and weather conditions
- What assistance is needed
- A description of the incident

A follow-up written report is required within 14 days and should include information required under SARA Title III, Section 304, in addition to the following:

- The cause and events leading to the release
- Measures taken to avoid reoccurrence

MARYLAND

Who to notify: Maryland Department of the Environment

Address: Hazardous and Solid Waste Management Administration
Maryland Department of the Environment
2500 Broening Highway
Baltimore, MD 21224

Phone: (410) 974-3551 (24-hour number; after hours allow the phone to ring 6 times for it to transfer to an agency that will contact the duty officer); if release is to the air ALSO call the Air Management Administration (410) 243-8700 after hours or (410) 243-8700 during business hours.

Reporting requirements: Notification is required for releases of ANY quantity of petroleum substance (within 2 hours) or hazardous material substance (immediately). Written follow-up reports are required within 10 days of completion of the cleanup.

MASSACHUSETTS

Who to notify: Regional Office of Department of Environmental Protection

Address: Massachusetts Department of Public Health
Right-to-Know Program
150 Tremont Street
Boston, MA 02111

Phone: During business hours call the number below that indicates the region in which the spill occurred.

Region	*City*	*Number*
Northeast	Woburn	(617) 935-2160
Southeast	Lakeville	(508) 946-2700
Central	Worcester	(508) 792-7653
Western	Springfield	(413) 784-1100

Emergency notification, training, and right-to-know — 617-727-7035

After business hours, call the State Police at (617) 566-4500

Reporting requirements: The Massachusetts Oil and Hazardous Materials List (MOHML) provides a comprehensive list of regulated substances, including those listed on CERCLA and SARA lists. These substances along with any substances that are flammable, corrosive, reactive, toxic, infectious, or radioactive must be reported in the event they are spilled or there exists a threat of their spilling in excess of their reportable quantities. A report of this type must be made within 2 hours of learning of the spill or spill threat. The following information should be given when reporting:

- Caller's name and phone number
- Location of the release
- Type and (approximate) quantity of the oil or hazardous material released
- Brief description of the incident
- Name and phone number of the owner and/or operator of the source of the spill if known
- Potential impact of the release on health, safety, public welfare, or the environment
- Cleanup measures taken or proposed

MICHIGAN

Who to notify: Department of Natural Resources

Address: Department of Natural Resources
Environmental Response Division
P.O. Box 30028
Lansing, MI 48909

Phone: In state only, 24 hours — (800) 292-4706
Out of state — (517) 373-9837

Reporting requirements: As per SARA Title III, Section 304.

MINNESOTA

Who to notify: Division of Emergency Management, State Response Center

Address: Office Emergency Response Commission
450 N. Syndicate, Suite 175
St. Paul, MN 55104

Phone: Dispatch Center — (612) 649-5451 or (800) 422-0798

Reporting requirements: As per SARA Title III, Section 304.

MISSISSIPPI

Who to notify: Mississippi Emergency Management Agency

Address: Mississippi Emergency Management Agency
P. O. Box 4501
Jackson, MS 39296-4501

Phone: Reporting Center — (800) 222-6362 (24 hours)

Reporting requirements: As per SARA Title III, Section 304 plus: ***NO*** exemption for transportation applies, and the state requires notification of any accidental release of oil — 1 drop on water or 1 pound on land — that results in exposure off-site.

Provide the information required under SARA Title III and include location of release, company name, and person reporting the release. Send a follow-up written report to the address above using state form MERC-III.

MISSOURI

Who to notify: Department of Natural Resources (DNR)

Address: Department of Natural Resources
Hazardous Substance Emergency Response Office
P.O. Box 176
Jefferson City, MO 65102

Phone: Reporting and Response Center — (314) 634-2436

Reporting requirements: As per SARA Title III plus these requirements:

- Any release of petroleum, including crude oil or any fraction thereof, natural gas, natural gas liquids, liquefied natural gas, or synthetic gas usable for fuel (or mixtures of natural gas and such synthetic gas), in excess of 50 gallons for liquids or 300 cubic feet for gases
- Any release of hazardous waste that is reportable under Sections 260.350 to 260.430 must be reported
- A release of a hazardous substance that requires immediate notice under Part 171 of Title 49 of the *Code of Federal Regulations*

If requested by DNR, a written report of the particulars of the incident shall be submitted. In addition to the information required under SARA Title III, the following information must be provided:

- The location of the emergency and directions to the site
- Names, addresses, and phone numbers of persons that may have information on the substances involved

- Actions taken and those actions that will be taken (and when they will be taken) to clean up the substance and end the emergency
- Any other pertinent information requested by the state

MONTANA

Who to notify: Disaster and Emergency Services Division

Address: ESD/DHES
Cogswell Building
Capitol Station
Helena, MT 59620

Phone: Reporting Center — (406) 444-6911 (24-hour number)

Reporting requirements: As per SARA Title III, Section 304. All releases of oil in any substantial quantity must also be reported.

NEBRASKA

Who to notify: Department of Environmental Quality

Address: Department of Environmental Quality
Emergency Response Unit
Box 98922
Lincoln, NE 68509-8922

Phone: Reporting Center — (402) 471-2186 (business hours)
State patrol dispatcher, 24 hours — (402) 471-4545

Reporting requirements: Immediate notification is required regardless of the quantity of an oil or hazardous substance release that occurs beneath the surface of the land, impacts or threatens waters of the state, or threatens the public health and welfare.

Immediate notification is required of a release upon the surface of the land of an oil in a quantity that exceeds 25 gallons or of a hazardous substance which equals or exceeds 10 pounds or its reportable quantity, whichever is less.

Notification is not required for releases of oil upon the surface of the land of 25 gallons or less that will not constitute a threat to public health and welfare, the environment, or a threat of entering the waters of the state.

All information known about the release at the time of discovery is to be included such as time of occurrence, quantity and type of material, location, and any corrective or cleanup actions presently being taken. The Department of Environmental Control may require interim reports until any remedial action is completed.

A written final report is required within 15 days after remedial action is completed. This report should contain the following information in addition to information required under SARA Title III:

- Location of release
- Person or persons causing and responsible for the release
- Cause of the release
- Environmental damage caused by the release
- Location and method of ultimate disposal of the oil or hazardous substance and other contaminated materials
- Actions being taken to prevent a reoccurrence of the release

NEVADA

Who to notify: Division of Emergency Management

Address: Emergency Management
2525 S. Carson Street
Capitol Complex
Carson, NV 89710

Phone: (702) 687-4240 (during business hours) and ask for the duty officer

(702) 687-5300 Highway Patrol (after hours) and ask for the DEM duty officer

Reporting requirements: As per SARA Title III, Section 304 and any petroleum release over 25 gallons. The follow-up written report is required within 30 days and a number given out during the initial call-in report must be included in the written report.

NEW HAMPSHIRE

Who to notify: Office of Emergency Management

Address: Office of Emergency Management
107 Pleasant Street
Concord, NH 03301

Phone: In state — (800) 852-3792
Answering service, 24 hours — (603) 271-2231

Reporting requirements: As per SARA Title III, Section 304.

NEW JERSEY

Who to notify: New Jersey Department of Environmental Protection (NJDEP)

Address: NJDEP - Division of Environmental Quality
Bureau of Communications and Support Services
ATTN: Bureau Chief
CN-027
Trenton, NJ 08625-0027

Phone: (609) 292-7172 (24-hour hot line)

Reporting requirements: Immediate notification is required at the hot line number listed above. NJDEP will ask a series of questions and explain what to do next. A follow-up written report is required NO LATER THAN 60 days following the incident. It shall include the same information as required under SARA Title III, Section 304.

NEW MEXICO

Who to notify: State Police

Address: Department of Public Safety
P.O. Box 1628
Santa Fe, NM 87504-1628

Phone: (505) 827-9126 or 827-9300

Reporting requirements: As per SARA Title III, Section 304.

NEW YORK

Who to notify: Department of Environmental Conservation (DEC)

Address: SERC
c/o NYS DEC
Bureau of Spill Prevention and Response
50 Wolf Road - Room 326
Albany, NY 12233-3510

Phone: NYS DEC Spill Hot Line (in state only) — (800) 457-7362

Reporting requirements: As per SARA Title III, Section 304.

NORTH CAROLINA

Who to notify: State Warning Point (Highway Patrol)

Address: North Carolina Division of Emergency Management
116 West Jones Street
Raleigh, NC 27603-13335

Phone: Hot Line Communications Center — (800) 858-0368
Office — (919) 733-3867

Reporting requirements: As per SARA Title III, Section 304.
Written report is required within 30 days of the incident.

NORTH DAKOTA

Who to notify: State Radio Communications

Address: Division of Emergency Management
Box 5511
Bismark, ND 58502-5511

Phone: Dispatch Center (in state only) — (800) 472-2121 (24 hours)

Reporting requirements: As per SARA Title III, Section 304.

NORTHERN MARIANAS

Who to notify: Northern Marianas State Emergency Response Commission

Address: Northern Marianas SERC
Office of the Governor
Capitol Hill
Saipan, C.M.
Northern Marianas Islands 96950

No other information was provided.

OHIO

Who to notify: Ohio EPA

Address: Ohio EPA
P.O. Box 1049
1800 Water Mark Drive
Columbus, OH 43266-0149

Phone: In state only — (800) 644-2917

Reporting requirements: As per SARA Title III, Section 304. Any extremely hazardous substance not covered by CERCLA is assigned a reportable quantity of one pound. Also, Ohio requires the following additional and/or more specific information during both phone and written reports:

When Reporting By Phone:

- Probable source of release
- Present and anticipated movement of released material
- Weather conditions
- Company or emergency personnel on scene
- Actions initiated/remediation procedure
- Time and date of release
- Time and date of discovery

When Reporting In Writing:

- What company official was the release reported to?
- What ID number was assigned by Ohio EPA during call?
- Who made verbal report to fire department, LEPC, and SERC?
- What is the location of incident
- Was any animal or vegetation damage observed?
- Was medical advice provided?
- What methods for detecting and monitoring were used?
- If release was airborne, what was the wind direction and speed?
- Was the public warned?
- What are the plans for prevention of future incidents?

Phone notification should be within 30 minutes of knowledge of release and the written follow-up report is required within 30 days.

OKLAHOMA

Who to notify: Department of Pollution Control (DPC)

Address: Department of Environmental Health Services (DEHS)
1000 N.E. 10th Street
Oklahoma City, OK 73117-1299

Phone: DPC (24-hour number) — (405) 271-4468

Reporting requirements: As per SARA Title III, Section 304. Written reports go to DEHS.

OREGON

Who to notify: Oregon Emergency Management Division

Address: Department of Environmental Quality
Regional Operations
10th Floor
811 S.W. 6th Avenue
Portland, OR 97204-1390

Phone: In state — (800) 452-0311
Out of state — (503) 378-4124

Reporting requirements: If a spill, release, or threatened spill/release in excess of the reportable quantity (according to Division 108 OAR 340-108-010) occurs, report it immediately. The spill need not be reported if ***ALL*** of the following are true:

- The spill or release is known to the party responsible for the material (the person owning or having control over the oil or hazardous material or their designated representative).
- It is completely contained.
- It is completely cleaned up without further incident, including fixing or repairing the cause of the spill or release.

The department may require the person responsible for a spill or other incident to submit a written report within 15 days. All aspects of the spill and steps taken to prevent a reoccurrence should be included. Any spills of pesticide residue (as defined by OAR 340-109-002) in excess of 200 pounds (approximately 25 gallons) must be reported.

PENNSYLVANIA

Who to notify: Pennsylvania Emergency Management Agency

Address: Pennsylvania Emergency Management Council
c/o Pennsylvania Emergency Management Agency
P.O. Box 3321
Harrisburg, PA 17105

Phone: (717) 783-8150 or (800) 535-0202; For releases during transportation, call 911 or the operator.

Reporting requirements: As per SARA Title III, Section 304, with the following additional requirements:

- During the phone-in report, include the name and phone number of the person making the notification and name of the person employed by the owner/operator who has the authority to supervise clean-up activities.
- In the follow-up written report, include actions taken to mitigate potential future incidents.
- Written follow-up reports must be submitted within 14 calendar days of the release.

PUERTO RICO

Who to notify: Solid Waste Department

Address: Puerto Rico Environmental Quality Board
Office of the Governor
Box 11488
San Juan, Puerto Rico 00910

Phone: (809) 725-5140 ext. 204 or 214; or (809) 722-1175; or (809) 767-8181

No other information was provided.

RHODE ISLAND

Who to notify: Department of Environmental Management

Address: SERC
State House Room 27
Providence, RI 02903

Phone: Dispatch office — (401) 277-3070
In state — (800) 498-1336

Reporting requirements: As per SARA Title III, Section 304.

SOUTH CAROLINA

Who to notify: SC Department of Health and Environmental Control (SCDHEC)

Address: SCDHEC
Bureau of Solid and Hazardous Waste Management
2600 Bull Street
Columbia, SC 29201

Phone: (803) 253-6488 (24 hours) or (803) 935-6321 (business hours); 911 or the operator for transportation incidents

Reporting requirements: As per SARA Title III, Section 304, and facilities should document the circumstances of the release and all communications.

SOUTH DAKOTA

Who to notify: Department of Environmental and Natural Resources (DENR)

Address: DENR
Ground-Water Quality Program
523 E. Capitol Avenue
Pierre, SD 57501-3181

Phone: (605) 773-3231 (after hours) or (605) 773-3296 (business hours)

Reporting requirements: As per SARA Title III, Section 304. South Dakota regulates several substances not normally covered under SARA Title III. South Dakota law requires reporting a known discharge of a regulated substance immediately if one of the following conditions exist:

- The discharge threatens or is in a position to threaten the waters of the state.
- The discharge causes an immediate danger to human health or safety.
- The discharge exceeds 25 gallons or causes a sheen on surface water or it exceeds any ground-water quality standards of chapter 74:03:15 or surface water quality standards of chapter 74:03:02.
- The discharge harms or threatens to harm wildlife or aquatic life.

Aside from the substances designated by SARA Title III, the following substances are regulated:

- Fertilizers as defined in SDCL 38-19-1(1), (2), (3), (12), (13), and (14), including fertilizer derivatives

- Pesticides as defined in SDCL 38-20A-1(1), (3), (4), (5), and (6), SDCL 38-20A-10; and those substances defined in SDCL 38-21-14(4), (5), (18), (19), (21), and (29), including metabolites and all active and inert ingredients of these substances
- Petroleum and petroleum substances, including oil, gasoline, kerosene, fuel oil, oil sludge, oil refuse, oil mixed with other wastes, refined or blended crude petroleum stock, and any other oil or petroleum substances
- Radiological, chemical, or biological warfare agents or radiological waste
- Hazardous wastes as described in 40 CFR Part 261, Subparts C and D (July 1, 1988)

Reports should include SARA Title III information and the following:

- Specific location of the discharge
- Type and amount of regulated substance discharged
- Responsible person's name, address, and telephone number
- An explanation of any response action that was taken
- List of agencies notified
- Suspected cause of the discharge
- Date and time of the discharge to the extent known
- Immediate known impacts of the discharge

TENNESSEE

Who to notify: Tennessee Emergency Management Agency

Address: Tennessee Emergency Management Agency
3041 Sidco Drive
Nashville, TN 37204

Phone: In state — (800) 262-3300, (615) 741-0001
Reporting and Dispatch out of state — (800) 258-3300

Reporting requirements: As per SARA Title III, Section 304.

TEXAS

Who to notify: Release to land or water — Texas Water Commission; Release from a pipeline — Texas Railroad Commission; Release to air — Texas Air Control Board

Address: Emergency Response Unit
Texas Water Commission
P.O. Box 13087
Austin, TX 78711

Railroad Commission
P.O. Box 12967
Austin, TX 78711-2967

Texas Air Control Board
12124 Park 35 Circle
Austin, TX 78753

Phone: Water Commission — (512) 463-7727
Railroad Commission — (512) 463-6832
Air Control Board — (512) 908-1876

Reporting requirements: As per SARA Title III, Section 304.

UTAH

Who to notify: Department of Environmental Quality (DEQ)

Address: DEQ Division of Environmental Response and Remediation
1950 West North Temple, 2nd floor
Salt Lake City, UT 84116

Phone: (801) 538-4100 (24-hour number)

Reporting requirements: As per SARA Title III, Section 304.

VERMONT

Who to notify: Vermont Emergency Response Management

Address: Vermont Department of Health
Division of Occupational and Radiological Health
Administration Building
10 Baldwin Street
Montpelier, VT 05602

Phone: In state — (800) 641-0681 or (802) 865-7730
Answering machine after hours

Reporting requirements: As per SARA Title III, Section 304 with follow-up written reports due no later than 7 calendar days after the incident.

VIRGIN ISLANDS

Who to notify: Environmental Health Department

Address: Virgin Islands Environmental Emergency Response Commission
Misky Center, Suite 231
Misky #45 A
St. Thomas, U.S. Virgin Islands 00802

Phone: (809) 774-3320

No other information was provided.

VIRGINIA

Who to notify: Department of Emergency Services

Address: Department of Environmental Quality
629 E. Main Street
Richmond, VA 23219

Phone: Office — (804) 762-4000
Reporting Spill Center — (804) 527-5200

Reporting requirements: As per SARA Title III, Section 304.

WASHINGTON

Who to notify: Emergency Management Division

Address: SERC
Department of Community Development
Emergency Management Division
P.O. Box 48346
Olympia, WA 98504-8346

Phone: SERC — (800) 258-5990
Duty officer and spill response — (206) 923-4964

Reporting requirements: As per SARA Title III, Section 304 for reporting to state level (SERC). The report should also include the reporting party, location, responsible party, cleanup status, and resource damage. Also, Washington has more stringent reporting requirements for the regional level and for reporting to the Department of Ecology (state and regional level). Refer to the following state regulations to ensure compliance with all applicable laws:

- State of Washington 1990 Oil Spill Law HB 2494
- Department of Ecology UST (Underground Storage Tank) Regulations Ch. 173-360-375 WAC, Nov. 1990
- Department of Ecology Model Toxics Control Act Cleanup Regulations Ch. 173-340-450 WAC UST Section, February 1991
- Department of Ecology Dangerous Waste Regulations Ch. 173-303-145 WAC (blue book), April 1991

WEST VIRGINIA

Who to notify: Office of Emergency Services (OES)

Address: West Virginia OES
Main Capitol Building
Room EB 80
1900 Kanawha Blvd East
Charleston, WV 25305

Phone: (304) 558-5380, 24 hours (an answering machine will answer after business hours)

Reporting requirements: As per SARA Title III, Section 304.

WISCONSIN

Who to notify: Division of Emergency Government (who will in turn contact the Wisconsin Department of Natural Resources and all other agencies necessary)

Phone: 24-hour hot line — (800) 943-0003

WYOMING

Who to notify: Wyoming Department of Environmental Quality (DEQ); SERC; LEPC

Address: Wyoming SERC
P.O. Box 1709
Cheyenne, WY 82003

Phone: DEQ (24-hour reporting center) — (307) 777-7781

Reporting requirements: As per SARA Title III, Section 304 and any release of petroleum to water or any release over ten barrels (420 gallons) of crude, petroleum condensate or produced water, or any release over 25 gallons of refined product must be reported to DEQ.

Index

D

E

F

G

H

U

V

W

X

Y

NOTES

NOTES

NOTES

NOTES

COMMENT SHEET

DATE ____________________ NAME ____

ADDRESS ____

ORGANIZATION REPRESENTED ____

CHAPTER TITLE ____ NUMBER ______________

SECTION/PARAGRAPH/FIGURE ______________________________ PAGE ______________

1. Proposal (include proposed wording or identification of wording to be deleted), OR PROPOSED FIGURE:

2. Statement of Problem and Substantiation for Proposal:

RETURN TO: IFSTA Editor
Fire Protection Publications
Oklahoma State University
Stillwater, OK 74078-0118

SIGNATURE ______________________________

Use this sheet to make any suggestions, recommendations, or comments. We need your input to make the manuals as up to date as possible. Your help is appreciated. Use additional pages if necessary.